ASE SCIENCE PRACTICE

teaching secondary CHEMISTRY

SECOND EDITION

EDITOR: KEITH S. TABER

Titles in this series:

Teaching Secondary Biology	978 1444 124316
Teaching Secondary Chemistry	978 1444 124323
Teaching Secondary Physics	978 1444 124309

The Publishers would like to thank the following for permission to reproduce copyright material:

Photo credits

p.8 Photo Researchers/Alamy; **p.13** *l* cartela – Fotolia; **p.57** Philip Johnson; **p.169** *l* 123idees – Fotolia, *r* dk/Alamy; **p.190** *l* Andrew Lambert Photography/Science Photo Library, *r* Paul Straathof, www.paulslab.com; **p.191** John Oversby; **p.254** Jean-Claude Revy, ISM/Science Photo Library; **p.271** John Boud/Alamy; **p.354** *tl* Nick Biemans – Fotolia, *tr* Vera Kuttelvaserova – Fotolia, *b* Vincent RUF – Fotolia; **p.355** *a* Holmes Garden Photos/Alamy, *b* Noam Graizer/Alamy, *c* F1online digitale Bildagentur GmbH/Alamy, *d* Bill Bachman/Alamy, *e* Powered by Light/Alan Spencer/Alamy.

Orders: please contact Bookpoint Ltd, 130 Milton Park, Abingdon, Oxon OX14 4SB. Telephone: (44) 01235 827720. Fax: (44) 01235 400454. Lines are open 9.00 – 5.00, Monday to Saturday, with a 24-hour message answering service. Visit our website at www.hoddereducation.co.uk

First published in 2012 by
Hodder Education,
An Hachette UK Company
338 Euston Road
London NW1 3BH

Impression number	5 4 3 2 1
Year	2016 2015 2014 2013 2012

Cover photo © BSIP, Laurent/Science photo library
Illustrations by Tony Jones/Art Construction
Typeset in 11.5 ITC Galliard by Pantek Media, Maidstone, Kent
Printed by MPG Books, Bodmin
A catalogue record for this title is available from the British Library

ISBN: 978 1444 124323

Contents

Contributors

Judy Brophy graduated in chemistry from Manchester University before training as a teacher at the Institute of Education. She taught science and chemistry for many years in inner city (11–18) London comprehensives, whilst taking an active role in the ASE. Judy specialised, successfully, in encouraging her A Level students, especially girls, to aim higher. As a visiting teacher at King's College London she passes on her experience to secondary science PGCE students.

Justin Dillon is professor of science and environmental education and Head of the Science and Technology Education Group at King's College London. After studying for a degree in chemistry he trained as a teacher at Chelsea College and taught in six inner London schools until 1989 when he joined King's. Justin is co-editor of the *International Journal of Science Education* and in 2007 was elected President of the European Science Education Research Association for a four-year term.

Philip Johnson taught chemistry up to A Level for thirteen years in 11–18 comprehensive schools before joining Durham University School of Education in 1992. He began researching into the development of students' understanding in chemistry while teaching in schools and continues to do so. His work is published in international science education research journals.

Vanessa Kind is Senior Lecturer in Education at Durham University and Director of Science Learning Centre North East. She has extensive experience of chemical education gained through teaching in London and Hull, and a previous lectureship at the Institute of Education, University of London. Vanessa was the Royal Society of Chemistry's Teacher Fellow 2001–2002. Her current research interests include pedagogical content knowledge for science teaching and post-16 chemistry education.

John Oversby has been a teacher of sciences and mathematics in Ghana and the UK for over 20 years, and a teacher educator at The University of Reading for 20 years. He is also an active member of The Royal Society of Chemistry and Chair of the ASE Research Committee. He has interests in modelling in science education. He has been the UK coordinator of a history and philosophy in science teaching project and is now international coordinator of a Comenius Climate Change Education network.

Keith S. Taber taught sciences, mainly chemistry and physics, in secondary schools and further education, before joining the Faculty of Education at the University of Cambridge, where he is mostly working with higher degree students. Whilst teaching, he undertook doctoral research exploring student understanding of the chemical bond concept. He was the Royal Society of Chemistry's Teacher Fellow in 2000–2001. He has written a good deal about chemistry and science education, and is editor of the journal *Chemistry Education Research and Practice*.

Kim Chwee Daniel Tan started his career as a chemistry teacher in 1990. He has been a faculty member of the (Singapore) National Institute of Education since 1998. He teaches higher degree courses as well as chemistry pedagogy courses in the pre-service teacher education programmes. His research interests are chemistry curriculum, translational research, ICT in science education, students' understanding and alternative conceptions of science, multimodality and practical work.

Georgios Tsaparlis is professor of science education at University of Ioannina, Greece. He holds a chemistry degree (University of Athens) and an M.Sc., and a Ph.D. (University of East Anglia). He teaches physical chemistry (including electrochemistry) and science/chemistry education courses. He has published extensively in science education, his research focus being on structural concepts, problem solving, teaching and learning methodology, and chemistry curricula. He was founder and editor (2000–2011) of *Chemistry Education Research and Practice* (CERP).

Elaine Wilson is a Senior Lecturer in science education and a Fellow of Homerton College at the University of Cambridge. Elaine was formerly a secondary school chemistry teacher who was awarded a Salters' Medal for Chemistry teaching. She now teaches undergraduates, secondary science PGCE students, coordinates a 'blended learning' Science Education Masters course and has helped set up a new EdD course. Elaine has received two career awards for teaching in Higher Education, a University of Cambridge Pilkington Teaching Prize and a National Teaching Fellowship.

Vicky Wong taught Science and Chemistry in the UK, Spain and New Zealand for ten years. She was the Royal Society of Chemistry Teacher Fellow in 2004–2005 and now works as an independent science education consultant running training courses for teachers, writing curriculum materials and undertaking research.

Introduction

Keith S. Taber

This book is part of a series of handbooks for science teachers commissioned by the Association for Science Education. This particular handbook is intended to support the teaching of chemistry at secondary level (taken here as ages 11–16 years), whether as a discrete subject or as part of a broader science course. The book has been written with a particular awareness of the needs of new teachers and of those teaching chemistry who would not consider it their specialism within the sciences. However, the book should prove to be of interest and value to anyone teaching chemistry topics at secondary level. The book has been written by a team of authors who collectively have a wide range of experience in teaching chemistry, supporting and developing teachers of chemistry, and undertaking research into teaching and learning in chemistry topics.

It is sometimes easier to characterise something by explaining what it is not. This book is not a chemistry textbook for teachers, although inevitably it discusses chemistry content as part of the process of describing and recommending approaches to teaching the subject. There are many good chemistry books available at various levels, and any teacher who is concerned about their knowledge and understanding in aspects of chemistry should first do some work to develop their own subject knowledge before considering approaches to teaching. So-called 'pedagogic content knowledge' will only be sound when we are building on subject knowledge that is sound (else we become very effective at teaching poor chemistry).

The handbook does not set out to act as a teaching guide for any particular curriculum or syllabus. We intend the book to be equally useful across different courses: whether the course is arranged as a set of traditional topics or organised in some other way, for example teaching concepts through the contexts of major areas of application of chemistry such as food, transport, fabrics, etc. The book is not tied to a particular stream or ability level of student. In some places chapters make explicit suggestions for differentiating between different groups of students, but you should always consider how the advice given here can best inform the teaching of your particular classes.

Nor does the book set out to be a manual for teaching chemistry in the sense of providing comprehensive coverage of all content that might potentially be included in a secondary chemistry course. Such a manual would inevitably be both voluminous and very quickly out of date as chemistry and chemistry teaching move on.

Ideas for effective chemistry teaching

The philosophy behind the present book is that it is more important to inform teachers about effective teaching approaches than to train them to apply specified teaching schemes in particular topics. Some of the recommendations we offer in the handbook are based on a good deal of classroom experience and draw on specific research into students' learning difficulties and effective innovations in teaching. The specific suggestions made for teaching these aspects of the subject can certainly be considered as the best research-informed advice currently available to teachers.

However, teaching is a contextualised process: students, facilities, curriculum requirements and so much more can make a difference to what counts as good teaching in a particular classroom on a particular day. So just as important as the specific suggestions for teaching particular concepts is the range of approaches adopted by authors across the chapters. Drawing upon the range of teaching and learning activities described here can inform the development of teaching that is both varied and responsive to the needs of particular students and classes. Although not everything that could possibly be covered is included in the book, the thinking behind the examples that are presented here offers the basis for developing effective teaching that can be adopted across chemistry teaching (and often well beyond).

Teaching about the nature of chemistry

In accordance with the philosophy outlined above, the reader will find a varied set of chapters included in the book. For example, features of the nature of chemistry as a science are highlighted in some chapters. Teaching about the nature of science (which has sometimes been referred to as 'how science works'), which recognises how all students need to understand science (i.e. should be scientifically literate) for their role as citizens (as consumers, voters, etc.), has become increasingly important. Such a perspective needs to inform all science teaching and readers may wish to think how the ideas highlighted in a number of the chapters here (such as the nature and role of models in developing understanding of the world) can be more widely applied.

Another key difference between the chapters is the centrality of the topics discussed. The first five chapters set out key ideas that are needed to understand any area of chemistry, whereas the other chapters stand alone to a much greater extent. It is also useful to note that while almost all of the chapters are clearly about chemistry, the chapter about Earth science overlaps strongly with several other sciences, and in particular draws on geology. Given the increasing importance of interdisciplinary work in science and – even more so – the importance of understanding the environment in science and in society more widely, this chapter reminds us that chemistry links with, builds upon, and feeds into, a wide range of scientific work.

A final difference between chapters that will be very obvious to readers is the extent to which practical work features. It is said that chemistry is a practical subject. This is certainly true, but – as a science – chemistry is just as much a theoretical subject: it is the interplay of theory and evidence that is at the heart of scientific work. Chemistry as a science is based upon observations that lead to the development of categories and the identification of patterns, and to ways of making sense of, and understanding, the phenomena. This involves the construction of models and theories, which can motivate empirical investigations that can then inform further rounds of theorising and experimentation. In particular, much of chemistry as a science can be considered to be about building models and selecting those that are useful to scientists despite inevitably having limited ranges of application. The notions of acids (Chapter 6) or oxidation (Chapter 7) certainly reflect this, being concepts that are the products of human imagination, designed to reflect the patterns found in nature and to have utility to chemists as tools for thinking, explaining, predicting and so supporting practical and technological work. Understanding chemical ideas in this way, as creative products of scientific work (rather than simply being descriptions of the way the world is), should help students make sense of how chemists have modified these concepts over time (as described in the case of acids in Chapter 6), and why sometimes we seem to operate with a range of not entirely consistent models (oxidation in terms of oxygen or electrons: Chapter 7; oxidation in terms of oxidation states: Chapter 9).

An activity which illustrates this general point in relation to particle theory (the topic of Chapter 2) is included in a publication available from SEP (the Science Enhancement Programme, details of which are given in the 'Other resources' section at the end of this introduction). The first of two group-work tasks in this activity 'Judging models in science' asks students to consider two types of particle models – particles like tiny hard billiard balls; particles as molecules with 'soft' electron clouds – and to consider which model better explains a range of evidence based on the observable properties of matter. Students will find that each model is useful for explaining some phenomena, but neither fits all the evidence – and of course both models are still found useful in science.

A key aspect of this Janus-faced nature of chemistry (looking to both the phenomena and the theories) is that chemistry is discussed in terms of the macroscopic/molar/phenomenal level and in terms of submicroscopic particle models which are used to make sense of those phenomena. These models exemplify the nature of science (offering an extremely powerful explanatory scheme for making sense of the nature of the material world), but are known to be challenging for students.

A central feature of the way chemistry is presented and discussed in classrooms is the set of representations (such as formulae and chemical equations) used. This is often seen as a third 'level' distinct from the molar and submicroscopic levels, but is more helpfully understood as a specialised language that allows us to shift between those two levels (see Chapter 3). Translating between observable phenomena, symbolic representations and theoretical models is a key part both of teaching, and learning, chemistry.

Practical work and 'experiments' in chemistry teaching

Given the nature of chemistry as a discipline, and of school chemistry as a curriculum subject, it is essential that students are introduced to the phenomena that the theories and models are meant to help us understand. Without experiencing these phenomena, there is little motivation for adopting the theoretical ideas. Within school science, practical work has long been seen as a key part of chemistry lessons in many countries (and especially in the UK). Yet research also suggests that a good deal of the practical work undertaken by students has limited impact on learning (or even on student motivation to continue studying the subject).

Readers should consider carefully whether practical work undertaken is going to be the most effective way of meeting the purposes of teaching. In some cases this will certainly be so. So the chapter on inorganic analysis, an area of secondary chemistry that is not always given as much attention as it once was, is largely developed around practical work that students can carry out. This makes sense, as the aims of teaching this topic include the ability to carry out analytical procedures and interpret bench observations to identify particular chemical species present. Other chapters vary considerably in the amount of laboratory practical work recommended. In some cases teacher demonstrations are recommended as being more effective ways of ensuring students can be helped to appreciate the chemical interpretations of observations

– as a teacher you can draw students' attention to the chemically relevant and away from the salient but incidental. In some topics alternative forms of engaging, active learning are suggested to help students learn concepts not easily appreciated from the laboratory work possible in school contexts. Ultimately, however, theory and practical experience both have important roles to play and need to be coordinated across secondary school learning.

One specific issue that has vexed me in this context is the use of the term 'experiment' to describe school science practical work. Very few chemistry practicals carried out in most schools have any genuine claim to be experiments (authentic investigations to test out hypotheses, rather than to illustrate what is already set out as target knowledge to be learnt), and using the term experiment loosely may undermine learning about the nature of science, when we know that students commonly use technical terms such as 'experiment', 'theory', 'proof' and so forth in very vague and inconsistent ways. However, within the culture of school chemistry the term experiment is often used for any laboratory practical activity, especially one that is undertaken by students (rather than being a demonstration). Some authors here have followed the common usage and referred to laboratory practical activities as experiments. This is fine in talk among teachers and technicians, but it may be wise to consider carefully which practical activities should be presented to students as experiments. It is probably best limited to those where they are genuinely testing some kind of hypothesis – where students are seeking to discover something they do not actually already know.

Ensuring safety during practical work

Many standard chemistry practicals have been carried out safely in schools for decades and there have seldom been serious accidents. Yet clearly there are particular potential hazards involved in some practical work. It is important to keep the potential risks of practical activities (whether observed by, or undertaken by, the students) in perspective, yet to remember that student safety must be your paramount concern. In this book, the margin icon shown here is used to alert you to particular safety issues that you should be aware of. Recommended activities are considered suitable for classroom use by the authors. However, assessing risk is not just about the activity, but also the people and the conditions. The same practical may be viable with some teaching groups and not others.

It is, therefore, very important to undertake careful risk assessments before deciding to do any practical activities in the classroom. Your risk assessment needs to consider the experience and skills of the person(s) carrying out the activity (teacher or students), how responsible the particular students are and the facilities available in the teaching room (presence of a fume cupboard, adequate ventilation, sufficient access to sinks, plenty of space for students to move around, support of teaching assistants, assistance of a qualified technician, etc.)

For any practical activity (including teacher demonstrations) the teacher should carry out a risk assessment. This might include:

- checking the model risk assessment supplied by their employer (this may be supplied by another agency, such as CLEAPPS or SSERC in the UK, to which the employer subscribes) and adjusting the model risk assessment as appropriate.
- referring to the hazards of the starting material and the products by consulting safety data sheets provided by the supplier, and adjusting the model risk assessment as appropriate,
- consulting more experienced teachers or experienced technicians.

As a result, the significant findings should be recorded and the appropriate control measures implemented. Whatever the outcomes of a risk assessment, including the risk of annoying hard-working technical staff, you should always be prepared to suspend or cancel laboratory work during a lesson if at any point you judge that student behaviour or some other factor makes continuing unwise.

The structure of the book

The book contains twelve chapters. The first three discuss the teaching of the most basic chemical ideas, which underpin all other chemistry topics. These are ideas that will need to be met early on in the secondary school, but will be developed in more sophisticated ways throughout the secondary years. Chapters 4 and 5 discuss other fundamental ideas which build upon those discussed in Chapters 1–3, and provide the basis for the level of theoretical exploration of chemistry expected of many upper-secondary level classes (as well as setting out the theoretical basis that is developed in post-secondary study). The next six chapters draw upon, and often assume knowledge of, these basic ideas, but need not necessarily be taught in the order they appear in the handbook. The final chapter has a slightly different flavour; it reflects upon how decisions are made about which chemistry topics should be a part of the secondary education for different groups of students. For readers in some contexts such decisions will be made centrally by curriculum authorities, but in many teaching contexts there may be important decisions to be made about which science topics should be part of a core curriculum, and which can be seen as either possible enrichment or as important for some, but not all, groups of students.

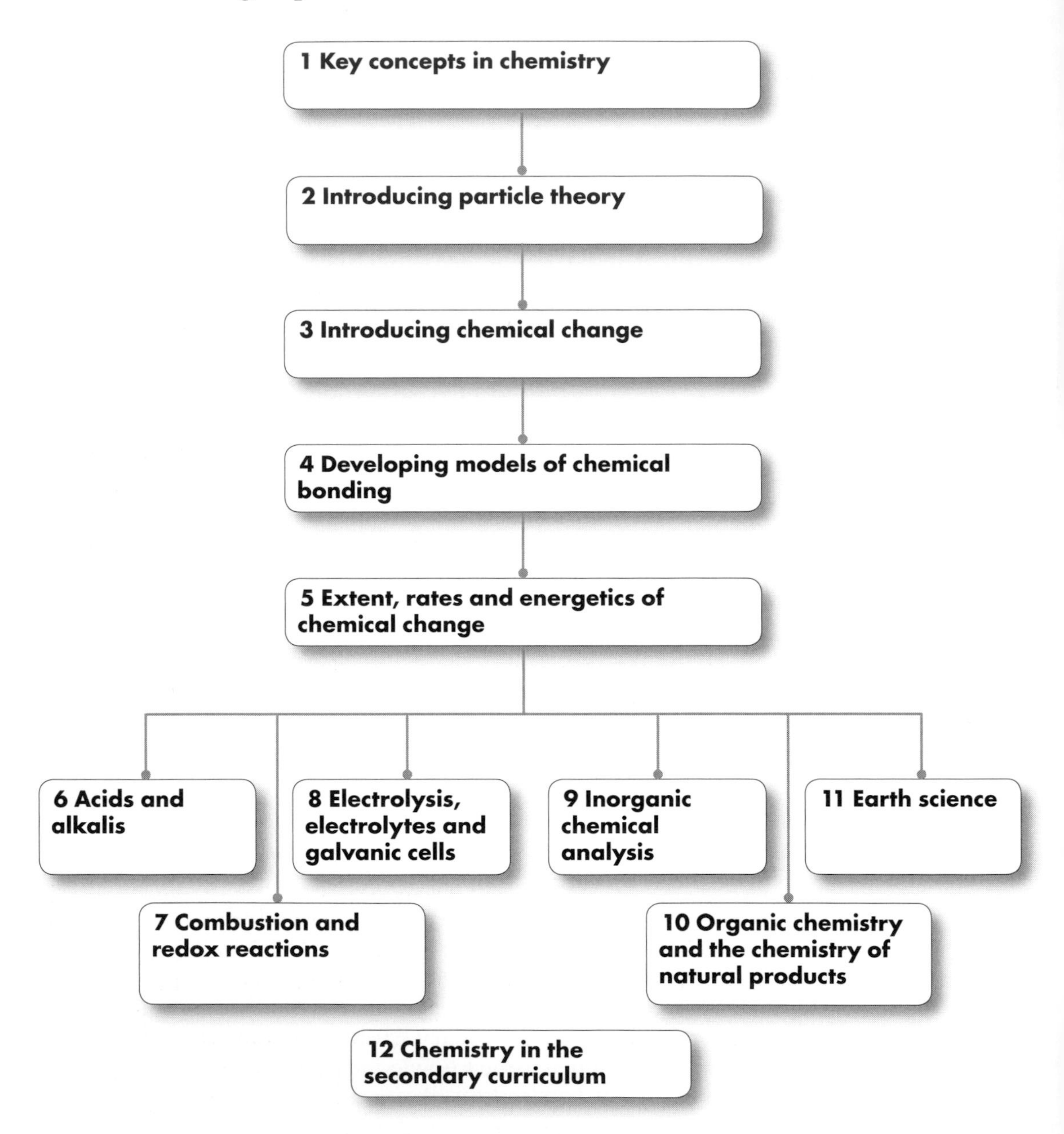

The structure of the chapters

As suggested above, the nature of the eleven topic-based chapters in this book reflects the diversity of topics found in a chemistry course, as well as the different teaching approaches judged worth recommending by the authors who have particular expertise in those topics. However, all of these chapters include a number of common features. Each chapter is divided into sections according to main subsections of the chapter topic, and each includes a schematic diagram of the chapter structure. Each chapter offers guidance on a route through the topic. These chapters also discuss the prior knowledge and relevant experience that students will bring to the study of the topic at secondary level, as well as highlighting the common learning difficulties and alternative conceptions that students are known to have in the topic. Although the authors here can offer guidance on the likely range of prior knowledge and common alternative conceptions to be found in many classes, each group of students is different, and – indeed – each learner is an individual with their own personal understanding of the world and their own way of interpreting what they are taught in chemistry classes.

This is something I am well aware of from my own work, where I have spent a good deal of time asking secondary-age students to tell me about their understanding of various science topics. There are some very common alternative ideas (alternative conceptions or misconceptions) that are found in just about any class. For example, it is very likely that in any upper secondary science or chemistry class you teach, at least some (and in many classes it will be most) of the students think that chemical reactions occur so that individual atoms can fill their electron shells. (If you cannot see what is wrong with that idea, then you may find Chapter 4 quite interesting.)

However, it is also the case that if you spend enough time asking students to tell you their ideas relating to science topics, you are likely to uncover some idiosyncratic notion they hold which is at odds with accepted scientific knowledge and which you have never heard suggested before. Student individuality and creativity is such that this is likely to remain the case even when you have been teaching chemistry for some decades! Some examples of student thinking about chemistry topics can be found on the ECLIPSE project website (full details of which are given in the 'Other resources' section at the end of this introduction), offering a taste of the diverse and often unexpected ways that different learners make sense of the topics they meet in school science lessons. (Note, the icon here is used in the margin where references are made to useful websites.)

As what students already think is a strong determinant of what they will learn in your lessons, it is worth spending some time exploring their ideas and in particular undertaking diagnostic assessment at the start of major topics to check on prior learning and elicit any strongly held alternative conceptions. In an ideal world teachers would have time to talk at length to each of their students. More realistically, ideas can be explored through open-ended classroom discussion that invites students' ideas in an accepting way and explores their consequences (and relationship with available evidence), before 'closing down' discussion to present the scientific models. This presentation of the 'scientific story' can then be done in ways that anticipate student objections and emphasise the reasons for the scientific explanations being adopted in the scientific community – especially where these explanations contradict students' own thinking. Classroom materials for diagnostic assessment have been published in many science topics and some are referred to in this volume.

The chapters here draw upon research evidence, although in common with other handbooks in the series, the chapters do not include in the text references to the research literature. However, recommendations for further reading and suggestions for resources likely to be found useful by teachers are listed at the end of chapters.

Keith S. Taber
Cambridge, 2012

Other resources

Books

The *ASE Guide to Secondary Science Education* (Martin Hollins, editor) offers a broad range of advice and information for those teaching science in secondary schools. ASE Publications.

An introduction to how students learn in chemistry and the nature and consequences of their alternative ideas, as well as a range of classroom resources to diagnose student ideas, can be found in *Chemical Misconceptions – Prevention, Diagnosis and Cure*. Taber, K.S. (2002). London: Royal Society of Chemistry. Volume 1: Theoretical background; Volume 2: Classroom resources.

To explore further ideas for teaching about the nature of science, readers might refer to the companion handbook in this series, *Teaching Secondary How Science Works*. Vanessa Kind and Per Morten Kind (2008). London: Hodder Education.

CLEAPSS provides support for practical work, and in particular health and safely information for school science. CLEAPSS is the source for such useful resources as the *Secondary Science Laboratory Handbook* and *Secondary Science Hazcards* (providing safety information and model risk assessments for handling chemicals). www.cleapss.org.uk

For a research-based discussion of effective practical work in school science see *Practical Work in School Science: A Minds-on Approach.* Abrahams, I. (2011). London: Continuum.

The group-work activity 'Judging models in science' is included in *Enriching School Science for the Gifted Learner.* Taber, K.S. (2007). London: Gatsby Science Enhancement Programme. Available from 'Mindsets Online': www.mindsetsonline.co.uk

Websites

An introduction to the common alternative ideas ('misconceptions') that students often develop in chemistry is provided in Kind, V. (2004). *Beyond Appearances: Students' Misconceptions about Basic Chemical Ideas* (2nd edn). London: Royal Society of Chemistry. Available at: www.rsc.org/Education/Teachers/Resources/Books/Misconceptions.asp

The Royal Society of Chemistry provides a wide range of resources to support chemistry teaching, which may be searched at: www.rsc.org/learn-chemistry

Examples of how students think about and explain chemistry (and other science) topics can be found at the ECLIPSE (Exploring Conceptual Learning, Integration and Progression in Science Education) project website: www.educ.cam.ac.uk/research/projects/eclipse

A range of resources for chemistry teachers recommended by colleagues on two email discussion lists (Chemistry-Teachers@yahoogroups.co.uk and Chemed-L@mailer.uwf.edu) are listed at: http://camtools.cam.ac.uk/access/wiki/site/~kst24/teaching-secondary-chemistry-resources.html

The magazine *Education in Chemistry*, published by the Royal Society of Chemistry, includes a regular feature called 'Exhibition Chemistry' offering 'ideas for chemistry demonstrations to capture the student's imagination'. These are accompanied by a video of the demonstration which can be viewed on the web: www.rsc.org/Education/EiC/topics/Exhibition_chemistry.asp

1 Key concepts in chemistry

Keith S. Taber

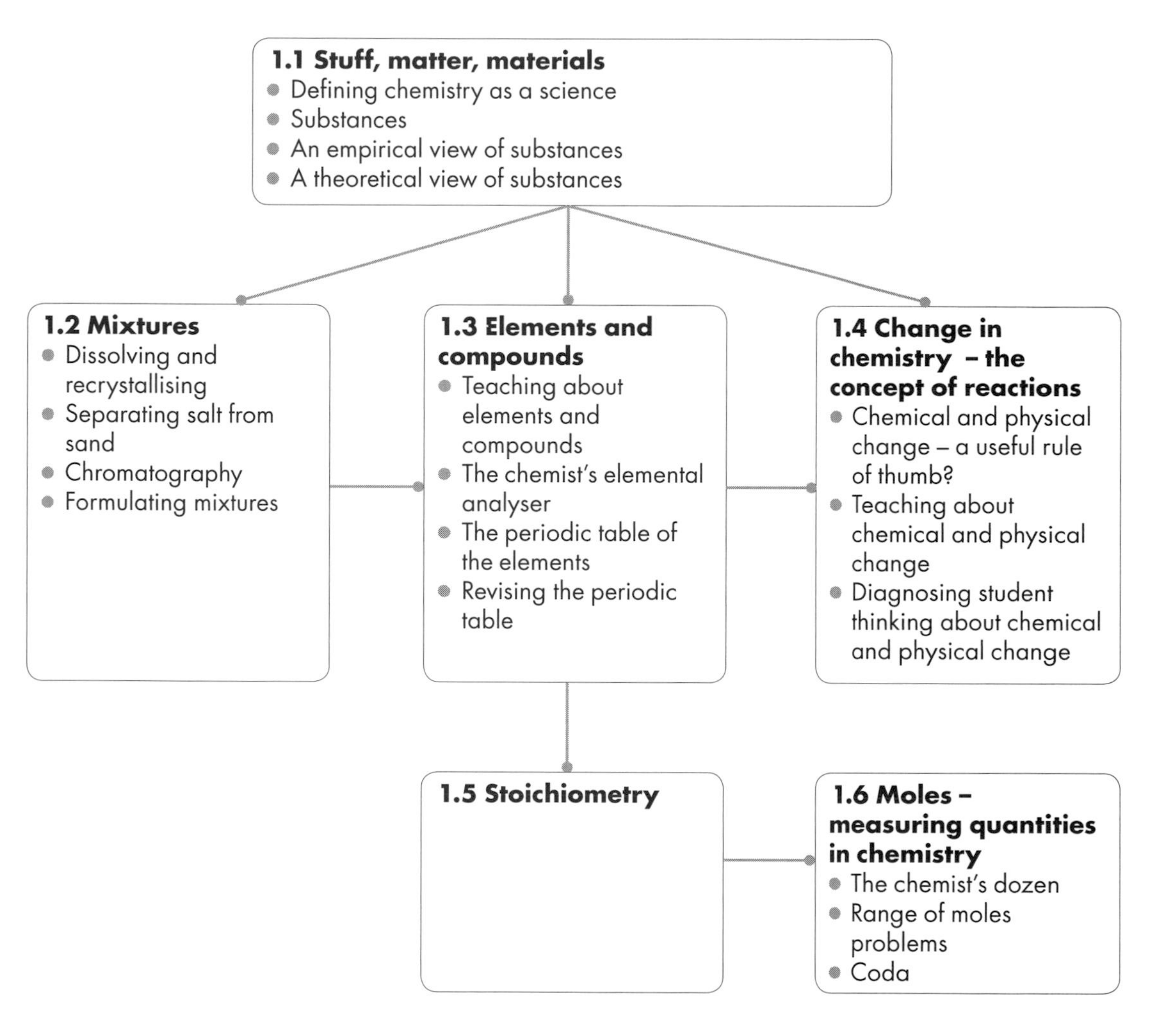

Choosing a route

This chapter will present key ideas that will be taught and developed through a spiral curriculum. Increasingly advanced treatments will be revisited throughout the secondary years in different contexts. This is important because many of the ideas met in chemistry are abstract, unfamiliar and even counter-intuitive. Learners therefore need time to come to terms with these ideas; to explore them and become familiar with them. Few students will be able to master these ideas when first meeting them, so it is important that the ideas are carefully introduced and then later reviewed and reinforced in a variety of contexts. Luckily, chemistry, as a subject, supports this teaching approach as many of the key ideas are relevant to teaching and learning across all topics. A balance will be needed between introducing new materials and revisiting previous teaching. It is easy to overload students' working memories, as they can only keep a limited amount of new information in their minds at any time. However, effective and meaningful learning will require students to relate teaching to their developing understanding of the subject. The key is to recognise that while ideas are still novel (and often somewhat strange) they will place a demand on the learner, but if they are regularly reinforced in various contexts, then over time these increasingly familiar ideas will shift from being an additional load on memory to acting as suitable support ('scaffolding') for new learning.

It is recommended, therefore, that after a major new idea is introduced (the distinction between chemical substances and mixtures of substances, say), you should look for opportunities to review the idea as often as possible over the next few weeks and months. Initially treat the reviews as if dealing with new material (for some students they will be received that way) and over time shift your approach to treating the ideas as taken for granted within the community of the chemistry class. Seeking regular formative feedback ('Jilly, can you remember what we called a substance with only one type of atom?'; 'Vijay, could you remind the class what we mean by a chemical reaction?') will provide guidance on how quickly such shifts are possible with particular classes.

Similar advice would be appropriate to many subjects, but in chemistry we have to deal with two particular complications that do not always apply in other subjects. As some of our key ideas are abstract and cannot be demonstrated directly, it is difficult to explain them clearly without reference to other equally abstract ideas. For example, consider the idea that a chemical change produces different chemical substances. To understand this statement, a student would already need to have a good grasp of the concept of chemical substances, so it would seem chemical substance needs to be introduced first. Yet understanding a chemical substance as something that retains its identity through phase changes (such as ice becoming water) to some extent requires one to already have some notion that such changes are not considered as chemical changes. Of course, a decision has to be made about which ideas should be considered most suitable as a starting point, but students will not be in a position to appreciate these concepts fully when first introduced. So learning in chemistry is iterative, involving some 'bootstrapping' of partially understood concepts, one upon another.

Perhaps this circularity can be avoided to some extent by defining a pure substance differently – in terms of structure at the molecular level. Yet learning about 'particle models', as we will see in this and the subsequent chapters, is challenging for students and we again run into the way ideas are intimately interlinked. Finding a simple 'particle' level definition of a pure substance that would apply unproblematically to all substances (such as neon, oxygen, water, common salt, sulfur, sugar and copper) might be a challenge for any teacher!

This chapter is organised, as it needs to be, as a linear presentation of topics. In one sense this is a logical sequence to follow in teaching, as it builds up the complexity of the ideas. However, while I would advise teachers to try to follow something like this sequence, it is more important to realise that whatever order is chosen, the effective teaching of these ideas will not be achieved in a single pass through.

1.1 Stuff, matter, materials

Previous knowledge and experience

Students will have had experience with materials as part of their primary education, as well as from their everyday experience of the world. Students are likely to be familiar in particular with the ideas of solids, liquids and (probably) gases, although their concepts here may be limited and imprecise (and talking about materials in this way, as being 'solids', 'liquids' and 'gases' may not be helpful, as will be explained in Chapter 2). Some may have met the particle model of matter and may be familiar with simple representations of the states of matter at a submicroscopic scale, though they are unlikely to have a good grasp of the actual scale at which these particles are considered to exist, nor a strong appreciation of the significance of the models.

A teaching sequence

■ Defining chemistry as a science

Definitions in science are notoriously unhelpful. They tend either to be very vague, too exclusive (i.e. seeming to omit things that should be included) or so technical that they are only useful to someone who already has a good understanding of what is being defined.

Chemistry is usually defined in terms of being about the nature, properties and structure of matter, or about the properties and interactions of different substances. While not inaccurate, such definitions are of limited value to students until they have already started to see 'matter' in chemical terms and to understand what chemists mean by 'substances'.

Chemistry is about the 'stuff' around us and about how we can think about this stuff in scientific terms. As a science, chemistry sets about analysing stuff in systematic ways and this often means working with simplifications and generalisations – at least as starting points and 'first approximations'. The science of chemistry involves building up a body of theory: a collection of principles, laws and models that can be used to make sense of, and so explain and predict, the properties of matter.

These are important points, as a student cannot be considered to understand chemistry in any depth unless she or he appreciates that, as a science, its central 'contents' are not the phenomena in nature, but the theoretical constructs people have developed to explain those phenomena. Most chemists are very interested in the phenomena themselves – we tend to be fascinated by the colour changes, the ability to produce smells and bangs and so forth. Students usually like this aspect of the subject, although for most the original fascination with smells and bangs is unlikely to last throughout secondary education if it is based purely on observing phenomena. What makes chemistry a science, and makes it a science that continues to fascinate students, is the ability to organise and explain the phenomena in terms of models of great explanatory strength; models that with some modification can be applied across the wide range of substances and reactions met in school science (and of course beyond).

■ Substances

One of the major simplifications adopted in chemistry is to focus on substances. This is a simplification because, in our normal environment, few of the materials we commonly come across are strictly 'substances' in the chemical sense. Figure 1.1 sets out the relationship between some key ideas in chemistry. So where 'matter' is a general term for stuff, we tend to use the term 'materials' for well defined samples of stuff that we can work with – glass, wood, sodium carbonate (washing soda), poly(ethene), diamond, sea water, paint, etc. From a technological perspective, these materials may have a similar status (different types of stuff that can be obtained, worked in various ways or used in different applications), but to a chemist they have rather different status. Materials may be pure substances like diamond or sodium carbonate, or mixtures such as air or paint. The use of the term 'mixture' is another simplification, as something like wood is more complicated – although it contains many substances, they are not simply mixed in a random way, but built into a complex structure. Some manufactured materials are also composite, such as 'fibreglass', which contains fibres of glass embedded in a polymer (plastic).

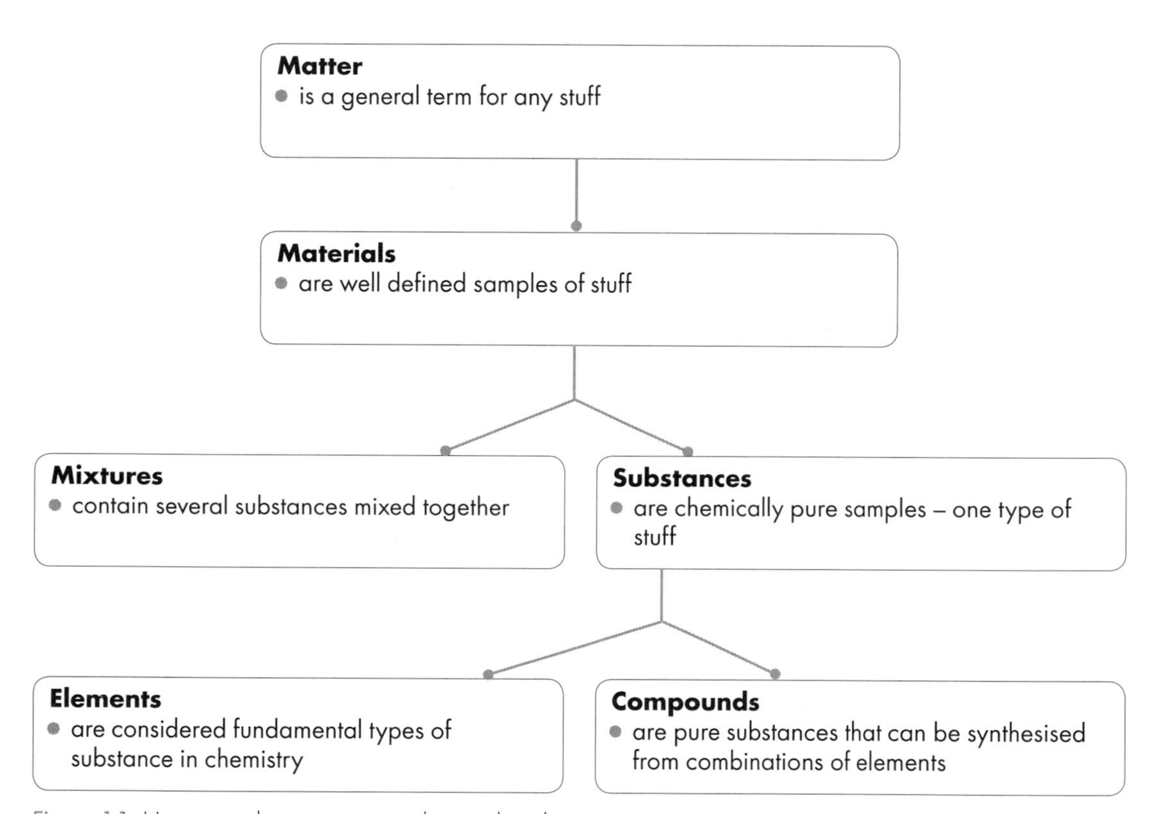

Figure 1.1 How some key terms are understood in chemistry

So most common materials in our environment, such as air, sea water, earth, wood and even steel, are not substances. This is a simple point but one which is not trivial for students. A key issue here in the minds of some students is the notion of 'natural' materials. For a chemist, natural products are those that derive from animal or vegetable sources, but are not considered to make up an intrinsically distinct type of stuff from other materials. (However, chemical terminology still retains vestiges of earlier thinking that living matter had some special vital essence, in our use of the terms 'organic' and 'inorganic'.)

For many lay people natural materials are considered to be intrinsically better (for example, safer) than 'synthetic' or man-made materials. The assumption seems to be that nature knows best, and man less so. From a scientific perspective, man is part of nature and any material that can be made by man is just as natural as anything secreted, excreted or extracted from a living organism. Indeed there are many berries, fungi, insects and amphibians that produce substances which are harmful or even lethal to people, whereas most synthetic products produced by chemists are subject to extensive safety testing before being allowed onto the market. Many natural products that were once difficult to obtain (for example, those requiring expensive processes to extract and purify tiny quantities of a substance present in living things) can now be synthesised much more effectively, and of course their chemical behaviour is unrelated to their origins.

As teachers, we need to be aware that many of our students may have absorbed at some level the notion of 'natural-good, synthetic-bad' and be prepared to challenge the supposition without ignoring or underplaying how many synthetic materials can be used to do harm, in weapons for example, and may bring significant environmental costs in manufacture or disposal.

A closely related idea is that of purity. When buying orange juice to drink, for example, we expect it to be 'pure' in the sense of just being material squeezed from oranges, and not including dead flies, sawdust or the farmer's finger nail cuttings. To assure the potential buyer of this, the manufacturer may well claim to be selling '100% pure orange juice', and in the context of selling and buying a drink this makes perfect sense.

However, students will need to be taught that no matter how pure our orange juice is in terms of only being juice from oranges, it is far from being a 'pure substance' in chemical terms. Orange juice is mostly water, but contains a wide range of other substances including fruit sugar, vitamin C, citric acid, various amino acids and flavonoids that make oranges taste different from lemons or grapefruits. Chemically, orange juice is a mixture of a lot of different substances, even though it is a natural product. A key distinction to be introduced and reiterated in teaching the subject, then, is that between materials which can be understood in everyday terms (orange juice is a different material from the glass, paper or ceramic cup we may drink it from) and the constituent substances that chemists analyse such materials into.

So the task of the chemistry teacher is to find a way to justify considering iron but not steel, methane but not petroleum, cellulose but not wood, sucrose but not honey, vitamin C but not orange juice, and so on, as substances. As this is a difficult distinction for those new to the subject, it is useful to have a wide range of examples that can be used when explaining the idea to the students.

However, the examples by themselves only seem persuasive to those of us who already appreciate the difference between materials *in general* and those *particular* materials that are substances. Students will have to have good reasons to see this as a meaningful and important distinction. There would seem to be two different approaches to thinking about what we mean by substances in chemistry: one of these is highly empirical and the other more theoretical. Both approaches offer challenges for the teacher but also a considerable opportunity to teach about the nature of science ('how science works').

■ An empirical view of substances

From a chemical perspective, materials are either pure samples of a single substance or a mixture of substances. When a material is a mixture, it can in principle be separated into its components. There are a number of common techniques that can be used to separate different classes of mixtures, and a mixture that can be separated by one separation technique will not necessarily be separated by another. For example, if sand is mixed with salt, it can be separated by dissolving (the salt) and filtering (Figure 1.2), but this process

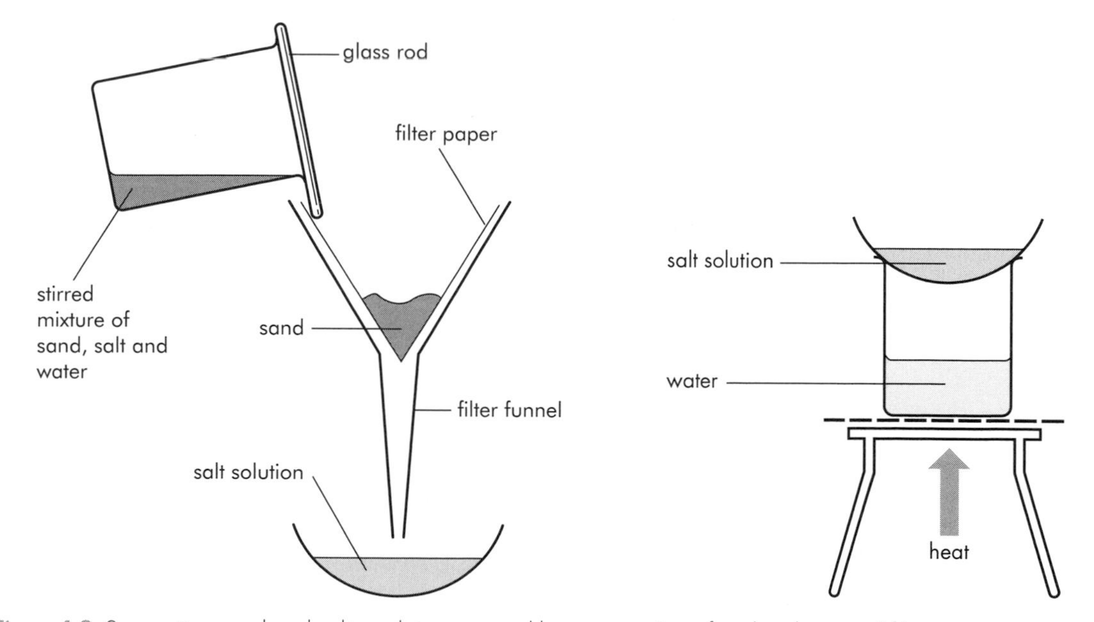

Figure 1.2 Separating sand and salt – salt is recovered by evaporation after dissolving and filtering

will not work if the sand is mixed with iron filings. Conversely, sand can be separated from iron filings by using a magnet, but this will not have an effect on a sand/salt mixture. Laboratory exercises in this area can easily become somewhat artificial: for example giving students a deliberately prepared mixture of sand and salt for them to separate. That is different from being able to take an unknown material, find out if it is a mixture and – if so – separate it into its components. That was the kind of challenge faced by Marie and Pierre Curie when they carried out their work identifying new chemical substances (the elements radium and polonium).

Figure 1.3 Marie Curie has the distinction of having been awarded Nobel Prizes for her contributions to both chemistry and physics

It is often possible to recognise a mixture because mixtures usually do not have a distinct temperature at which they melt/freeze or boil/condense (see Chapter 2). Having identified a material as being a mixture, it is then possible to subject it to the battery of separation techniques available to the chemist. So, for example, consider a liquid that was considered to be a mixture (because a sample had been found to boil over a range of temperatures). It may be that if the liquid is heated, one component of the mixture will evaporate, leaving a solid residue (for example, this would happen with salt solution). However, it is also possible that all of the mixture would boil off. If a fractional distillation apparatus were set up (Figure 1.4) and the components of the mixture were found to have very different boiling points, it would be possible to separate them by carefully collecting condensate from vapour produced at different temperatures. If, however, the components have similar boiling temperatures, separation by this technique will prove difficult. Perhaps another technique, such as a form of chromatography, might separate the components in such a case.

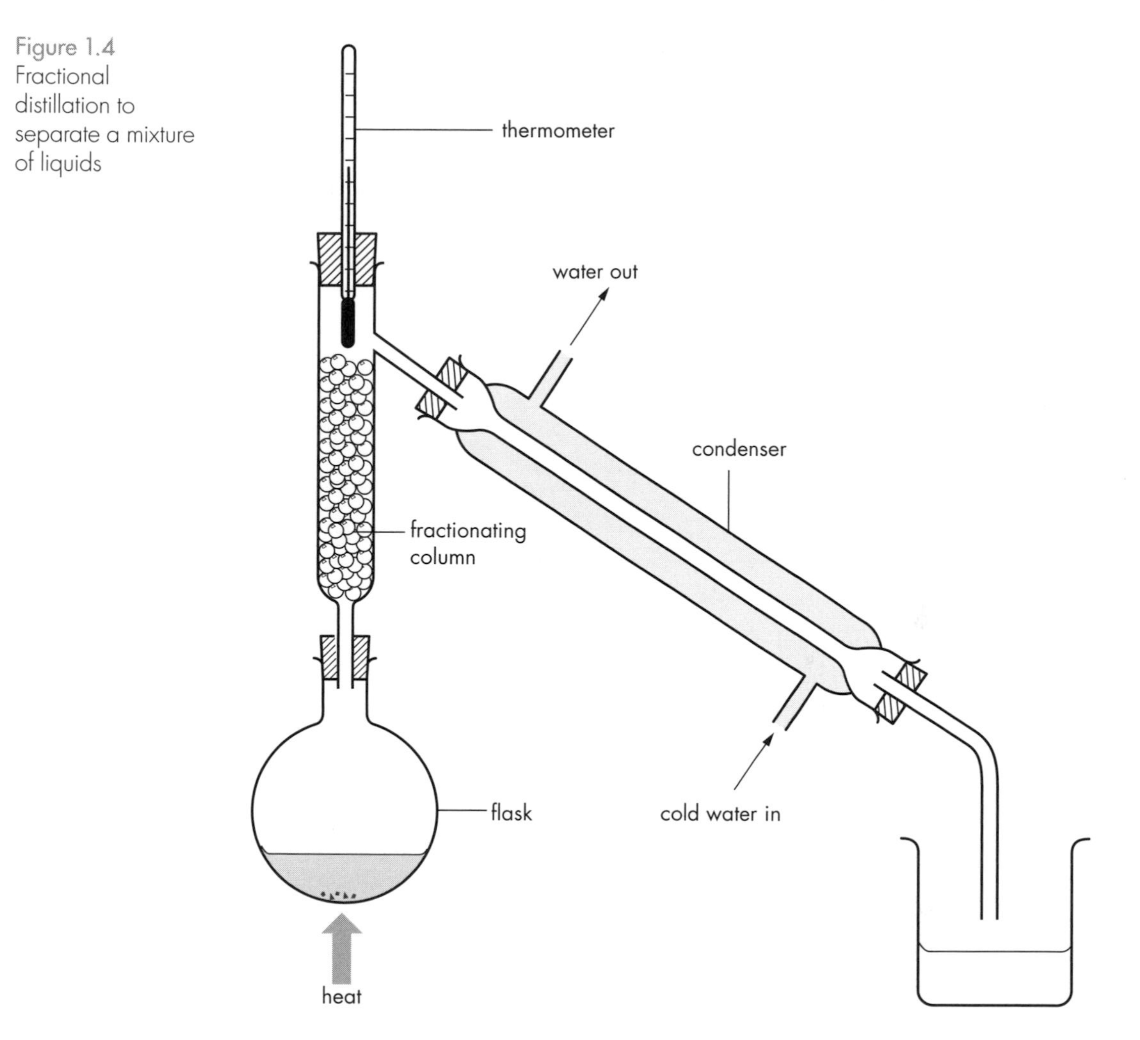

Figure 1.4 Fractional distillation to separate a mixture of liquids

■ A theoretical view of substances

From a theoretical perspective, we would say that a single substance is one that has a homogeneous chemical composition. The problem is that many mixtures, such as air, sea water, orange juice and bronze, often appear uniform enough. We say they are homogeneous mixtures.

Once students have learnt about basic particle theory (Chapter 2) and then progressed to learning about how many substances consist of molecules or ions (Chapter 3), it becomes much easier to communicate the chemical notion of a (pure) substance. Distilled water, but not sea water, consists just of water molecules; sodium chloride consists of the same repeating pattern of sodium and chloride ions throughout the crystal; copper consists of a repeating lattice of copper ions with associated electrons. In contrast, in brass, both copper ions and zinc ions are present, but there is no regular repeating pattern such as is seen in the representation of sodium chloride (Figure 1.5a). It is clearly possible to show students diagrams of materials as chemists imagine them at the scale of molecules and ions to illustrate the distinction.

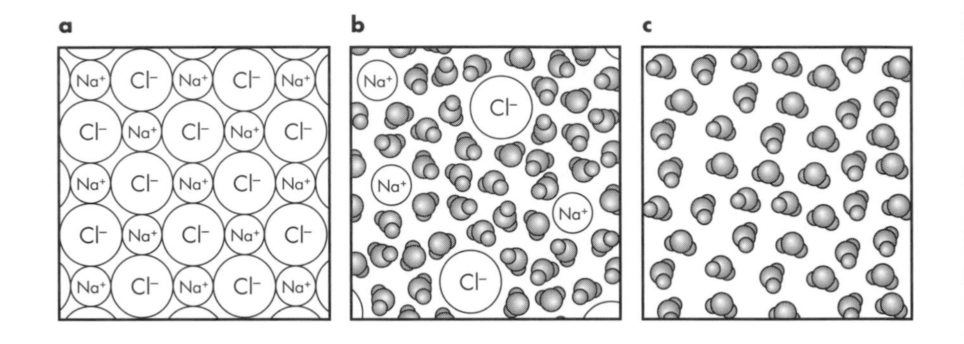

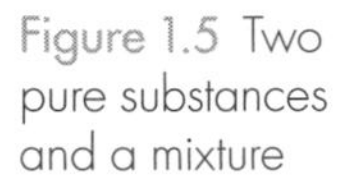
Figure 1.5 Two pure substances and a mixture

Figure 1.5 illustrates this point with the example of salt and water. To the teacher, the salient points to note about these images are likely to be clear: water is a pure substance because it consists only of one type of entity (water molecules, part (c)); sodium chloride is a pure substance as there is a constant composition (the alternation of Na^+ and Cl^- in part (a)) throughout the material. The solution (part (b)) shows the sodium and chloride ions mixed up with molecules of water. To the student, the distinction between pure substance and mixture may be less clear: the order in NaCl is not present in water (because it is in the liquid state, not the solid state) and – unlike water – sodium chloride does not comprise one type of entity at this submicroscopic scale.

The images in Figure 1.5 are, of course, representations of models that we use in chemistry, and such representations offer challenge and opportunity. The challenge is in part that (as discussed below) molecules and ions are not quite like things we can draw as discretely bounded lumps; in part the difficulty of showing three-dimensional, dynamic scenes in a flat image; and in part that in using introductory models we tend to need to ignore complications. So part (c) of Figure 1.5 shows that water contains only one type of particle, water molecules. Actually, even pure water contains a very small proportion of hydrogen and hydroxyl ions, $H^+_{(aq)}$ and $OH^-_{(aq)}$, which are important for some of its properties (Chapter 6). Yet chemists do not consider that this fact makes water a mixture rather than a single substance and it is not a complication we would wish to introduce when students first meet these ideas.

The opportunity here is to recognise that the limitations of our models actually reflect an important aspect of chemistry as a science. One of the key ways in which science works is through the development of useful simplifications. It is not 'wrong' or 'untrue' to say that water only contains water molecules: it is a useful 'first-order' approximation that is appropriate for many purposes. When students are encouraged to think about representations in this way, then the later introduction of increasingly sophisticated models, more suitable for some purposes, need not be seen as contradicting what students have previously learnt.

1.2 Mixtures

Previous knowledge and experience

Students will have much experience of mixtures of objects as well as examples of sets of the same type of thing in their everyday life. An approach building upon familiarity with such everyday examples is recommended.

A teaching approach

A good approach is to start with some models of mixtures at a level that all the students in the class can appreciate.

For example, some glass jam jars can be set up with various contents:

- steel ball bearings of the same size and appearance
- marbles of the same size and appearance
- marbles of different sizes and colours
- marbles of the same size, but different colour inserts
- a mixture of ball bearings and marbles of a single size and appearance
- a 'fruit and nut' mixture, such as sultanas, currants, peanuts and brazil nuts
- mixed nuts
- peanuts (unsalted and unspiced)
- salted peanuts
- honey-coated banana chips.

Clearly, these are just suggestions and the actual examples may be varied. You may wish to add additional examples or in large classes have two or three jars of each example. The jars should be labelled with numbers or letters rather than descriptive names. Students should work in small groups (pairs or threes) and should attempt to complete a table to show which jars they think contain mixtures. Importantly, students should be asked to agree in their groups before recording an answer. Where the group does not initially agree, or is not sure, they should discuss the example and try to come to an agreement. Warn students that they may be asked to give reasons for their decisions. The focus on dialogue is important both because research suggests such talk supports learning and also because part of the role of the science teacher is to introduce learners to the nature of the 'discourse' of science, which involves examining evidence, presenting and discussing ideas, developing agreements and seeking to persuade others of your ideas. While the teacher and textbook are likely to be better informed about chemistry than the students, and part of teaching involves presenting and justifying currently accepted scientific models, too much authoritative teaching (the teacher telling, the students listening) gives a poor impression of what doing science is really like. Effective science teaching therefore involves a balance between developing students' thinking and argumentation skills, and helping them learn accepted scientific ideas. This is especially important in chemistry where students will meet a progression of models during their time in school classes, as they need to appreciate that scientific models are human inventions that sometimes have limited ranges of application, and in some circumstances need to be replaced by more sophisticated thinking.

After the class has completed the exercise, hold a plenary session where you explore the decisions students made and their reasoning (especially in any cases where groups did not agree). There is clearly room for some disagreement about what counts as being similar enough not to be considered a mixture (this can be used to reinforce the point about scientists using models and useful simplifications). If students have not noticed that the peanuts in one jar are salted, point this out, and ask if it makes a difference to their classification.

In the case of the banana chips, point out that the contents are not just banana, as these are honey-coated banana chips. Ask students whether that makes a difference to their decision. I suggest this latter example because of an experience I once had in a 'health food' shop. On a label I read the words 'a mixture of honey-coated banana chips'. Being a science teacher, I asked the shop assistant what the honey-coated banana chips were mixed with. The shop assistant did not understand the basis of the question, but I concluded that the shop was selling honey-coated banana chips mixed with other honey-coated banana chips – not a good model of a mixture to a chemist. This point is worth emphasising to the class when they meet particle diagrams of compounds. It is quite common for students to think that if there is more than one element represented, they are dealing with a mixture, even if there is only one type of unit (molecule) present (Figure 1.6).

Figure 1.6 Mixtures? Both the honey-coated banana chips and the compound are best understood as having one basic type of unit.

This also raises an important issue about scale. The jars with the marbles and other everyday objects are mixed at a scale where the individual entities can be observed directly. Below, a similar activity will be suggested with mixtures of substances. Some of these can also be recognised as mixtures under close observation (sand and salt, for example); but others can only be understood to be mixtures from the perspectives of a particle model, where the particles being mixed are much too small to be directly observed. Salt solution, for example, does not look like it is a mixture of different things, but can be separated into salt and water by a suitable separate technique.

For those jars that are considered to contain mixtures, ask students whether they could be changed into something that was not a mixture. This would mean using a separation technique. In these examples, this could be quite crude, as tweezers or even fingers could do the job. You might want to start demonstrating this process just to reinforce the idea. With some groups you might decide to select a couple of students who like to be doing things with their hands to separate the marbles and ball bearings into two different jars. Asking them to do this by transferring between jars with tweezers might add a little challenge! With other groups, simply making sure they recognise that such a separation is possible will suffice. Modelling a process such as filtering would be possible for some mixtures, using a suitable gauge garden sieve.

The next activity moves from using a model to actually considering materials, so use boiling tubes sealed with bungs or corks (rather than everyday containers like jam jars) to indicate that you are looking at the kind of examples of interest to chemists.

Students are again asked to judge which samples are mixtures, but this time they are asked to consider examples such as:

- sand mixed with salt
- sugar mixed with salt
- salt solution
- black ink
- air
- iron filings mixed with sulfur powder (flowers of sulfur)
- iron filings mixed with copper turnings
- water.

There is an opportunity here for a range of practical work that will allow students experience of using basic laboratory equipment. As always with practical work in science, it is important to ensure students' minds are working as well as their hands – and so they are thinking about the scientific ideas behind the activities. This means differentiating activities for particular classes, or even for particular groups within classes. Ideas could include those detailed below.

■ Dissolving and recrystallising

Showing that different salts have different solubilities in water is a useful exercise. Students can be asked to modify a simple 'add solid to water, stir and observe' activity to increase the challenge.

- Can they provide a quantitative measure of solubility?
- Can they see whether water temperature affects solubility?
- Can they see whether the amount of material dissolved changes over time?

A given amount of solvent will dissolve a specific amount of a particular solute at a certain temperature. Often, though not always, solubility increases as temperature increases. It may sometimes take a noticeable time for a solution to become 'saturated'.

Recrystallisation is used as a purification technique. Here the challenge may be to make crystals from an initially fine-grain (powder) sample, by using a dropping pipette to add samples of solutions to a watch glass or Petri dish, and waiting for evaporation. Using hand lenses or low-power microscopes can add to the excitement of making crystals.

■ Separating salt from sand

This activity (Figure 1.2 on page 7) involves a sequence of steps: adding water to the mixture; stirring; filtering; evaporation of the solvent from the solution. When I carried out this activity with first-year secondary students, a DART (Directed Activity Related to Text) was employed to ensure students thought about the logic of the sequence of steps. The DART consisted of a set of labelled diagrams showing the steps and their end states. Groups had to cut out the images and sequence them correctly before starting the practical work. A similar process could be completed as a whole-class activity on an interactive white board, with the final agreed sequence left on display while the students work.

■ Chromatography

The theory of chromatography is not normally studied at secondary level, but the basic idea is straightforward (and some students are likely to ask about the mechanism). Chromatography relies on the variations in 'affinity' between different substances. Although various materials may be used, the principle is the same in different forms of chromatography. In a very basic set-up, a liquid is allowed to soak into and move up a paper rectangle onto which a sample has been spotted. The degree to which the sample moves will depend upon how the substance interacts with the paper and the fluid. If the sample is a mixture, then some components may adhere less strongly to the paper, or dissolve more readily in the liquid, and so move further than others.

The name *chroma*tography refers to the colours that are often seen, although the technique works regardless of whether the sample is coloured or not. This probably seems an obvious point to readers, but will not be obvious to some students. When direct observation does not allow us to see the substances being separated, more sophisticated techniques are needed. In 'GCMS' (gas chromatography–mass spectrometry), the mixture is separated when a vaporised sample passes through a long column with a carrier gas, before the components are identified by mass spectroscopy (a technique discussed later in the chapter). The most advanced forms of chromatography are very 'high tech', with gas–liquid chromatograms costing tens of thousands of pounds. However, separating the dyes in coloured inks using filter paper is a traditional school practical (Figure 1.7) that most students enjoy and which shows that we cannot always tell what is a mixture simply by inspection.

Good separation of coloured inks is usually possible with alcohol-based solvents (such as butan-1-ol), but these are harmful and require good ventilation (they are also flammable). Acceptable results can sometimes be obtained using water as the solvent, which is more suitable for younger students. However, this does mean finding suitable pens to use in advance, as many permanent inks will not be appropriate.

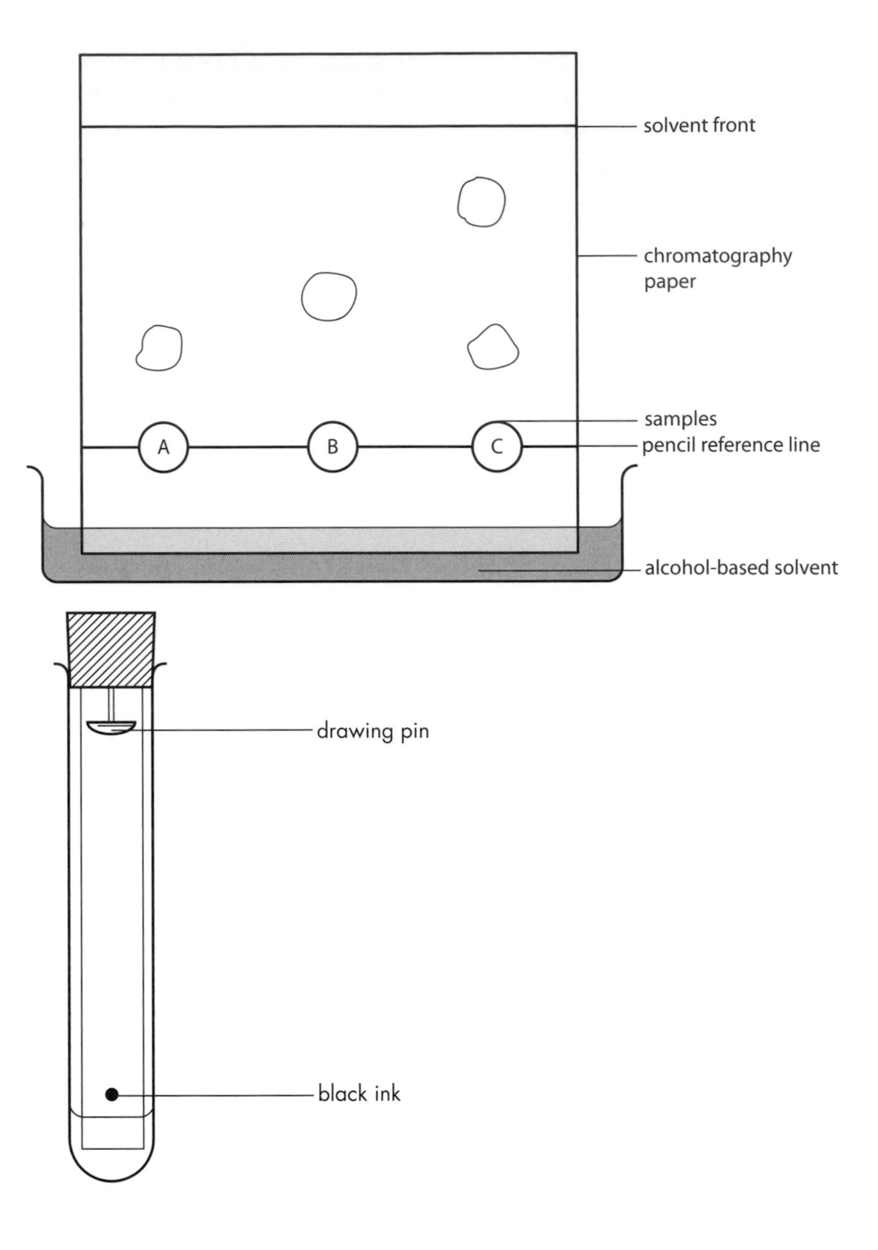

Figure 1.7 Paper chromatography

Teachers can set up a 'forensic' context for practicals such as this, with a sample of ink from a crime scene and samples from several suspects. While this may seem artificial, students often enjoy identifying the 'offender' and it is a good example of how chemistry is used in forensic science.

■ Formulating mixtures

Practical work to demonstrate the separation of mixtures reflects an important aspect of chemistry, but could give the false impression that chemistry is all about breaking down and analysing, when an equally important role of chemistry is in producing new materials. Some of this involves synthesising new compounds, but chemists also play a very import role in formulating new mixtures suitable for particular purposes. Many cleaning materials, cosmetics, adhesives, paints and varnishes, medicines and foodstuffs, for example, can be considered as designed mixtures – where chemists have carefully researched the effects of mixing various ingredients in different proportions to find the composition with most suitable properties for a particular purpose. With a carefully chosen range of potential ingredients, it would be possible to organise investigative work to explore how the qualities of (for example) a skin cream depend upon the formulation used.

1.3 Elements and compounds

The ancients referred to air, earth, water (along with fire and the ether) as 'elements' because they considered these substances basic. We still hear the air we breathe and the water we drink referred to as 'elements' (for example, in some advertising campaigns), as this seems to tap into something in the popular psyche. However, in chemistry, the term 'element' is reserved for a limited number of substances. In Figure 1.1 (page 5), two major distinctions were shown:

1 between materials that are mixtures and others that are substances
2 between substances that are elements and those that are compounds.

This is potentially a very difficult idea for students to grasp – if substances are pure, single types of stuff, then how can some of them be compounded from others? Figure 1.8 overleaf revisits Figure 1.1, but acknowledges this issue. The distinction beneath the dashed line is likely to make more sense to students once they have learnt about atoms, ions and molecules.

Indeed the dashed line in Figure 1.8 highlights one of the reasons that learners find the element concept a challenge. The modern scientific notion of element is a kind of hybrid concept, containing at least three somewhat distinct facets. This is shown in Figure 1.9, which represents the three overlapping meanings of 'element'.

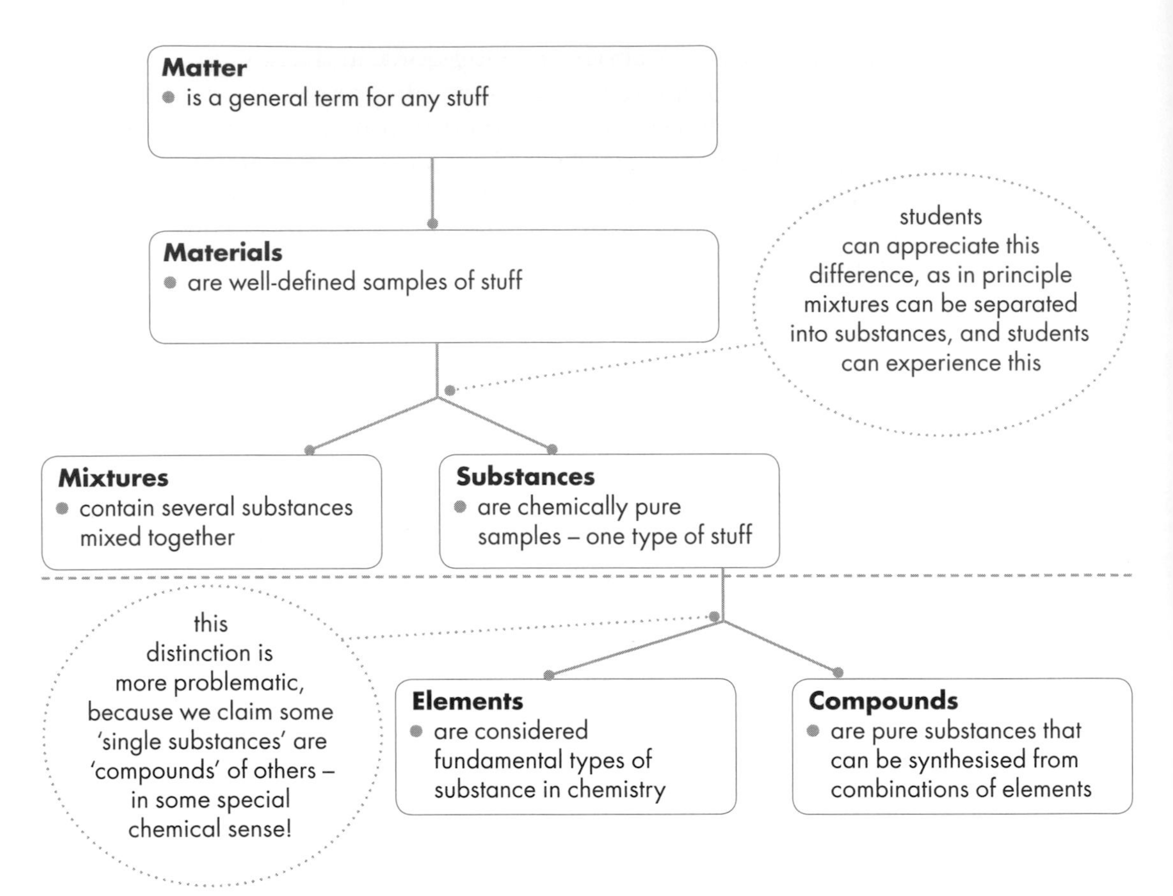

Figure 1.8 The distinction between elements and compounds is important, but may be challenging for students.

Figure 1.9 Three facets of the 'element' concept

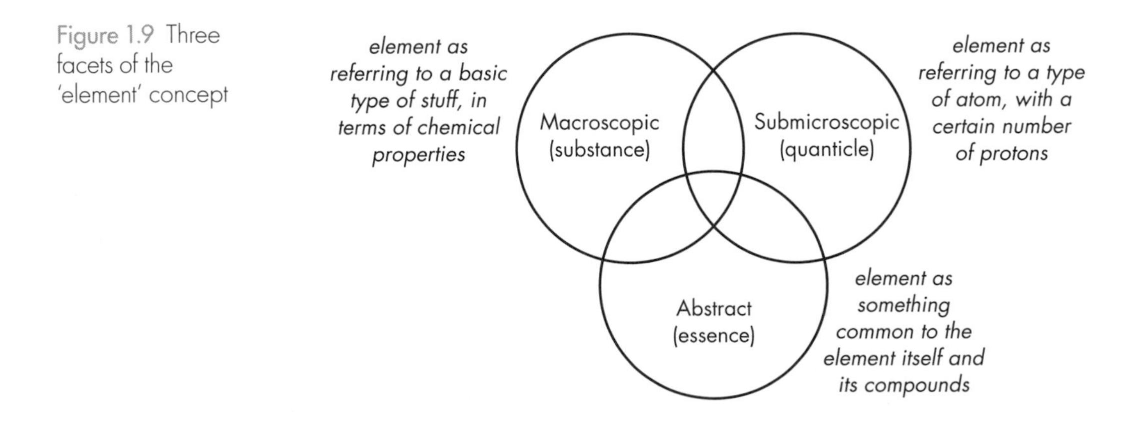

So here we are talking about an element as a basic kind of substance. In terms of chemical properties, an element is a substance that (unlike a compound) cannot be converted to any more basic substance by chemical means. However, the periodic table of the elements is often presented in terms of properties of atoms and in particular the distinction in terms of atomic number, in other words the number of protons in a nucleus defining the element (hydrogen: 1; helium: 2; lithium: 3; etc.). We are working here with two complementary meanings for the idea of an element, one at the (macroscopic) level of phenomena we can demonstrate to students (substances and their reactions), the other deriving from a theoretical model in terms of conjectured submicroscopic entities ('quanticles'; read more about these in section 1.5).

However, there is also a sense in which an element is considered to be present, in a virtual or potential sense, within its compounds. This use is more common among French-speaking chemists; in the English-speaking world we normally consider it quite inappropriate to suggest that sodium is somehow present in sodium chloride or hydrogen in water. Yet, of course, chemical formulae ($NaCl$, H_2O, etc.) tell us that the compounds somehow 'contain' the elements.

These distinct meanings of the same term are unfortunate and can lead to confusion among students. For an element (as a substance) has a set of properties that would distinguish it from either a collection of its atoms or from one of its compounds. So the substance sodium has quite distinct properties from a large number of discrete sodium atoms, as well as having different properties from the other substances that are its compounds: sodium oxide, sodium chloride, etc.

An additional complication, and one that arguably is best put aside when first teaching about elements, is the existence of allotropes: where samples of the same element (in terms of the atoms present) can exist in different physical forms with different properties. So, for example, solid samples of carbon can exist as diamond or graphite (or indeed other forms) at the same temperature and pressure. These allotropes of carbon have very different physical properties (hardness, electrical conductivity). As another example, common oxygen (dioxygen, O_2) and ozone (O_3) are both gaseous forms of the element oxygen (both contain only atomic nuclei with eight protons), yet they have different chemical properties.

Diamond and graphite are certainly different materials, although both are forms of the same element, carbon, and so are generally considered to be the same substance. Yet if dioxygen and ozone have distinct chemical properties, it could be argued that they are actually different substances, although both forms of element number 8, oxygen. This shows the limitations in the set of concepts and demarcations that chemists use to describe and classify the different substances found in nature. This complication is unhelpful when introducing the key ideas of substance and element, but when it is raised (by a particularly astute student) it offers a context for discussing an important aspect of the nature of science. Our chemical concepts – element, substance, acid, oxidising agent, etc. – are human constructions designed to help us make sense of the patterns we find in nature. Some of these human constructions fit unproblematically upon the patterns we observe, but nature is often too subtle and nuanced for our simplest classifications and definitions to always work.

■ Teaching about elements and compounds

The distinction between elements and compounds is one of the most fundamental ideas in chemistry, so it is important that students are able to appreciate it. However, there is no immediate way to distinguish elements from compounds, as both are single substances. Pure samples of either elements or compounds may be reactive or inert and will give the sharp pattern of phase transitions expected of single substances. That is to say, unlike a mixture, they will change state at a sharp melting/boiling temperature (Chapter 2). So, unlike the distinction between single substances and mixtures, there is no simple way of demonstrating whether a substance is an element or a compound.

Having some sealed tubes with labelled samples of different elements and compounds can illustrate this well. Avoid materials such as alkali metals and phosphorus that must be kept in a protective medium as this could confuse younger students.

The common definition that an element cannot be changed into anything simpler by chemical means needs to be treated carefully. It is correct as long as we consider an element to be 'chemically simpler' than a compound – but to a student that may seem a rather circular argument. The notion that there are some substances that are more basic, and which can combine to give all the other substances, is fine in principle. Yet the historical development of this area of chemistry shows just how much evidence and argument was needed to establish our modern understanding of the elements.

Most students will accept the basic principles here and there may be a temptation to simply present the idea of chemical elements as if it is unproblematic. However, it is probably better that students appreciate something of just how much hard work was involved in establishing this basic idea: that the careful and difficult experiments and measurements of a good many scientists over a long period of time led slowly to our modern understanding. If time allows, it may be worth showing some pictures of early chemists at work in their (often makeshift) laboratories, to show that much that we take for granted in science today was once cutting edge and the basis of intense debate. This can give students a better feel for what working in science is like than simply presenting the outcomes of previous scientific work as in the sanitised form in which ideas are often reported in textbooks.

This is again an area that will benefit from an iterative treatment during the secondary years. Once students have progressed to learning about atoms (Chapters 2 and 3), it becomes possible to shift to a definition of an element as a substance with only one kind of atomic core ('kernel') present. This ignores isotopes, of course, and so this is an area where it is useful if students have started to develop a feel for how we are using models in chemistry (Table 1.1).

Table 1.1 Complications of defining an element

Way of thinking about elements	Comment
Most basic kind of substance	Could only be useful if we have a criterion for what makes one substance more basic than another.
A substance that cannot be broken down into anything simpler by chemical means.	Again, this would only be useful if we already know what we mean by simpler.
A substance that contains/is made up of only one type of atom.	Only useful once students know about atoms. Few elements (the noble gases) actually contain discrete atoms in their structure. Does not acknowledge isotopes.
A substance that contains only one type of atomic core.	Acknowledges that most elements exist with (covalent, metallic) bonding which means they do not contain atoms as such. This is a more complex idea, though. Does not acknowledge isotopes.
A substance that contains only one type of atomic nucleus.	Avoids the issues of whether there are atoms in most elements. However, students need to be aware that the nuclei are balanced by electrons in the structure. Does not acknowledge isotopes.
A substance that contains atoms/nuclei with the same number of protons.	This definition allows for isotopes, but is more abstract than referring to 'one type of atom'.

Table 1.1 presents some of the advantages and disadvantages of different ways of defining the notion of an element. Clearly a parallel list could be drawn up for compounds. The most useful approach to take will depend to some extent on the class and in particular on how much relevant prior knowledge students bring to the lesson. Ideas based on notions of atoms, or subatomic particles, can clearly only be introduced after students have learnt about atoms and atomic structure, respectively.

The existence of isotopes, versions of an atom with the same number of protons but a different number of neutrons, adds a level of complication. Different isotopes have atoms/nuclei that are in this sense different, but are distinguished by a property which is not considered central to defining the element. Generally, samples of an element will contain a mixture of different isotopes and so the relative atomic mass (the mass of an atom of the element relative to a standard, usually an atom of carbon-12, i.e. with 12 nucleons) quoted for the element usually reflects this. For example, the relative atomic mass of chlorine is about 35.5, reflecting a roughly 3 : 1 mixture of chlorine-35 (17 protons, 18 neutrons) and chlorine-37 (17 protons, 20 neutrons). Isotopes (like allotropes, discussed above) represent complications that teachers need to decide when to introduce, and this may mean initially leaving them aside until students are confident with the more basic ideas.

However, it should be recognised that even the explanations which do not draw upon atomic ideas are abstract. So to consider a compound as a single substance which can be broken down into more basic substances is going to be challenging for many learners. A possible teaching model here might be a jar of peanuts still in their shells (i.e. groundnuts). This can represent a single substance, as the jar only contains one kind of object. However, it is possible to process the nuts to break up these objects and separate them into kernel and shells. However, care is needed in using such an analogy to ensure that students appreciate that the kernels and (now broken) shells represent elements that were joined into a compound. Both the kernel and shell are actually made of complex materials and students need to appreciate that they are being used as components of a model.

A diagnostic task to explore students' understanding of the distinctions between – and their ability to recognise – elements, compounds and mixtures in diagrams showing molecules (entitled 'Elements compounds or mixtures') can be downloaded from the Royal Society of Chemistry website (see the 'Other resources' section at the end of this chapter). You may wish to use this to elicit students' thinking, for example if you wish to check the understanding of students who should already have mastered these ideas. Many upper secondary level students are likely to demonstrate confusion or misunderstanding of these key distinctions, indicating the need to review this topic before continuing to more advanced material.

■ The chemist's elemental analyser

Of course chemists do have a device that can determine whether a substance is an element or a compound, and this is the mass spectrometer. The mass spectrometer will break up a molecule (whether of a single substance or from a mixture) to produce discrete atomic ions, which can be detected by their different masses (or, technically, the mass to charge ratio, but most ions produced have unit charge). In many applications of mass spectrometry, molecules in a sample are broken into various fragments that collectively give clues to the overall structure. However, if the sample is treated so that it is fully atomised, then the spectrum produced shows the range and relative numbers of different atoms present. So atomic mass spectrometry acts as an elemental analyser.

The theory of the technique is beyond most secondary-age students and is a 'sixth-form' topic, but the existence of a machine that can decompose a compound and show that at some level it contains more than one kind of component could be useful when first introducing the idea of elements and compounds. Indeed, just as with a technique like chromatography (page 15), it can be used to illustrate the possibility of separation long before students are ready to appreciate how it works (Figure 1.10).

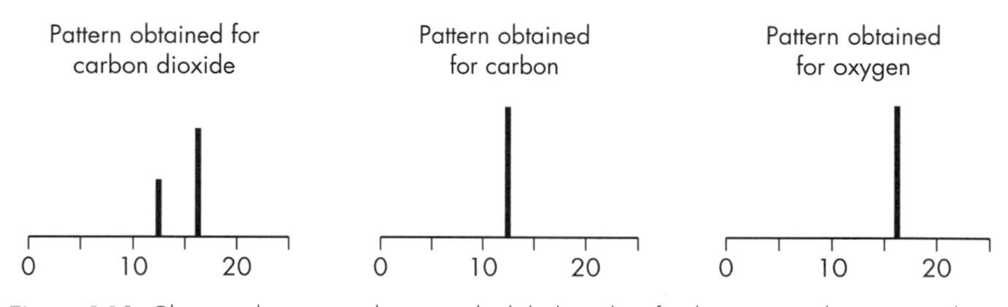

Figure 1.10 Chemists have a technique which helps identify elements and compounds. Carbon dioxide is a single substance, but can be shown to 'contain' the elements carbon and oxygen.

I would suggest that although the black-box approach to using mass spectrometry from early in secondary school need not be problematic in itself, it is important for students to realise we are talking about something here which is not just another technique to separate the components of mixtures. Rather it should be seen as chemically changing a single substance that is a compound into its elements. Mass spectrometry relies on physical separation techniques using electric and magnetic fields, but before this can happen, a sample has to be chemically decomposed. So mass spectrometry is a technique that decomposes compounds (a chemical step) so that they can then be physically separated.

Of course, Figure 1.10 again ignores the issue of isotopes. There is nothing wrong in using this kind of simplification, but I would advise telling students that there are some complications which are being ignored because they only become important when chemists need to look at things in more detail. It is useful to make a teaching point by being explicit whenever you use a simplification or model, both because it avoids students feeling they have been misled later, and because it is important for students to realise that science proceeds through the development and testing of various models, representations and theories. This is 'how science works'.

Mass spectrometry can be revisited later in the school, once atomic structure and isotopes are studied, when more complex diagrams showing isotopic composition can be considered. Even at that point, there is no need for students to know the details of how the technique works, although the basic physics required for a qualitative understanding is usually taught at upper secondary level.

The great advantage of talking to students about mass spectrometry is that it provides a basis for accepting the idea of elements as something special and for identifying when we are dealing with them. This can allow the teacher to set students simple practical work that can then be discussed in terms of elements and compounds without the circularity noted above.

For example, the electrolysis of water, using the Hoffman voltameter apparatus, could be followed by presenting diagrams of what the mass spectra of the water and the gaseous products would look like – what chemists would find if they tested the three different substances (the water and the gases collected at each electrode) in their 'elemental analyser' (Figure 1.11).

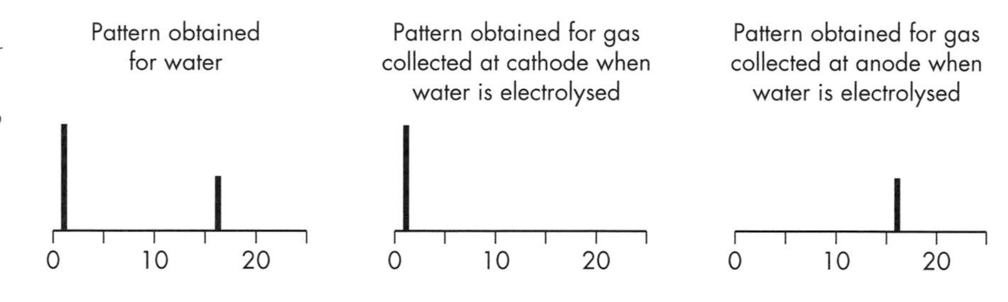

Figure 1.11 Electrolysis of water converts a compound into two elements.

As another example, consider the common practical where weighed magnesium is heated in air (usually in a crucible to minimise the loss of the powdery product and to avoid exposure to the bright visible and ultraviolet radiation emitted in the reaction) to demonstrate that the product has a greater mass than the reacting magnesium. The practical is often included in a topic on burning, sometimes as an illustration of counter-evidence to the phlogiston theory – the historical idea that burning is the release of something (called phlogiston) found in flammable materials, and responsible for that property. This was a focus of major debate in the development of modern chemical ideas, and is sometimes used as an example of how science proceeds ('how science works').

Combustion is a reaction with oxygen; common examples of burning lead to an apparent loss of material (as the products are often formed as gases such as carbon dioxide), but magnesium forms a solid oxide. Students could carry out this practical work, but you are advised to have some suitable sample results available, as inexperienced hands may well find the weighed product less massive than the magnesium they started with.

In a plenary session, students' ideas on what is going on in the experiment, and what the measurements may be taken as evidence for, can be invited. Before closing down discussion, and explaining current scientific thinking (which should include making sure students realise that measured changes in mass are due to not weighing everything present before and after a reaction), the class can be shown the results chemists get when they test magnesium, and the powder obtained by burning magnesium, in their elemental analyser (Figure 1.12).

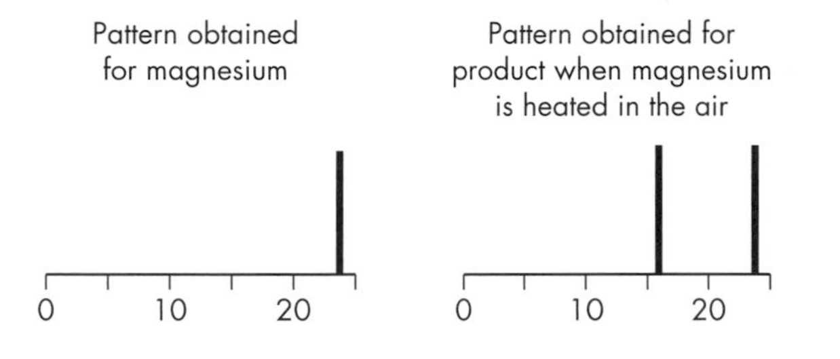

Figure 1.12 The product from burning magnesium includes a second element.

If the idea of the elemental analyser is already familiar from earlier work, then students should be able to suggest from Figure 1.12 that:

- magnesium is an element – one type of substance
- the powder produced in the reaction is made up of two basic substances
- one of the substances is the original magnesium.

You can also tell students that the powder has been found to be a single substance, as it has a precise melting temperature, so as it 'contains' two elements it must be a compound.

The elemental analyser is a flexible idea that can be introduced in many contexts to reinforce the distinction between elements and compounds. It can help us with the rather abstract notion of how a compound can be said to contain elements. This can be a major problem for students. Sodium is a dangerous metal and chlorine is a nasty gas that was used to kill people in war. If sodium chloride is a compound that contains sodium and chlorine, then surely – many students think – it should exhibit some of their horrible properties. The sodium in the compound should react with water and the chlorine should attack the respiratory systems of those who come in contact with it.

This would be true if sodium chloride were a mixture, but a compound is a single substance that has its own properties quite unlike those of its 'constituent' elements. Sodium chloride does not contain the substances sodium and chlorine, and so they are not present to exert their properties. Too much 'salt' is bad for our blood pressure, but we will not be badly burned or choked by it and a little is important in our diet. Only if sodium chloride is subjected to the extreme conditions of the elemental analyser can it be broken down to show that in a sense it is 'made of' the elements sodium and chlorine.

The periodic table of the elements

This discussion of the sense in which a compound contains its elements reflects the ambiguity in the core chemical concept of the 'element' (Figure 1.9, page 18). Chemists use the term element to refer to both the basic substances themselves and something more abstract that might be thought of as their 'essence', and which can be considered to be present in samples of the elemental substance and its compounds. What are actually – physically – considered to be present in both these contexts are nuclei with particular proton numbers; again this is something that can only be discussed with students after they have learnt about atomic structure.

When chemists were developing an understanding of the elements, and attempting to identify them and distinguish them from compounds (from which they were sometimes very difficult to extract), they realised that one thing that seemed to distinguish elements from each other was how much of one element would react with a certain amount of another element. By considering mass ratios involved in various reactions, chemists slowly started assigning masses to the different elements. It became clear that they varied widely, but that there were patterns in how much of one element would react with a certain amount of another (you can find out more about this in the section entitled 'Stoichiometry', page 35), which were easier to understand once mass were assigned to the different atoms so that the ratio of atoms could be considered.

It was eventually recognised that the relative mass of the atoms was not the crucial feature, but something that varied along almost the same sequence as mass: what is now called the atomic number. Figure 1.13 shows a modern arrangement of elements in the periodic table, showing both the atomic numbers and the approximate relative masses of the elements.

Figure 1.13 A modern version of the periodic table of elements

Period	Group 1	Group 2											Group 3	Group 4	Group 5	Group 6	Group 7	Group 0
1					1 H 1 hydrogen													4 He 2 helium
2	7 Li 3 lithium	9 Be 4 beryllium											11 B 5 boron	12 C 6 carbon	14 N 7 nitrogen	16 O 8 oxygen	19 F 9 fluorine	20 Ne 10 neon
3	23 Na 11 sodium	24 Mg 12 magnesium											27 Al 13 aluminium	28 Si 14 silicon	31 P 15 phosphorus	32 S 16 sulphur	35.5 Cl 17 chlorine	40 Ar 18 argon
4	39 K 19 potassium	40 Ca 20 calcium	45 Sc 21 scandium	48 Ti 22 titanium	51 V 23 vanadium	52 Cr 24 chromium	55 Mn 25 manganese	56 Fe 26 iron	59 Co 27 cobalt	59 Ni 28 nickel	63.5 Cu 29 copper	65 Zn 30 zinc	70 Ga 31 gallium	73 Ge 32 germanium	75 As 33 arsenic	79 Se 34 selenium	80 Br 35 bromine	84 Kr 36 krypton
5	85 Rb 37 rubidium	88 Sr 38 strontium	89 Y 39 yttrium	91 Zr 40 zirconium	93 Nb 41 niobium	96 Mo 42 molybdenum	99 Tc 43 technetium	101 Ru 44 ruthenium	103 Rh 45 rhodium	106 Pd 46 palladium	108 Ag 47 silver	112 Cd 48 cadmium	115 In 49 indium	119 Sn 50 tin	122 Sb 51 antimony	128 Te 52 tellurium	127 I 53 iodine	131 Xe 54 xenon
6	133 Cs 55 caesium	137 Ba 56 barium		178.5 Hf 72 hafnium	181 Ta 73 tantalum	184 W 74 tungsten	186 Re 75 rhenium	190 Os 76 osmium	192 Ir 77 iridium	195 Pt 78 platinum	197 Au 79 gold	201 Hg 80 mercury	204 Tl 81 thallium	207 Pb 82 lead	209 Bi 83 bismuth	209 Po 84 polonium	210 At 85 astatine	222 Rn 86 radon
7	223 Fr 87 francium	226 Ra 88 radium		261 Rf 104 rutherfordium	262 Db 105 dubnium	263 Sg 106 seaborgium	262 Bh 107 bohrium	Hs 108 hassium	Mt 109 meitnerium	* 110	111	112						

139 La 57 lanthanum	140 Ce 58 cerium	141 Pr 59 praseodymium	144 Nd 60 neodymium	147 Pm 61 promethium	150 Sm 62 samarium	152 Eu 63 europium	157 Gd 64 gadolinium	159 Tb 65 terbium	162 Dy 66 dysprosium	165 Ho 67 holmium	167 Er 68 erbium	169 Tm 69 thulium	173 Yb 70 ytterbium	175 Lu 71 lutetium
227 Ac 89 actinium	232 Th 90 thorium	231 Pa 91 protactinium	238 U 92 uranium	237 Np 93 neptunium	244 Pu 94 plutonium	243 Am 95 americium	247 Cm 96 curium	247 Bk 97 berkelium	251 Cf 98 californium	252 Es 99 einsteinium	257 Fm 100 fermium	258 Md 101 mendelevium	259 No 102 nobelium	260 Lw 103 lawrencium

key

atomic mass
symbol
atomic number
name

The periodic table is iconic and there are many variations available. Indeed a quick web search shows adoptions of the periodic table format to systematise desserts, rock bands and cartoon characters! More significantly it can be appreciated that there is no single correct form of the periodic table, but rather that it is a type of model used to organise chemical ideas and information – and so different versions are especially useful for different purposes. This again reflects the nature of chemistry as a science: scientists develop models to help them understand aspects of nature. Different versions of the periodic table can best reflect different aspects of the patterns among the elements that chemists have discovered in nature. For example, some versions of the periodic table are more aligned with the chemical properties of the elements, whereas others are organised according to the electronic structures of the atoms of the elements. That the same basic arrangement of periods and groups fits both of these very different considerations reflects how well atomic theory helps explain chemical processes observed in the laboratory.

The background to the periodicity of the elements can be demonstrated by plotting charts of various properties against atomic number. This can be done by hand, but could also be a useful ICT-based activity, where data from a spreadsheet can be used to produce various types of chart. An important teaching point is the relative merits of using line graphs compared with bar charts. As there is no meaning to interpolating between elements (there cannot be an element between element 15 and element 16 in the way there can be a time between 15 s and 16 s, for example), line-graph formats are useful for highlighting trends – such as ionisation energy changes down a group – as long as it is appreciated that atomic number is not a continuous variable.

Traditionally, this type of activity has been reserved for more advanced levels of study (where the patterns are considered in terms of underlying theories, for example to explain patterns in ionisation energies). However, such an activity could be useful as a means of linking this topic with the nature of chemistry as a science. For example, a class of upper secondary students could be divided into a number of groups, each of which is given the task of plotting a different property against atomic number and considering whether there is any evidence for considering that property to repeat in a periodic pattern. Different groups could make brief presentations, before a class discussion to synthesise ideas. Such an activity could allow students to practise scientific argumentation and focus on the relationship between ideas and evidence in science. Unlike many school activities, the evidence in this case is complex and does not lead easily to a clear conclusion, showing what an intellectual achievement the original development of the periodic table was for chemists who did not yet know about atomic structure.

■ Revising the periodic table

A diagnostic task based on completing a concept map for the periodic table ('The periodic table') can be downloaded from the Royal Society of Chemistry website (find out about this in the 'Other resources' section at the end of this chapter). Concept mapping is a useful technique for developing students' metacognitive (learning to learn) skills as well as being potentially open ended and so inviting more creativity than many tasks we set in chemistry classes. Research suggests that students are often very poor at linking up the different ideas they meet in class for themselves. This can be quite problematic in chemistry: a subject that is quite diverse in nature, but which is strongly integrated through central concepts and key theoretical ideas. Concept mapping encourages students to look for links between different ideas.

The diagnostic task is likely to be particularly useful:

- as a revision task after students have been taught about the periodic table or
- to check on the understanding of students who should have covered these ideas earlier in their secondary chemistry careers and to determine whether their prior knowledge is sufficient or whether the topic needs to be revisited.

1.4 Change in chemistry – the concept of reactions

Chemical changes, also known as reactions, bring about a change of substance. Consider the example discussed earlier of magnesium burning in the air:

magnesium + oxygen → magnesia

In this process we start off with magnesium. This is, in chemical terms, a single substance. It is a metal that has a shiny appearance when clean, conducts electricity, melts at 650 °C and burns with a bright white flame. Air is a mixture of gases, but the reactive component that is present in quite a high concentration is oxygen. Oxygen is a colourless, odourless gas that can be liquefied (at normal pressures) below about –219 °C (its boiling temperature is 54.36 K). After the magnesium has burned in air, there is no magnesium left. If we carried out the reaction in a sealed container, with just the right amount of oxygen present to react with the amount of magnesium used, then there would also be no oxygen left afterwards. These two substances have 'disappeared' in the reaction. However, a new substance has been produced. Magnesia (chemical name, magnesium oxide) is a white powder, quite different from either magnesium or oxygen. It is an insulator used as a refractory material (when high temperatures are needed), as its melting temperature is 2852 °C (3125 K) and it is hygroscopic – that is, it will absorb water from the atmosphere. It is also used as an indigestion remedy. The new substance, the product, was not present before the reaction and the reactant substances no longer exist. However, the total amount of material has not changed (3 g of magnesium will react with 2 g of oxygen to give 5 g of magnesia). The elemental analyser (Figure 1.14) shows us that the original substances are in a sense present in the product.

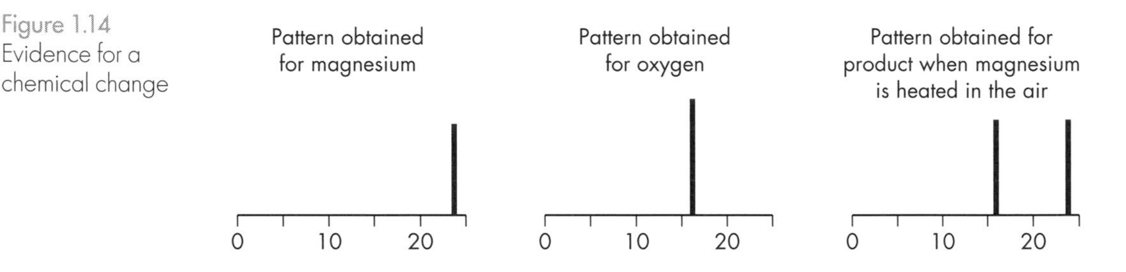

Figure 1.14 Evidence for a chemical change

■ Chemical and physical change – a useful rule of thumb?

A distinction that is often introduced in school chemistry is between physical and chemical changes. After a chemical change, we have a different substance or substances than before. After a physical change, we have the same substance in a different state or phase. So if ice is warmed it will melt, and if the water obtained is heated, it will boil to give steam:

$$\text{ice} \rightarrow \text{water} \rightarrow \text{steam}$$

$$H_2O_{(s)} \rightarrow H_2O_{(l)} \rightarrow H_2O_{(g)}$$

Now ice, water and steam have some very different properties and can be considered different materials. However, scientifically they are different states of the same chemical substance: hydrogen oxide (or, rather undemocratically, just 'water'). These changes – ice melting and water boiling – are not chemical changes (Chapter 3). Yet, to a novice student, such changes may seem just as dramatic as some chemical reactions. So appearances do not always give us a clear indication of whether a change is chemical or not.

Traditionally, a number of criteria have been used to contrast chemical and physical changes:

Table 1.2 Some criteria used to distinguish chemical and physical changes

Chemical change	Physical change
A new substance (or substances) produced	Same substance(s) before and after the change
(Usually) involves a large energy change	(Usually) does not involve a large energy change
Irreversible	Reversible
Involves breaking of bonds	Does not involve breaking of bonds

The first criterion listed is the most fundamental and is generally clear cut as long as the substances present before and after the change are known. If a new substance has been produced, it will almost certainly have different melting and boiling temperatures than the original substance. The other criteria are much more dubious. Some chemical changes involve a great deal of energy being released, such as the example of burning magnesium in air, or even require a considerable energy input, such as the example of the electrolysis of water. However, other reactions may not obviously involve large energy transfers, for example when the enthalpy and entropy changes more or less cancel each other out (Chapter 5). The rusting of iron is a chemical reaction, but usually occurs so slowly that it is not apparent whether the process involves much energy transfer.

It is important to be careful with language when discussing energy changes: energy is always conserved and if we talk about a reaction 'producing' energy, it might seem to suggest that the energy has been created in the reaction. The companion volume in this handbook series on Teaching Secondary Physics includes a chapter setting out an approach to teaching about energy, and it is probably sensible if this approach (based around energy being transferred between different stores via pathways) is adopted in all science classes in the school. From this perspective, the reactants may be considered as a chemical store of energy. When an exothermic reaction takes place, some of the energy from that store is transferred through a pathway (heating) to a different store (in terms of the increased thermal energy of the now warmer products and their surroundings).

Generally speaking, physical changes are more readily reversible than chemical changes. However, again this is not a very definitive criterion. The idea that chemical reactions tend to either 'go' or not is a useful approximation, but there are many examples of reactions that can be readily reversed (see Chapter 5). In principle, all reactions involve equilibria of forward and reverse reactions, and can be reversed by changing the conditions sufficiently. When hydrogen and oxygen are exploded, it takes a pedant to claim that there is also a process of water molecules being converted into oxygen and hydrogen molecules as the reaction proceeds, which means the reaction will continue for ever. Technically such a claim may be true, but for all practical purposes the explosion reflects a reaction that very quickly goes to completion.

One technique that can be used to separate iodine from sand is to warm the mixture gently in an evaporating basin, over which is placed an upturned beaker or funnel. The iodine will sublime – turn to vapour – before recondensing on the cold glass, separated from the sand. The same technique may be used if ammonium chloride is mixed with the sand. In both cases the separation is achieved because sand (which has a high melting temperature) is mixed with another substance in the solid state that is readily changed into a vapour by warming, and then readily recovered as a solid sample when the vapour is in contact with a colder surface. There are reversible changes involved in both cases:

$$\text{solid iodine} \rightarrow \text{iodine vapour}$$

$$\text{ammonium chloride} \rightarrow \text{ammonia} + \text{hydrogen chloride}$$

In the first case, the process involves only changes of state: evaporation and condensation – collectively called sublimation. However the second case involves one substance (a salt) changing to two other substances. To a student seeing these changes demonstrated, there would be little basis to infer that one is (usually considered as) a chemical change, but not the other.

(It is worth noting that iodine, ammonia and hydrogen chloride are all hazardous substances – and if this demonstration is attempted, a proper risk assessment is needed.)

The final criterion in Table 1.2 concerns whether bonds are broken and made during a change, and this can only be meaningful for students once they have learnt about particle models of the submicroscopic structure of matter (Chapter 2). In a chemical change, there will be the breaking of bonds that hold together the reactants, and the formation of new bonds in the products. However, we have to be careful here what we mean by 'bond' (this is discussed further in Chapter 4).

When ice melts and water boils, 'intermolecular' forces between molecules are disrupted and this includes the breaking of hydrogen 'bonds'. However, when people talk about bond breaking in the context of chemical and physical changes, they tend to mean strong chemical bonds such as covalent, ionic and metallic bonds (Chapter 3). Yet even this is not clear cut. When metals evaporate or are boiled, metallic bonds are broken, although the vapour is not normally considered a different substance. When elements such as carbon and phosphorus undergo phase changes relating to allotropy, there is breaking, and forming, of bonds, which might suggest these changes are chemical and that the different forms of the same elements should be considered different substances. As suggested above, the status of different allotropes of the same element introduces a complication that is probably best avoided when first introducing the key ideas of chemistry.

(Allotropes occur where different structures are more stable under different conditions of temperature and pressure. So in conditions deep in the Earth, the most stable form of carbon is diamond. However at the Earth's surface, diamond is less stable than graphite. Diamonds are 'meta-stable' which means that although they theoretically change to graphite after being brought to the surface, this is a very slow process. Chemistry tells us that diamonds are not 'forever' – but this need not worry us mortals.)

A particularly tricky case occurs when we dissolve materials to form solutions, especially materials with ionic bonding (Chapter 4). Dissolving tends to involve small energy changes, and to be readily reversible, and is generally considered a physical change. However, to dissolve an ionic compound such as sodium chloride (table salt), the strong ionic bonds between the sodium and chloride ions have to be overcome (and new bonds must form between the ions and solvent molecules). This would seem to suggest that dissolving can be a chemical change according to the criterion of bond breaking and formation (Table 1.2).

■ Teaching about chemical and physical change

It appears that the distinction between chemical and physical changes is a rather messy one, with no clear criteria to help students understand the difference. Some chemistry teachers avoid the distinction completely and consider that it is not useful. However, some school curricula and examination specifications include the topic, so it cannot be avoided in all instances. It makes little sense to teach chemical and physical changes as any kind of absolute distinction, as this would seem to be unsupportable in terms of the chemistry and will ultimately involve pointless rote learning and/or be a source of frustration for students. However, the idea that a chemical change involves changes in substances is a key idea which should be introduced early in school chemistry and reiterated in suitable contexts as students develop sufficient background knowledge to understand the principle.

Talking about changes of state as being physical changes (because despite apparent material differences, the same substance is present after the change) is also a central teaching point (Chapter 2). The distinction is worth emphasising in these terms whenever suitable contexts are met. Ideas about energy changes and reversibility can be introduced (especially if specified in the curriculum) but should be presented as 'rules of thumb', that is as heuristics that are often useful, but which can mislead us. Similarly, when students have learnt about different types of chemical bond, the extent to which it is helpful to think about bond breaking/formation as being indicative of chemical change can be explored. With many students this might best be limited to another useful rule of thumb, that breaking and making of strong chemical bonds is often associated with chemical changes, but can sometimes occur without a chemical change.

For older students ready for a challenge, this topic can form the basis of a useful group discussion task. However, the focus should be the extent to which the criteria suggested in Table 1.2 are useful in making a distinction. The task would not be about coming to the right answer, but rather be based on using evidence (from their knowledge of chemistry or resource material) to argue a case for their position. In this way a rather unsatisfactory topic becomes the basis for practising scientific argumentation, reviewing knowledge of specific chemistry and learning something about the nature of science. (Table 1.2 can be considered to present a model of changes studied in chemistry, which – like all models – has limitations.)

■ Diagnosing student thinking about chemical and physical change

A diagnostic task to explore students' understanding of the distinction between physical and chemical change ('Changes in chemistry') can be downloaded from the Royal Society of Chemistry website (more information can be found in the 'Other resources' section at the end of this chapter). You may wish to use this before teaching to elicit students' prior knowledge, for example if you wish to check the understanding of a class that should already have mastered these ideas. Alternatively, you may wish to set this as an exercise before exploring with a class the limitations of this particular model. Students may be interested to know that when chemistry teachers completed this task, they did not all agree on their answers on some items!

1.5 Stoichiometry

A key pattern in chemical reactions, alluded to above, concerns the constant ratios found in most chemical reactions. (Most, because sometimes the same reactants may give different products: iron and oxygen, for example, can form FeO, Fe_2O_3, Fe_3O_4 or a mixture depending on conditions.) This is a key point worth emphasising to classes. Some students will appreciate this very quickly, but with many classes it is worth working through sufficient examples to be sure the students are comfortable with the idea that reacting masses can be multiplied by any arbitrary factor, as long as the substances represented in the reaction equation are treated the same. So, in our example of magnesium and oxygen reacting, considered earlier in the chapter:

3 g of magnesium will react completely with 2 g of oxygen to produce 5 g of magnesium oxide.

6 g of magnesium will react completely with 4 g of oxygen to produce 10 g of magnesium oxide.

9 g of magnesium will react completely with 6 g of oxygen to produce 15 g of magnesium oxide.

12 g of magnesium will react completely with 8 g of oxygen to produce 20 g of magnesium oxide.

24 g of magnesium will react completely with 16 g of oxygen to produce 40 g of magnesium oxide.

60 g of magnesium will react completely with 40 g of oxygen to produce 100 g of magnesium oxide.

600 g of magnesium will react completely with 400 g of oxygen to produce 1 kg of magnesium oxide.

3 kg of magnesium will react completely with 2 kg of oxygen to produce 5 kg of magnesium oxide; etc.

In general terms,

1.5 X of magnesium will react completely with X of oxygen to produce 2.5 X of magnesium oxide.

Although the ratios found in different reactions are constant, they are not always this simple:

1 g of hydrogen will react with 8 g of oxygen to form 9 g of water.

23 g of sodium will react with 35.5 g of chlorine to give 58.5 g of sodium chloride.

100 g of calcium carbonate will decompose on heating to give of 56 g of calcium oxide and 44 g of carbon dioxide.

The existence of such ratios can be explained by the models chemists use of the structure of matter at the submicroscopic level, and indeed is part of the reason chemists initially adopted such ideas in an instrumental way: that is, as ideas that worked as useful tools for thinking about chemistry, but which might not reflect an underlying reality. Over many years these ideas were found to provide a core set of models that could provide unifying frameworks for making sense of chemistry. Today most chemists consider molecules, atoms, protons, electrons and so forth to be real objects: but it is important to remember that these ideas are a set of theoretical models, even if a very useful and successful one. I am not suggesting these entities do not exist, but that our scientific models of these entities are subtle and still being developed, and the mental models that most of us have of them are at best partial and approximate versions of the best descriptions science can currently offer.

It is certainly very important to teach these ideas as theoretical, because although the models are successful and central to modern chemistry, it is not helpful if students think our models of atoms and molecules are precise realistic descriptions. Certainly the models introduced at secondary level fall somewhat short of this. As just one example, the notion that atoms contain 'shells' of electrons should not be taken to imply either that there is any kind of physical shell which contains the electrons (as some students assume), nor that the electrons in a shell can always be considered as equivalent. Students who select chemistry as a subject for further study will soon run into problems if they develop fixed ideas along these lines. It is much better to teach that atoms often behave as though they have electrons arranged in shells, but to warn students that scientists have found this to be a simplification. This approach provides students with a more authentic understanding, avoids over-commitment to the model that might impede more advanced learning, and better reflects the nature of chemistry as a science.

Teaching basic 'particle' theory will be discussed in the next chapter (Chapter 2). The key ideas needed here to see how this theory explains stoichiometry include:

1 Matter is quantised: that is, at a submicroscopic level, matter consists of myriad discrete bits. We often refer to these bits as 'particles' although this is an analogy with familiar bits of matter like salt grains or specks of dust. The quanta of matter are at a much smaller scale and also have strange properties that are not like familiar particles – sometimes two of them can be in the same space, for example. Electrons and neutrons are both used in diffraction experiments – beams of electrons or neutrons will be scattered when directed at suitable target materials, offering indications of the structure of the target materials. This shows

that under suitable conditions these 'particles' show wave properties when they interact with materials in ways quite unlike how we expect classical particles to interact. If we use the term 'particles' with students, we need to make sure we are clear about what we mean. (Students have sometimes been found to be confused about whether salt and sugar grains are examples of these 'particles'.) I will refer to these quanta of matter as 'quanticles' to emphasise the difference.

2 There are a small number of basic types of quanticles (i.e., particle-like entities at the submicroscopic scale) of interest in chemistry. The latter qualification is useful because physicists will talk about a large zoo of different particles – some of which are highly unstable under normal conditions and can only be produced in very specialised (and expensive) high-energy colliders. In terms of a model which is useful for teaching secondary chemistry, it is usually enough to know about protons, electrons and neutrons.
3 Protons, neutrons and electrons are usually found clumped together (except at very high temperatures) and it is these clumps that are often the level of 'quanticle' most useful for discussing what is going on in chemistry. Protons and neutrons are bound together in nuclei by what is known as the strong nuclear force, and electrons are attracted to nuclei because of their opposite electrical charges – electrons are negatively charged and protons in the nuclei are positively charged. The clumps that form are usually neutral (molecules) or nearly neutral (ions) – simply because forces exist between oppositely charged quanticles, attracting them together. Under most important conditions, ions tend to be found either in neutral lattices or surrounded by other quanticles (so in aqueous solution, ions are surrounded by a sheath of water molecules that are attracted to the ion).
4 The electrons do not get attracted right into the nuclei (luckily, as when that happens you get a neutron star, where no chemical substances or normal materials can exist), but are associated with one or more nuclei according to some rather complex rules.

Table 1.3 overleaf outlines the key quanticles we commonly talk about in explaining chemistry. It is worth remembering that quanticles are not like familiar particles: they are more like a hybrid of a particle and a wave. The entries in the table are organised in terms of 'clumpiness'. So protons, neutrons (collectively termed nucleons) and electrons can all be considered as single quanta of matter in chemistry. The nucleus is a clump of nucleons. The atomic core is a clump containing a single nucleus, surrounded by electrons, and is the largest type of clump that is generally unchanged in chemical processes. Molecules, atoms and ions are larger clumps which are modified in chemical processes and which tend to be characteristic of particular substances.

Table 1.3 Quanticles – theoretical 'wave–particle' objects used by chemists to describe and explain the submicroscopic structure of matter

Quanticle	Description	Extension notes
Electron	Considered a fundamental particle; has a negative charge (–1); is much less massive than a nucleon	The electrical charge is about 1.6×10^{-19} C in SI units. Electrons have inherent angular momentum ('spin'). The mass of an electron is about 9.1×10^{-31} kg. This is so much less than the mass of a proton or neutron it is often considered negligible
Nucleon	A collective term for the particles found in nuclei, in other words protons and neutrons	Made up of three quarks. Mass is 1.7×10^{-27} kg – often called one atomic mass unit
Proton	A positively charged entity that is attracted to other nucleons by the strong nuclear force	Made up of three quarks. Mass is 1.7×10^{-27} kg – often called one atomic mass unit
Neutron	A neutral entity that is attracted to other nucleons by the strong nuclear force	Made up of three quarks. Mass is 1.7×10^{-27} kg – often called one atomic mass unit
Nucleus	The clump of protons and neutrons at the centre of an atom	Nuclei are unchanged by chemical processes, although some are unstable and undergo radioactive decay
Atomic core	A nucleus and any 'shells' of electrons that can be considered to be fully associated with that nucleus – that is electrons not in the valence or outermost shell	In some countries this is called an atomic kernel. In most chemical changes, the atomic core (kernel) remains unchanged (whereas there are changes in the arrangements of outer electrons)
Molecule	A neutral entity comprising one or more atomic cores and an outer layer of electrons that electronically cancels the nuclear charge	Note – in common use, the term molecule is sometimes reserved for species with two or more atomic cores: not 'monatomic molecules' such as He, Ne etc.
Atom	A neutral species with one nucleus; it has the same number of electrons as protons	Most atoms have outer (valence) electron shells that are not symmetrical and atoms are rarely found under normal conditions. The exceptions are the inert gases
Ion	A charged entity with one or more atomic cores	Ions that are commonly found have particular valence shell electron arrangements
Simple ion	An ion with a single atomic core surrounded by a shell of electrons	Simple ions differ from atoms because they have either too many or too few electrons to cancel the nuclear charge
Molecular ion	An ion with several atomic cores surrounded by an outer 'layer' of electrons	Molecular ions differ from molecules because they have either too many or too few electrons to cancel the nuclear charges

Table 1.3 uses the alternative terms valence shell and outer shell. It is important to be aware that some students may only pick up on synonymous terms when they are explicitly acknowledged by the teacher. This can be a problem, for example if the teacher uses one term, and the student textbook an alternative. I recall one student who became confused when he moved between teachers – where one referred to valence shells, and the other to outermost shells – as he did not appreciate these were simply alternative terms for the same thing. This type of confusion can readily be avoided if we are aware when students are likely to meet common synonyms.

Stoichiometry is largely explained in terms of the electronic configurations of the atoms of the elements, as stable species are those where the atomic cores are well shielded from other species by the outer 'layer' of electrons. This tends to happen in ions and molecules, or sometimes atoms, with particular patterns of valence electrons (Chapter 3). This leads to atomic cores combining in fixed ratios. The evenness (symmetry) of the pattern of electron density is more important than the neutrality of the species, so ions such as Na^+ and O^{2-} are found in common materials, whereas the atoms Na and O are not. However, it is important to realise that the stability of ions is only possible because they are usually found in neutral lattices or surrounded by solvent molecules; isolated ions are usually less stable than the corresponding atom.

So, for example, magnesia has the formula MgO because magnesium forms an ion, Mg^{2+}, and oxygen an ion, O^{2-}, both of which have a symmetrical pattern of charge and which can be stabilised by being formed into an MgO lattice (a great many Mg^{2+} alternating with the same number of O^{2-}) which is neutral overall. (Although other ions can be formed, such as Mg^{3+}, O^- etc., these structures are too unstable to be readily stabilised.) The reason why magnesium and oxygen most commonly form these particular ions (as well as why, for example, a molecule of ammonia has three hydrogen atomic cores and only one nitrogen atomic core) can be understood in terms of their electronic configurations. This is explained further in Chapter 3.

Models of the structure of the atom that scientists find useful vary considerably in complexity, but in introductory chemistry it is useful to think that the electrons around a nucleus are arranged in shells. The elements in successive groups across a period of the periodic table reflect increasing numbers of electrons in a shell, and the breaks for a new period reflect the starting of a new shell. (This is, alas, a simplification, as only the first two shells fill completely before a new shell is begun. So, in the third shell, only eight of the maximum of 18 electrons are in place before the fourth shell is used, in the first element of period 4, potassium.) Versions of the periodic table that represent the electronic structures of the atoms of the elements (Figure 1.15) can be useful in appreciating the stoichiometry of chemical compounds.

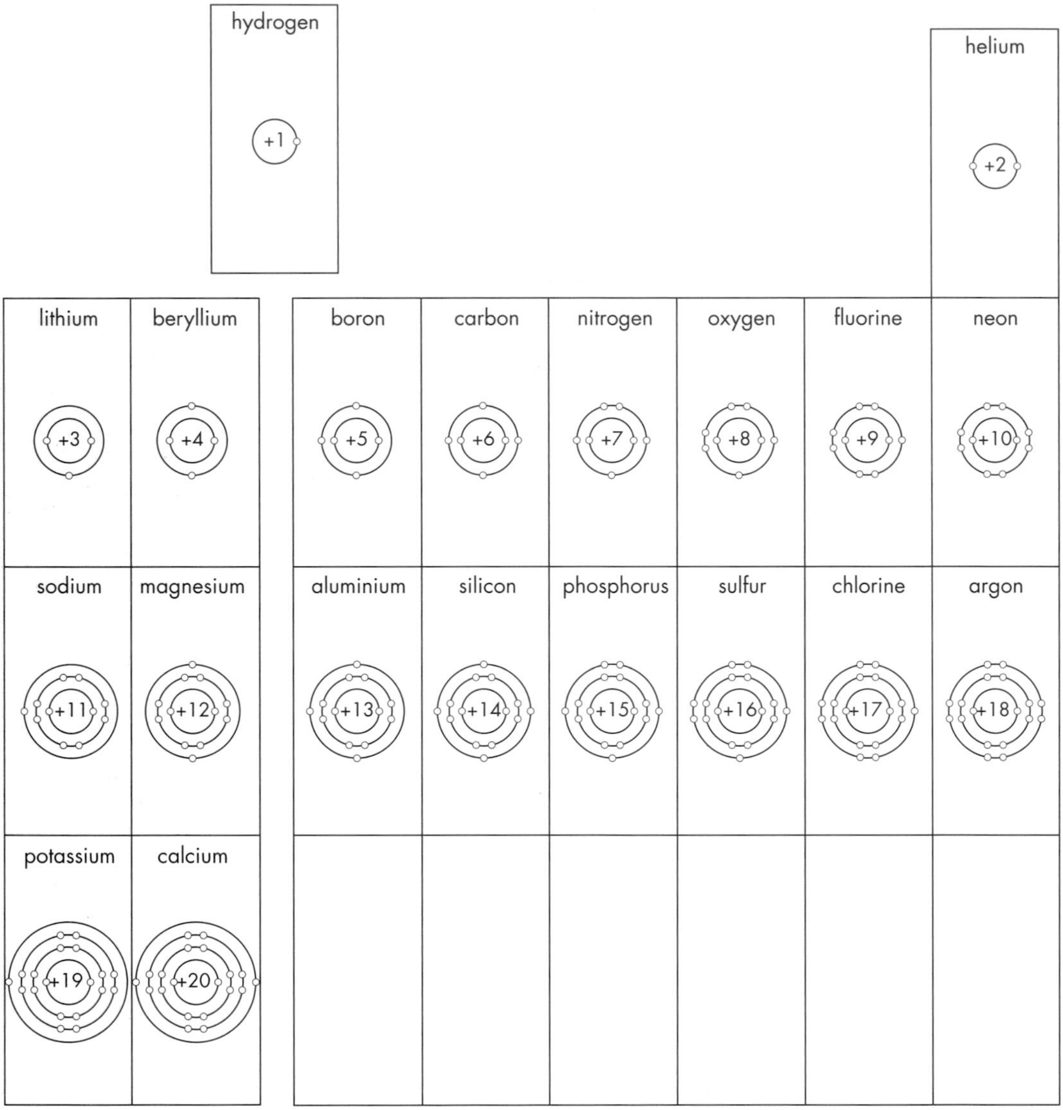

Figure 1.15 A section of the periodic table in terms of electronic structure of atoms

1.6 Moles – measuring quantities in chemistry

The phenomenon of stoichiometry – that chemical reactions involve precise mass ratios of reactants, leading to precise mass ratios of products – is understood in terms of the models of the substances being comprised of quanticles, such as molecules and ions, at the submicroscopic level. The ratios themselves relate to how different substances can be understood in terms of the composition of those substances at the submicroscopic level.

So we have just seen that magnesia has the formula MgO because it comprises equal numbers of Mg^{2+} and O^{2-} ions. However, although the ion ratio is 1 : 1, this does not mean that equal masses of magnesium and oxygen react: rather we have seen that 3 g of magnesium will react completely with 2 g of oxygen. This is because the nucleus of a magnesium ion is more massive than the nucleus of an oxygen ion. Each ion of magnesium, element number 12, has a nucleus containing 24 nucleons (12 protons and 12 neutrons), and each ion of oxygen, element number 8, has a nucleus containing 16 nucleons (eight protons and eight neutrons). The 1 : 1 ratio of ions in this case relates to mass ratios of 24 : 16 (which happens to simplify nicely to 3 : 2 in this particular case). In reality, because of the presence of isotopes, a small proportion of these ions will have different numbers of neutrons in their nuclei and so the precise values are not quite so neat!

The same principles apply to other examples, although the numbers are not always so convenient. So calcium fluoride has the formula CaF_2, as the common ion of fluorine is F^- and that of calcium is Ca^{2+} (Figure 1.15). Therefore the crystal lattice of CaF_2 contains twice as many fluoride ions as calcium ions (as it will be neutral overall). Calcium is element 20 and the ion (ignoring isotopes) has mass number 40 (20 protons, 20 neutrons); whereas for fluorine the ion (again ignoring isotopes) has mass number 19 (9 protons, 10 neutrons). The mass ratio for reacting calcium and fluorine is 40 : (2 × 19 =) 38. So 40 g of calcium will react with slightly less, 38 g, of fluorine (and 20 g of calcium will react with 19 g of fluorine; 10 g of calcium will react with 9.5 g of fluorine, etc.).

Using mass ratios allows us to scale up or down the amounts of reacting materials as much as we like (40 tonne of calcium will react with 38 tonne of fluorine; 40 μg of calcium will react with 38 μg of fluorine, etc.). However, it is sometimes useful to have a standard way of talking about how much material we are using, that allows us to shift directly from talking about individual ions or molecules to talking about laboratory scale amounts of material. For this, chemists scale to a unit of chemical amount known as the mole.

Consider the example of the reaction between hydrogen and oxygen in which two molecules of hydrogen will react with one molecule of oxygen:

$$2H_2 + O_2 \rightarrow 2H_2O$$

The reacting mass ratio for hydrogen and oxygen therefore depends on two factors: the relative molecular masses of oxygen and hydrogen and the ratio of molecules that react together.

As the mass of electrons is so small compared with the masses of protons and neutrons, and as neutrons are only very marginally heavier than protons, the reacting masses of substances follow very closely from a consideration of where the nucleons (neutrons and protons) are in the reacting substances. Consider the case of hydrogen and oxygen reacting (Table 1.4).

Table 1.4 How reacting mass ratios depend on where the nucleons are

Substance	Hydrogen	Oxygen	Water
Molecules involved in reaction	$2H_2$	O_2	$2H_2O$
Nuclear composition	Two molecules of hydrogen, each with two nuclei containing one nucleon each $(2 \times 2 \times 1)$	One molecule, with two nuclei, each containing 16 nucleons $(1 \times 2 \times 16)$	Two molecules, each with three nuclei, containing 1, 1 and 16 nucleons $(2 \times \{1 + 1 + 16\})$
Total number of nucleons	4 nucleons	32 nucleons	36 nucleons

Literally, this tells us that 3.64×10^{-30} kg of hydrogen (two molecules) reacts with 29.12×10^{-30} kg of oxygen (one molecule) to produce 32.76×10^{-30} kg of water (two molecules). However, because mass is quantised, this also tells us that any reaction of hydrogen and oxygen will involve multiples of these masses: 2000 molecules of hydrogen will react with 1000 molecules of oxygen; 2 000 000 molecules of hydrogen will react with 1 000 000 molecules of oxygen; 2×10^{24} molecules of hydrogen will react with 1×10^{24} molecules of hydrogen, and so forth.

■ The chemist's dozen

This is the context in which the idea of the mole, as a measure of the 'amount of substance' is used in chemistry. A lot of students find the idea of the mole difficult and this is presumably due to the combination of two factors: the abstract nature of the concept and the expectation that students will apply maths in a chemical context.

The important ideas to get across to students are:

1 that we can explain stoichiometry (constant reacting ratios) in terms of the masses of the particular molecules and ions involved in the reactions concerned, or the 'relative' formula masses if we allow for different isotopes; however,
2 that it is not very practical in the laboratory, as chemists operate with samples that contain billions and billions of molecules or ions and need to measure out samples in a way they can easily manage (not by counting billions of molecules or ions that are far too small to be seen!).

So in most reactions, a useful way of measuring reactants is in terms of mass: how many grams or kilograms, of substance are being reacted. (For fluids, we may measure volumes of reactants.) At the bench level, the gram is a more suitable starting point, so the mole is based on masses in grams that reflect the number of nucleons present in a sample. So for hydrogen (two nucleons per molecule), one mole would be 2 g; and one mole of oxygen (32 nucleons per molecule) is 32 g. As two molecules of hydrogen react with each molecule of oxygen, two moles of hydrogen (4 g) are required to react with one mole of oxygen (32 g). Thus the reacting masses in grams reflect the nucleon ratios in Table 1.4.

So the mole can be understood as based upon a multiplier like a dozen (12), a score (20) or a gross (144), only a lot larger. Indeed, as each nucleon only weighs 1.7×10^{-24} g, it turns out that the multiplier (known as Avogadro's number) is about 6.0×10^{23} – somewhat bigger than a standard dozen! So a mole of a substance, no matter which substance, contains 6.0×10^{23} formula units: there are 6.0×10^{23} H_2O molecules in one mole of water (formula H_2O) and 6.0×10^{23} of each of the ions Na^+ and Cl^- in one mole of sodium chloride (formula NaCl). Because the formula for calcium fluoride is CaF_2, one mole of calcium fluoride contains 6.0×10^{23} Ca^{2+} ions and twice that many (1.2×10^{24}) F^- ions.

Table 1.5 How reacting masses depend on where the nucleons are

Substance	Hydrogen	Oxygen	Water
Molecular formula	H_2	O_2	H_2O
Relative formula mass	2	32	18
Stoichiometric ratio (from the reaction equation)	2 H_2	(1) O_2	2 H_2O
Reacting mass ratio	$2 \times 2 = 4$	32	$2 \times 18 = 36$
Reacting masses (for reacting one mole of oxygen)	4g	32g	36g
Reacting masses (to generate one mole of water)	2g	16g	18g

The strength of describing the amount of substances involved in reactions in terms of moles (abbreviated to mol) is that it can be scaled up or down by any amount:

$$
\begin{array}{ccccc}
2H_2 & + & O_2 & \rightarrow & 2H_2O \\
2 \text{ mol} & : & 1 \text{ mol} & : & 2 \text{ mol} \\
2 \times 2\text{g} = 4\text{g} & : & 32\text{g} & : & 2 \times 18\text{g} = 36\text{g} \\
10 \text{ mol} & : & 5 \text{ mol} & : & 10 \text{ mol} \\
10 \times 2\text{g} = 20\text{g} & : & 5 \times 32\text{g} = 160\text{g} & : & 10 \times 18\text{g} = 180\text{g} \\
0.1 \text{ mol} & : & 0.05 \text{ mol} & : & 0.1 \text{ mol} \\
0.1 \times 2\text{g} = 0.2\text{g} & : & 0.05 \times 32\text{g} = 1.6\text{g} & : & 0.1 \times 18\text{g} = 1.8\text{g}
\end{array}
$$

We could carry out similar analyses for other examples, such as the formation of magnesium oxide or calcium fluoride, covered earlier in the chapter. The relative formula mass is based on the common formula for a substance, even if it is not molecular. So magnesium oxide, a lattice containing an extensive array of ions, is said to have a relative formula mass of 32, although it does not contain discrete Mg–O units (see Chapter 4).

Students seem to vary in how they find it best to work out chemical ('mole') calculations (perhaps related to different 'learning styles' or 'thinking styles'). Some prefer to use verbal arguments, some do the algebra and some use graphical schematics of various kinds. In general, it is found that learning of scientific ideas is supported when multiple forms of representation are used. It would seem sensible to model different approaches and allow students to find something they are comfortable with. This could be introduced in a dialogic form by setting a task that students are asked to work on in pairs or small groups, and then asking each group to explain their approach to the class. Invite comments on the strengths and weaknesses of each approach. You can add any key approaches not suggested by the class. Given that many students struggle when asked to work using a particular formalism or set of rules preferred by the teacher, this could be a good use of time: it will show students that there are different ways of thinking about mole problems and indicate that several different approaches can be used to get the right answer. (They are all equivalent at a fundamental level of course.) This can encourage them to think about the logic of what they are doing, rather than trying to learn an algorithm that will fail as soon as they have a slightly different form of task to complete. This could also encourage them to value understanding over rote learning of rules of thumb.

This approach also reinforces how in chemistry there are often alternative ways of representing the same information. Research suggests that students tend to make better progress when they have a repertoire of alternative representations they can draw upon to do their chemical thinking. Finally, if this seems a rather brave way of approaching using maths in chemistry classes, it might be worth talking to the maths staff. In many schools this general approach of getting small groups to work on problems and then share and discuss alternative methods is a core feature of maths lessons. (For information on using dialogic group work in science lessons, refer to the epiSTEMe website detailed at the end of the chapter).

■ Range of moles problems

At secondary level, students are usually only asked to address moles questions of limited complexity. However, for students who progress in the subject, the demand of problems can get quite high.

As well as calculating the masses of reactants or products, as in the example above (where the complexities of the mass ratios can vary according to the reaction, of course), students can be asked to solve problems with gases (where the volume of a gas at particular temperature and pressure depends upon the number of moles present, i.e. following Avogadro's principle) and solutions (where the amount of solution is measured by volume and a conversion according to molarity, the number of moles per unit volume of solution, has to be made). Avogadro's law tells us that the molar volume of any gas is the same under any stated conditions. In practice this is an approximation, but a good one for many gases under a wide range of conditions, which makes it a very useful principle. (The law applies well when a gas approximates to an ideal gas, i.e. where the volume of the molecules themselves are insignificant compared to the volume the gas occupies, and when the molecules do not significantly interact with each other.)

As with most of the topics discussed in this chapter, the mole is a fundamental idea in chemistry, which is best introduced with fairly simple examples, then regularly revisited over an extended period of time. In terms of introducing the topic, it is worth noting that the sheer range of possible examples (any chemical equation can be used as the basis; any feasible value for reacting masses can be used) means that such a topic lends itself to ready differentiation for students. Where some students may struggle with examples involving simple ratios and masses chosen to give simple whole number answers, others may soon be ready for more demanding examples with complex mole ratios and arbitrary choices of reacting masses to be calculated to several decimal places.

■ Coda

This initial chapter has introduced a range of key concepts, most of which are abstract and are known to be challenging for many learners. However, these topics are central to chemistry and essential for a modern understanding of the subject that goes beyond simple description of substances and their reactions.

Individually, the notions of elements and compounds, of theoretical models of structure at submicroscopic level, of chemical change, of chemical equations and stoichiometry, and of the mole, are likely to be found difficult by many students. Despite this, however, as these ideas become familiar, they also become mutually supporting as they build to give a coherent way of thinking about chemistry. This is a way of thinking that supports the wide range of chemical explanations that has made chemistry the successful science it is – developing new materials to make life healthier, safer, more comfortable and more entertaining. Teaching these ideas is not the work of a small number of discrete lessons, but should rather be the basis of teaching over the secondary years, being regularly reinforced with new examples in new contexts. In this way, what are initially odd and abstract ideas to be learnt and remembered can become a set of familiar thinking tools that link together to turn chemistry from being an amazing but mysterious subject of colours, smells and bangs into a coherent way of making sense of the material world.

Other resources

Websites

The Wellcome Foundation has a collection of images that can be used in teaching under a Creative Commons licence. The collection includes drawings of Lavoisier at work and of Priestley's chemical apparatus.
http://images.wellcome.ac.uk

A number of useful tools about the elements and the periodic table can be found on the Royal Society of Chemistry's website. Look for Online Chemistry Resources.
www.rsc.org/learn-chemistry

The University of Nottingham's *Periodic table of videos*, a series of short films discussing different elements, may be found at:
www.periodicvideos.com

Many examples of the ways in which secondary school students understand and think about chemistry concepts are reported on the ECLIPSE (Exploring Conceptual Learning, Integration and Progression in Science Education) project website:
www.educ.cam.ac.uk/research/projects/eclipse/

A number of diagnostic assessment tasks relevant to this chapter are included in *Chemical Misconceptions – Prevention, Diagnosis and Cure*. Taber, K.S. (2002). London: Royal Society of Chemistry. Volume 1: Theoretical background; Volume 2: Classroom resources. The individual classroom resources may be downloaded from the Royal Society of Chemistry's website:
www.rsc.org

Incorporating effective group discussion tasks in secondary science and mathematics was a focus of the epiSTEMe (Effecting Principled Improvement in STEM Education) project. Information about this project is available at the project website:
http://www.educ.cam.ac.uk/research/projects/episteme/

Books

Information about analytical techniques may be found in: Faust, B. (1997) *Modern Chemical Techniques*. London: Royal Society of Chemistry.

2 Introducing particle theory

Philip Johnson

2.1 Melting and solidifying
- Melting points
- Defining a substance

2.2 A 'basic' particle model
- The solid and liquid states
- Introducing the basic particle model

2.3 Boiling and condensing
- Predicting the gas state
- Boiling water
- Boiling points
- 'Gases'

2.4 Dissolving
- Recognising and explaining dissolving
- Solutes in the liquid and gas states
- Empty space in the liquid and solid states
- Intrinsic motion and the liquid state
- Solubility

2.5 Evaporation into and condensation from the air
- Evaporation and boiling
- Temperature and energy
- Explaining evaporation below boiling point
- Factors affecting the rate of evaporation
- Reconciling boiling and evaporation
- Condensation from a water–air mixture

2.6 More on the gas state: pressure, weight and diffusion
- The gas state and pressure
- Mass and weight of the gas state
- Diffusion involving the gas state

2.7 What, no 'solids, liquids and gases'?
- Three types of matter?
- The concept of a substance is important

Why teach particle theory?

Chemistry is all about substances. Substances can be involved in three kinds of change: change of state, mixing and chemical change. Our descriptions of phenomena are in terms of substances. When a lump of lead changes to a runny liquid, it is still said to be the same substance. When a sugar crystal dissolves in water, the substance sugar is still thought to be there. The substances wax and oxygen change to the substances water and carbon dioxide in a candle flame. Why do such descriptions make sense? Liquid is very different from solid, so how can liquid lead be the same substance as solid lead? The crystal disappears, why not the sugar? How is it that substances can cease to exist with new ones created in their place?

These descriptions owe their sense to particle theory, where a substance is identified with a particle. Observations are theory led, and particle theory at varying levels of sophistication determines the way chemists speak and think about the material world. Therefore, if we want students to really understand chemistry, we need to develop their understanding of particle theory. At entry level, a 'basic' model referring to the particles of a substance is good enough for changes of state and mixing. At greater resolution, identifying a substance with an 'atom structure' (which atoms are bonded to which) accommodates chemical change. This chapter confines itself to the introduction of a 'basic' particle model, leaving atoms to Chapter 3.

Previous knowledge and experience

■ The concept of a substance

The following picture of students' thinking is based on extensive research. To the uninitiated, what chemists mean by a substance is not at all obvious. (Chapter 1 discusses this at length.) Most materials in everyday experience are mixtures of substances; few are relatively pure samples of substances. Moreover, everyday thinking assigns identity to a sample of material by its history rather than its current properties. The focus is on where something has come from and what has happened to it. Although the idea of mixing is readily appreciated, materials 'as found' are not necessarily categorised as either substances or mixtures of substances. For example, there would be no distinction between milk (a mixture of substances) and sugar (a substance, sucrose) as ingredients. Rust is still thought to be iron, but in a different form (or a mixture of iron and something else); the preoccupation is with its origin. Whereas, from the chemist's point of view, rust is a new substance in its own right because of its properties; the substance iron no longer exists.

Students are generally mystified by the gas state and are very far from thinking that 'gases' are even material like that which presents in the solid or liquid states. Difficulties with the gas state are not surprising. Historically, the idea of a gas being a substance was not established until the early eighteenth century with Black's work on carbon dioxide. Recognising gases as substances was an important precursor to Lavoisier's experiments with oxygen and his founding of modern chemistry. Much of school chemistry involves one or more reactants and/or products in the gas state. Understanding the gas state is pivotal in understanding chemistry.

The concept of a substance is something that needs to be learnt and goes hand in hand with particle theory. Without substances, particles have no identity. The distinction between pure samples and mixtures is also crucial; there are no 'milk' particles in the same sense as there are sugar (sucrose) particles. (Note that we should not talk about 'pure substances' since the term is misleading. The question of purity relates to the sample not to the idea of a substance. For example, in this beaker is there only water? Substances are just substances – there are not pure and impure substances.)

■ Alternative particle models

The 'basic' particle model poses challenges. To the eye, matter appears to be continuous and imagination is needed to think in terms of extremely small, discrete particles. Initially, some students may construct an image of particles embedded in continuous matter. Unfortunately, textbooks sometimes show such images and talk about particles 'in' a solid/liquid/gas, which could lead students astray. Here, the particles are not the substance, they are extra to it. Ideas of particle movement are consistent with this model since movement will be determined by the state of the continuous matter (for example, particles can move around 'in' a liquid). Identifying the particles with the substance helps to avoid such misconceptions, for example, 'sugar particles', 'water particles' and 'oxygen particles'.

Those students who do see the particles as being the substance often start by attributing macroscopic properties to the particles. For example, students may think that a copper particle is hard, malleable, conducts electricity and is copper coloured. It takes time to appreciate that the physical nature of the particles is irrelevant, with the characteristic properties of the states determined only by movement and spacing, not by what individual particles are like.

The different kinds of movement associated with the different states are not so problematic for students. However, the intrinsic nature of the movement is more demanding, especially for the solid state. The most challenging aspect of the model is the notion of empty space. Logically, if the particles are the substance, there must be nothing between the particles – anything of a material nature would be more particles. However, students are inclined to think there is something in between. Many will say 'air', although what is meant is often very vague.

Choosing a route

The chapter overview presents a sequence which develops understanding in a cumulative progression where each step builds on previous steps. It seeks to avoid introducing too many ideas at once but also ensures that the key ideas involved in an event are not overlooked. What is necessary and sufficient for a coherent account at the desired level of understanding is the aim.

Assuming students do not already hold the concept of a substance, melting behaviour is an accessible first step. Melting at a precise temperature indicates a pure sample of a substance and the temperature identifies which substance. The 'basic' particle model can then be introduced to explain melting and why different substances have different melting points. Once established for the solid and liquid states, the model is used to predict the gas state (and boiling). This approach recognises that students will not know what a gas is and, importantly, demonstrates the predictive function of scientific models.

The basic model is then used to explain dissolving, which includes looking at diffusion within the liquid state. To account for evaporation into and condensation from the air below boiling point, ideas of energy distribution are introduced. In terms of understanding the nature of models, this illustrates development of the model to accommodate further phenomena without undoing the main ideas. To tidy up we return to the gas state and consider pressure and weight and some well known diffusion experiments for illustrating intrinsic motion which draw on other ideas to differing degrees. Finally, we consider the advantages of introducing particles within a substance-based framework over a 'solids, liquids and gases' framework.

To a large extent, the sequence is governed by the logical structure of the content. However, the basic model could be consolidated for the solid and liquid states in the context of dissolving before moving on to the gas state. Each stage is now considered in more detail.

2.1 Melting and solidifying

Melting points

There are many aspects to melting, at varying depths of understanding. To begin with, introduce the idea of a melting point as when something is 'hot enough to melt'. As a class experiment, students could be challenged to find the melting point of candle wax by the method shown in Figure 2.1. (Note that a sample of candle wax is not just one substance but, within the limits of the experiment, there is a precise enough melting point. Candle wax has the advantages of being familiar and readily cut to convenient size. Octadecanoic acid (stearic acid, melting point 70 °C) or 1-hexadecanol (cetyl alcohol, melting point 56 °C) are alternatives.)

Discussion should draw attention to the independence from sample size – if melting temperature depended on amount, the experiment would not be worth doing. (Of course, the time it takes for a sample to melt does depend on amount, for example snowflake versus iceberg.)

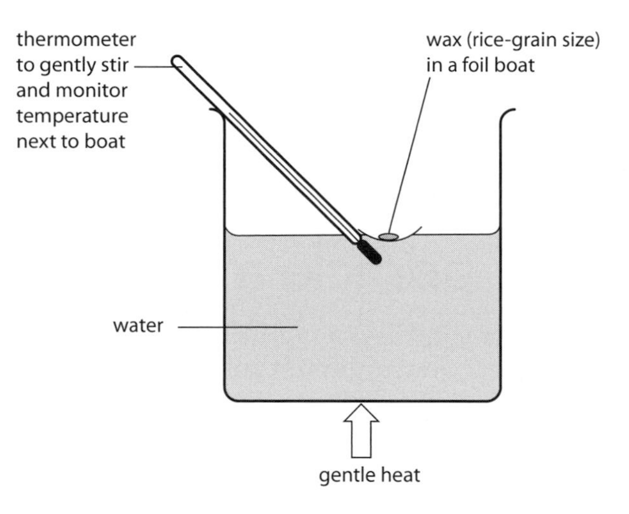

Figure 2.1 Finding the melting point of candle wax

Demonstrations of melting lead (melting point 328 °C) and sodium chloride (melting point 801 °C) can follow. (For the latter use large crystals from rock salt and three Bunsen flames.) The temperatures cannot be measured, but draw attention to the distinct change from solid to runny liquid which is characteristic of a precise melting point.

Defining a substance

A sample of a substance can now be defined as something which has a precise melting point, and solid and liquid can be defined as two of the states a substance can be in. Figure 2.2 highlights melting point as a 'switching' temperature. On heating, a substance melts when it reaches this temperature. On cooling, a substance solidifies on falling to this temperature.

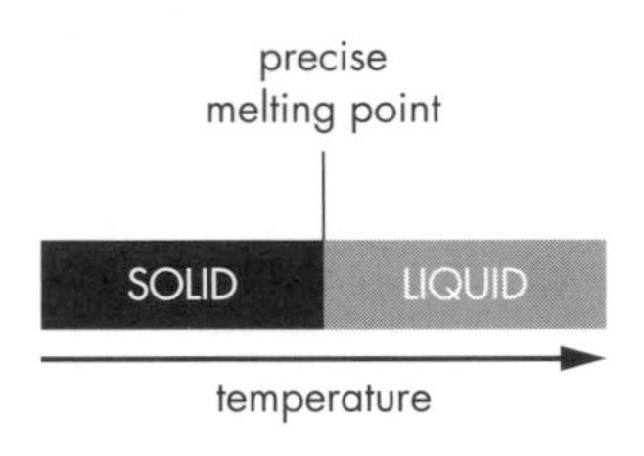

Figure 2.2 A substance has a precise melting point.

Many students think the temperature needs to fall well below melting point before solidifying occurs; perhaps the result of an association of 'freezing' with 0 °C. (This is why the word 'solidify' is preferred to the term 'freeze' in the language of chemistry.) More of the story is revealed in heating/cooling curve experiments and their interpretation in terms of latent heats. However these are difficult ideas going well beyond requirements at this stage. Thinking that solidifying takes place at just below melting point represents significant progress for most students and is a position to build on later. Other points for later are the absence of a precise melting point for mixtures (such as chocolate) and, after chemical change, the decomposition of some substances before melting (such as calcium carbonate).

2.2 A 'basic' particle model

Particle theory involves a number of ideas which work together to form a model. The minimum core for a 'basic' particle model is:

- A sample of a substance is a collection of identical particles; the particles are the substance.
- The particles hold on to each other; the 'holding power' is different for different substances.
- The particles are always moving in some way – they have energy of movement. Heating gives the particles more energy of movement – they are more energetic.

'Holding power' is preferred to talk of forces since the strengths of forces depend on the distances between particles, and distances change. Holding power belongs to the particle and does not change. However, holding power applies only to particles of the same substance – it cannot be used to predict the strength of hold between particles of different substances.

The solid and liquid states

The state of a substance depends on the balance between holding power and energy of movement (which depends on the temperature of the sample). For the solid state, the holding power can keep the particles close together in fixed places and the particles can only vibrate. For the liquid state, the holding power keeps the particles close together but not in fixed places. The particles move around from place to place. The melting point of a substance depends on the holding power – the stronger the hold, the higher it is.

In diagrams, we recommend using shapes other than circles for substance particles. This has the advantage of making it easier to show the disorder of the liquid state. With circles there is a tendency to move particles too far apart. Furthermore, if circles are reserved for atoms, the distinction between the substance and sub-substance levels of theory is emphasised.

Figure 2.3 The same substance in the solid and liquid states

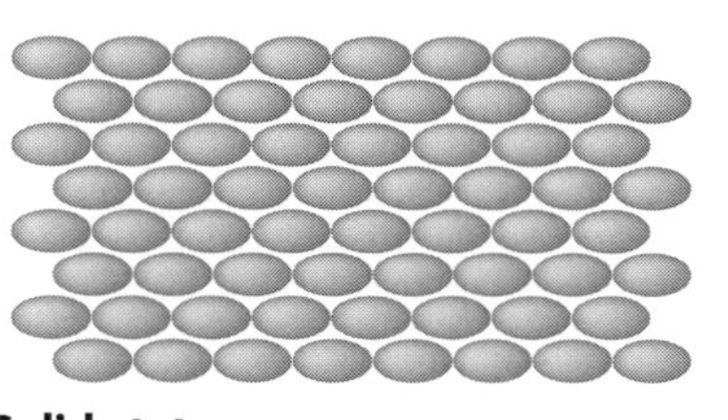

Solid state
The particles hold themselves close together.
Each particle is in a fixed position.
The particles are vibrating.
There is an ordered arrangement.

Liquid state
The particles hold themselves close together.
The particles move around from place to place, randomly.

Introducing the basic particle model

With respect to understanding the nature of models it should be noted that the phenomenon of melting of itself is not direct evidence for the existence of particles. The model can be presented as a useful way of thinking. As noted in Chapter 1, everyday discourse often uses 'particle' to describe small yet visible pieces of material and it is important to emphasise the special scientific meaning within the particle theory. A small drop of water ($0.05\,cm^3$) contains the order of 1.7×10^{21} water particles (molecules). Employing terms such as 'grains', 'pieces', 'bits', 'lumps', 'specks', 'droplets' and 'globules' in other contexts helps to make the distinction. Appreciating the absolute scale of particle size is a challenge to all of us. However, a sense of being extremely small is good enough to operate with the model.

The change in movement – from fixed places to moving around – is the key point in explaining the change from solid to liquid. To counter notions of individual particles having a state, emphasise that the particles themselves do not change – individual particles do not melt. Point out that we have not said what individual particles are like, physically. All we can say is they are not like anything we know. They behave like rubber balls but they are not made of rubber! It is also worth noting that the distances between the particles hardly changes. For melting ice, the particles actually move closer together. Moving further apart does not explain the change from solid to liquid. In considering the balance between 'hold' and 'energy', students have a tendency to think of the hold weakening rather than increased energy overcoming the hold enabling the particles to move around. (The change in average distance apart is very small and hence the average strength of forces changes little. At times, moving particles will be very close and if not energetic enough will be held in place.)

■ Limitations of an energy-only explanation

An energy explanation of melting is a limited view which does not account for ice melting at temperatures down to –12 °C in the presence of salt. Like all changes, a more complete explanation of melting concerns entropy which considers the arrangement of energy and particles. At melting point, the decrease in entropy of the surroundings (due to loss of energy supplied for the melting) is equal to the increase in entropy associated with the change from solid to liquid situation. However, ice in the presence of salt melts to form a solution, and solutions have higher entropy than pure liquid. Therefore, the increase in entropy from ice to salt solution is greater than the increase for ice to water. The greater increase compensates a greater decrease in entropy of the surroundings and so ice melts at a temperature below 0 °C (a lower temperature means a bigger change in entropy for a given change in energy). Of course, this treatment is very advanced, but students will know about salt on icy roads and may well ask. We need to acknowledge that there is more to understand about melting. For now one could say that it is easier for ice to melt to form part of a solution than a pure liquid. When measuring melting points it is important to have pure samples of substances.

2.3 Boiling and condensing

Predicting the gas state

Having established the basic particle model for the solid and liquid states, students could be invited to speculate about the continued heating of substances in the liquid state. Imagine heating a drop of water in a sealed plastic bag (with all air excluded) to higher and higher temperatures. What might happen to the particles? With little or no prompting some students will suggest that the particles could move apart because they have too much energy for the hold to keep them together. Objects held together by Velcro® can be used in an analogy. The ability to hold – the Velcro – does not go away but with enough energy the objects will separate. Students could imagine themselves wearing suits made from the two sides of Velcro arranged in a chequered pattern. They would stick together as opposite patches matched up, but with enough energy they could move apart. On losing energy, the Velcro would be able to keep them together again. (Patches of Velcro glued to polystyrene or wooden balls make an effective model.)

What is observed if the particles move apart? Students may not be so sure about this but here is an experiment worth trying! Since heating water in a plastic bag is not so easy, inject a small volume of water (0.05 cm^3) into a sealed gas syringe preheated to around 150 °C instead. On the bench out of its oven, the syringe retains a high enough temperature long enough to give satisfactory results. Explain that the small amount of water will quickly heat up on contact with the hot glass. The plunger moves out quickly, the liquid water disappears (some bubbling may be seen) and inside the syringe is clear (Figure 2.4). Practice is recommended since the volume of gas is very sensitive to the volume of liquid water, and speed is of the essence. Reasonably consistent volumes around the 100 cm^3 mark are achievable, but back-up syringes are advisable in case another go is necessary. On cooling, condensation appears as the plunger moves in. (The plunger might not go right back in because some air may intrude through the pierced seal.)

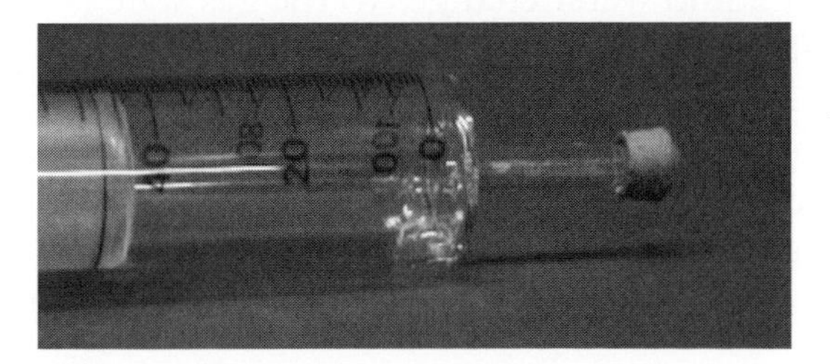

Figure 2.4 Water changing to the gas state

There are many important points to discuss. When particles move apart the gas state is created (Figure 2.5). The particles are still water particles so this is water in the gas state. Draw attention to the huge change in volume. If the particles close together only take up 0.05 cm^3, what is between the particles when they are apart in the gas state? Many students will want to say air (it looks like air inside the syringe), but remind them that there is only water in the syringe – where could air have come from? We are forced to conclude there is nothing between the particles – empty space. Some students may suggest that the particles expand. This is consistent with the observations but point out that we would have to explain why particles can expand. It is simpler to say the particles do not change and scientists always prefer the simplest model that works. For solid to liquid the key change is a change in movement (fixed positions to moving around); for liquid to gas the key change is a change in spacing (from close together to apart). Ideally, students will not have heard about water being made of hydrogen and oxygen since this can be a source of confusion. If they have, some will say the gas is oxygen and/or hydrogen, because these are known to be 'gases'. Again, point out that we do not need a more complicated explanation (and tests show it is still water).

Figure 2.5 The gas state

Gas state
The particles do not hold themselves together. Particles are apart, moving freely in all diections. There is a lot of empty space (nothing) between the particles.

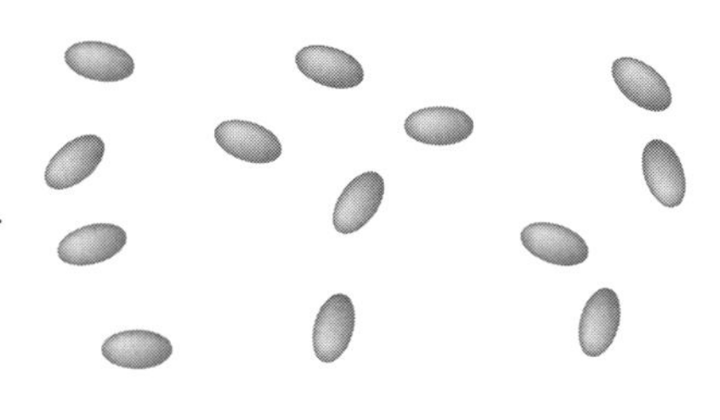

Boiling water

The syringe experiment helps students to understand the more complex event of a beaker of boiling water. What are the big bubbles? Previously, most will have said 'air' but now they know water in the gas state looks like air. The parallel with the syringe experiment can be made. At the bottom of the beaker, a small amount of liquid water changes to the gas state (some particles move apart) to form a bubble. Draw attention to the mist above the boiling water. Is this water in the gas state? No! Water in the gas state cannot be seen. The mist is where the gaseous water has condensed to form small droplets of liquid water. To corroborate, generate another syringe of clear gaseous water, remove the cap and expel the gas to give a small puff of mist on cooling in the air.

Boiling points

Students could measure the boiling point of water; the boiling points of other substances such as ethanol (79 °C) and propanone (56 °C) could be demonstrated (avoid naked flames, heat on a hot plate in a fume cupboard). As with melting, place the initial focus on the idea of boiling point as a 'switching' temperature. On heating, a substance changes from the liquid to the gas state when it reaches this temperature. On cooling, a substance changes from the gas to the liquid state when it falls to this temperature (it condenses). Again, full discussion of temperature–time graphs associated with the change of state could be left for later. However, the steady temperature during boiling is very noticeable and could be explained in terms of needing energy to change from liquid to gas. This is why the liquid does not change to gas all at once (and a parallel could be drawn with melting). The substance and its states chart can now be completed as in Figure 2.6.

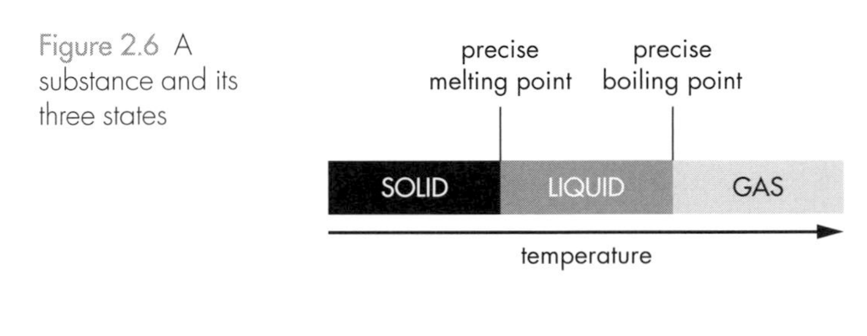

Figure 2.6 A substance and its three states

'Gases'

Students are now in a position to understand what 'gases' are. These are substances with boiling points below room temperature and therefore in the gas state at room temperature. In effect, they have already boiled to the gas state. Boiling points of these substances are low because the holding powers of their particles are very weak. At room temperature, the particles are energetic enough to be apart. Like water in the gas state, nearly all gases are clear and colourless. Looking the same hardly helps with the notion of being so many different substances. Putting a burning splint to samples of nitrogen (splint goes out), oxygen (splint burns brighter) and methane (catches alight) shows they are very different. Introducing a jar of chlorine shows that the gas state can have colour, but is still clear. It is worth stressing the clarity of the gas state to counter any confusion about mists and smokes being gases. The gas state is clear because particles are separate from each other and we cannot see individual particles.

■ Air

Air can now be understood as a mixture of substances that are all above their boiling points. Translating the percentage composition into particles is a useful consolidation exercise. Out of every 10 000 particles, 7800 are nitrogen, 2100 are oxygen, 93 are argon, four are carbon dioxide and three are various others. Given all the discussion in the media about carbon dioxide, students will be surprised to find out there are so few particles of it. However, proportionality is the point. Putting more carbon dioxide into the atmosphere makes a significant increase because there is not much to start with.

2.4 Dissolving

Dissolving provides a context to strengthen and develop students' understanding of the basic particle model, but also to recognise its limitations.

Recognising and explaining dissolving

For a focal event, students could watch a single, large crystal of common salt (sodium chloride) dissolve in water on gentle stirring. The crystal's gradual disappearance is explained by salt particles mixing in with water particles. We can see a crystal because there are lots of particles together in a lump. We cannot see individual salt particles dispersed among the water particles so the solution looks clear (nor can we see individual water particles, just the water as a whole). The crystal has dissolved but the salt particles are still there. Suspensions of fine powders are more challenging. Many students will say powdered chalk (calcium carbonate) dissolves on stirring because it spreads throughout the water turning it white. For some students it is worth spending time on what a powder is – they could make one by grinding a lump of chalk in a pestle and mortar. Emphasise that each tiny piece of chalk is still made of billions and billions of particles. On close inspection of the suspension, the tiny pieces of chalk are visible and eventually settle to the bottom. The pieces of chalk have not dissolved. From this, clearness emerges as the key criterion for recognising dissolving (as with the gas state). Copper sulfate could be introduced to show a clear but coloured solution. Comparing the results of passing solutions and suspensions through filter paper reinforces their difference and provides another opportunity to apply the particle model. (Make sure the filter paper grade is fine enough.)

Solutes in the liquid and gas states

Lest students think dissolving only applies to solutes in the solid state, those in the liquid and gas states (at room temperature) should be included. For the liquid state, ethanol, glycerine, olive oil and volasil are suitable examples to test in water. The first two dissolve readily, the last two do not – all can be used by students. (Although olive oil is a mixture, it all behaves in the same way.)

For a solute in the gas state, injecting 1 cm^3 of water into a gas syringe of ammonia (100 cm^3) is an effective demonstration. The plunger moves in quickly at first, slowing until all of the ammonia has dissolved in the water. Explaining these observations tests students' understanding of the gas state. How can 100 cm^3 of ammonia dissolve in only 1 cm^3 of water? The key point is that most of the 100 cm^3 is nothing – empty space. We know that the particles of 100 cm^3 of water in the gas state make a small drop when together in the liquid state. So, 100 cm^3 of ammonia in the gas state is equivalent in amount to a drop or a small crystal. A drop or small crystal dissolving in 1 cm^3 of water is not such a surprise.

It is worth emphasising that the original state of a solute has no bearing on the nature of the solution. Particle diagrams for solutions of sugar, ethanol and ammonia look the same save the different solute particle shapes (Figure 2.7). A dissolved substance does not have a state as such.

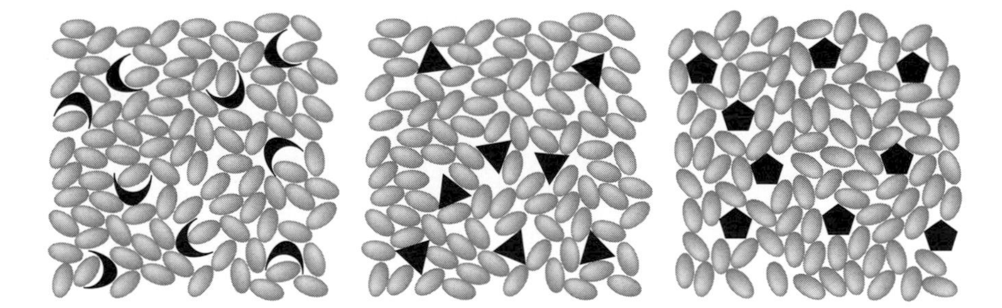

Figure 2.7 Particle diagrams of sugar, ethanol and ammonia solutions

Empty space in the liquid and solid states

Volume and empty space can be covered in the context of liquid state solutes. When 50 cm^3 of ethanol and 50 cm^3 of water are mixed, the total volume is around 96 cm^3. Even though particles in the liquid state are close together, there will be pockets of space where the shapes do not fit together snugly, especially since they are moving around. The particles of another substance could go into some of this space and vice versa. Mixing pasta shells and lentils makes a good analogy. By the same argument, there will be pockets of empty space between particles in the solid state too. Students could be challenged to find out if volume is 'lost' with solid solutes.

Intrinsic motion and the liquid state

To target the idea of intrinsic motion, students could be asked to predict and then observe whether a crystal of salt dissolves without stirring. This also moves thinking on to the mechanism of the change. Even for still water, individually, the particles are moving around. Bombarding water particles knock salt particles out of the crystal, gradually dismantling it. Using a coloured crystal like copper sulfate or potassium manganate(VII) shows how the particles of the dissolved substance spread out due to the intrinsic random motion of the liquid state (a classic diffusion practical).

■ Why does stirring speed up dissolving?

Stirring has a big effect on the rate of dissolving. Why? Most students will suggest that stirring gives the particles more energy. This is a plausible explanation. However, more energetic particles would mean a higher temperature and stirring does not increase the temperature of the water measurably. Stirring moves batches or swathes of particles around but it does not increase their individual, random movement. So, why does stirring speed up dissolving? The diffusion experiment with a coloured crystal gives a clue. Without stirring, a high concentration of solute particles builds up next to the crystal. Through random collisions, some particles are knocked back to the crystal and rejoin. Therefore, the rate at which the crystal actually disappears depends on the difference between the rate of particles 'leaving' and rate of particles 'rejoining'. Stirring moves 'dissolved' particles away from the crystal and hence reduces the rate of rejoining (so long as the incoming solution has a lower concentration). Using hypothetical numbers can help to make the point. For example, if the rate of leaving is 100 particles per second and the rate of rejoining without stirring is 60 particles per second, the rate of dissolving is 40 particles per second. On stirring, the rate of leaving will still be 100 but the rate of joining will reduce to, say, 20. The rate of dissolving will be 80 particles per second. You may feel this is too advanced for your students. However, the idea of rejoining is not so difficult. Importantly, it counters the teleological thinking that tends to creep into explanations: particles have no intentions. Furthermore, the notion of a two-way process lays important groundwork for understanding dynamic equilibrium at a later stage.

Solubility

The basic model explains dissolving but it cannot deal with solubility. There is no simple explanation for why up to 203 g of sucrose will dissolve in $100\,cm^3$ of water, compared to 36 g of sodium chloride and only 0.1 g of calcium hydroxide (all at room temperature). All are saturated solutions but saturation has got nothing to do with filling up spaces. Similarly, there is no simple explanation for the different effects of temperature on the solubility of different substances, with some increasing (such as potassium nitrate), some decreasing (carbon dioxide) and some almost unaffected (sodium chloride). Entropy is needed to understand solubility. However, solubility is a topic that cannot be ignored in early secondary school chemistry and this is a case where the limitations of our model should be openly acknowledged. Our model is not wrong, but other ideas are needed to tell the whole story of dissolving and this will have to wait. Meanwhile, viewing saturation in terms of equal rates of 'leaving' and 'rejoining' does accommodate the wide range in solubilities but does not explain why this arises at such different concentrations (down to effectively zero) and the differential effect of temperature.

Another aspect of solubility is the use of different solvents. For example, why will sodium chloride dissolve in water but not in propanone, whereas olive oil will dissolve in propanone but not in water? Developing the basic model with notions of different ways of holding and compatibility between ways gives some account. So, sodium chloride particles hold on to sodium chloride particles and propanone particles hold on to propanone particles, but sodium chloride and propanone particles do not have a good hold for each other and will not mix. On the other hand, sodium chloride and water particles do have a hold for each other. A general idea of types of hold prepares the ground for the more detailed look at structure and bonding covered in Chapter 3.

2.5 Evaporation into and condensation from the air

Evaporation and boiling

From everyday experience, students will know that water evaporates at temperatures well below boiling point. The basic particle model can deal with the overall disappearance – water particles become mixed in among air particles – but it cannot explain how this can happen. Moreover, to explain boiling we said particles were energetic enough to overcome the hold and move apart. If water must be at 100 °C to boil, how can particles escape from each other when less energetic at much lower temperatures? There seems to be a serious contradiction here which undermines our model rather than merely exposing its limitations as was the case with solubility. Bringing in ideas of energy distribution resolves the problem.

Temperature and energy

The idea of energy distribution is best introduced with the gas state. Without going into details of momentum, students can appreciate that energy is exchanged in collisions (snooker is a useful analogy). The total amount of energy within the sample stays the same but individual particles at any one time have different amounts of energy. We can say that temperature relates to the *average* energy of the particles. Much can be done with a simplified idea of low-, medium- and high-energy particles. At higher temperatures there are more high-energy particles and fewer low-energy particles. The same arguments apply to the liquid and solid states (energy is exchanged between neighbouring vibrating particles). It is worth emphasising that individual particles do not have a temperature in the same way that they do not have a state.

Explaining evaporation below boiling point

Evaporation is explained as follows (Figure 2.8). High-energy water particles that happen to be at the surface of the sample, and that happen to be moving in an outward direction, will be able to overcome the hold and move into the air. The escaped water particles mix with the air particles (here, we will not distinguish air itself as a mixture of substances). At the same time, high-energy air particles collide with other water particles (those of medium and low energy) at the surface. Energy is transferred in the collisions so the water particles gain enough energy to escape. In effect, the water particles are 'knocked out'. In this way all of the water particles will eventually escape from the saucer and mix in with the air particles.

Viewing the event in terms of energy and temperature, the initial loss of high-energy water particles (that can leave on their own) will cause the temperature of the remaining sample to drop (the average energy will be lower). Energy will then transfer from the room to the remaining water in the saucer (through collisions with high-energy air particles). This helps to maintain the number of particles with enough energy to escape, and so the particles of the sample gradually disperse among the air particles. Although not as obvious as heating with a Bunsen, the room is providing the energy to maintain the evaporation of the water.

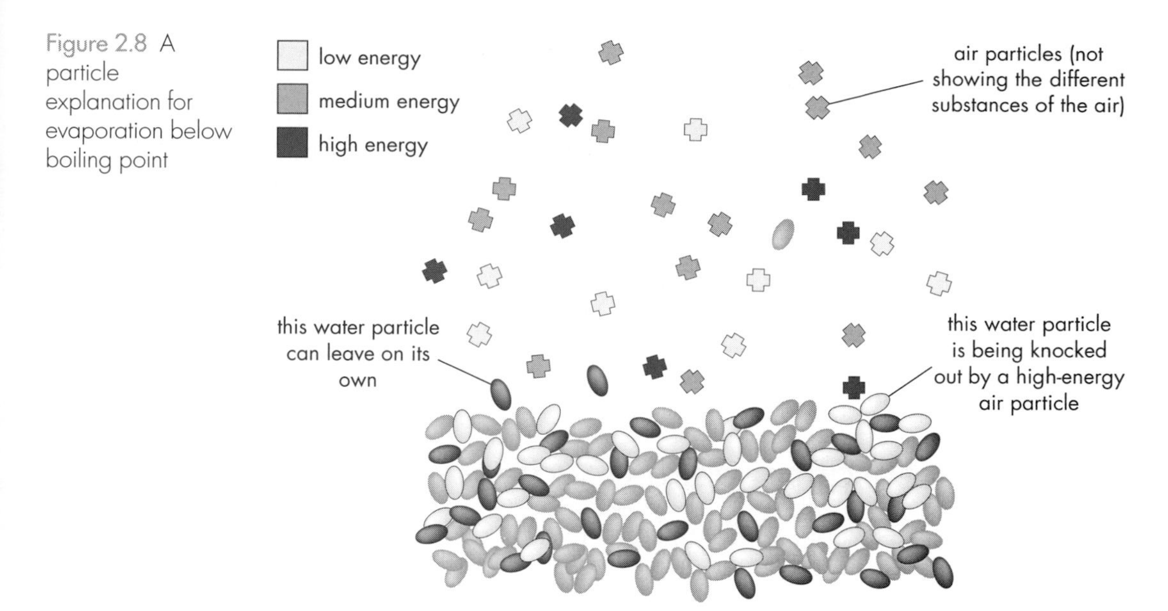

Figure 2.8 A particle explanation for evaporation below boiling point

Factors affecting the rate of evaporation

The model easily accounts for the effects of surface area, temperature and a breeze on the rate of evaporation. The action takes place at the surface and so with a larger surface area, more can happen at once. At higher temperatures there are more high-energy particles about. The effect of a breeze parallels the effect of stirring on dissolving. In the random movement, water particles could return. The actual rate of evaporation is the difference between the rate of leaving and rate of rejoining. A breeze sweeps away water particles, thus reducing the rate of rejoining. (Note that applying ideas of energy distribution refines our previous explanation of dissolving. Only the high-energy solvent particles 'knock' particles away from a crystal (and higher-energy crystal particles are more easily knocked out). Rate of dissolving increases with temperature because more particles have high energy.)

Reconciling boiling and evaporation

Having explained evaporation below boiling point with our developed model, we should reconsider boiling. It is worth highlighting the differences. Evaporation takes place at any temperature (between melting and boiling) and there are no bubbles. Boiling takes place at a particular temperature and there are bubbles of water in the gas state. For evaporation, the particles leave one by one from the surface to form a mixture with air particles. We cannot see individual particles leaving and at any temperature there will be some high-energy particles for this to happen. For boiling, a bubble is a pure sample of water in the gas state that forms inside the liquid water (where it is being heated). It takes more than one particle to form a bubble. This means there have to be enough high-energy water particles able to overcome the hold at the same time. There are only enough high-energy water particles when the temperature reaches 100 °C. Our improved model gives a better account of why boiling takes place at a particular temperature and can also explain why boiling point depends on external pressure. This will be covered later in the chapter.

Condensation from a water–air mixture

The appearance of condensation on a cold object is a perplexing event for many students. They know the liquid is water, but are not sure where it comes from. Many students think the water goes straight up to the clouds after evaporating rather than occupying the air all over (most water cycle diagrams do depict it going straight up). Our explanation of evaporation shows how water particles can mix into the air and this opens up the possibility of some water particles being a normal part of air. However, a little more thinking with the model is required.

Although escaping water particles have high energy, all will not retain this status. On mixing and colliding with air particles, a distribution of energies among the water particles will re-establish. The distribution will match that of the air particles since the whole collection of particles is at one temperature. In that case, when low- and medium-energy water particles meet up (as they will through random movement) one could ask why they do not cluster together to form a droplet of water in the liquid state (they will not have enough energy to escape their hold). This could happen, but a high-energy particle (water or air – most likely air because there are more air particles than water particles) could also knock the water particles apart (as with evaporation). The outcome will depend on the relative rates or chances of the two competing processes. The chances of water droplets being broken apart will depend on the number of high-energy particles, which depends on the temperature. The chances of water particles joining will largely depend on the concentration of water, and also the temperature since this affects the proportion of low- and medium-energy particles within the concentration. In a clear air–water mixture, like normal room temperature air, the rate of breaking apart is greater than the rate of clustering and so droplets do not form. On cooling the mixture, the rate of breaking apart decreases. When the rate of breaking goes below the rate of clustering, droplets of water will form. The temperature at which this happens will depend on the concentration of water.

In explaining the appearance of condensation, the emphasis is usually on temperature but concentration is an equally important factor. For example, when a jar is placed over a lighted candle, condensation forms despite the glass being warmed. This is because the flame gives out water and the concentration inside becomes very high. On lifting the jar the condensation quickly evaporates because the concentration falls – there is no change in temperature.

■ Could evaporation and condensation be explained without ideas of energy distribution?

If ideas of energy distribution are felt to be too intricate for some students, a simpler treatment of water–air mixtures could be given. Evaporation could be explained in terms of bombarding air particles knocking out water particles. Importantly, this explains why evaporation happens below boiling point. Water particles then become mixed with air particles, with the air particles keeping them apart. The more energy the air particles have, the better they are at keeping the water particles apart. Therefore, on cooling there comes a point when the water particles cannot be kept apart and condensation forms. How cold the mixture has to be for condensation to form depends on the concentration of water in the air: the more water, the less cold it needs to be.

2.6 More on the gas state: pressure, weight and diffusion

Our arguments for water–air mixtures also apply to pure samples of substances in the gas state. Higher-energy particles prevent lower-energy particles from clustering. Where the holding power is very weak, only the very lowest-energy particles would cluster and these are far outnumbered by the rest.

The gas state and pressure

Gas state pressure is readily tackled by the basic particle model. Bombarding particles exert a force on the walls of their container. The total force on an area depends on how strong the collisions are and the number of collisions at any one time. For a given sample of gas, the average strength of collisions depends on the average speed, in other words, temperature. The number of collisions will depend partly on the average speed but also on the number of particles in a given volume (concentration). Therefore pressure increases by either raising the temperature or squashing to a smaller volume.

Pressure considerations could be addressed in some of the events we have looked at. For the 'drop of water in hot syringe' experiment, one could ask why the plunger stops moving and explain that this is when the pressure of gaseous water inside equals atmospheric pressure outside. For a beaker of boiling water, the bubbles form when there are enough high-energy particles to give a pocket of gas with a pressure equal to atmospheric pressure. From this we can predict that water will boil at a lower temperature if the external pressure is reduced. On injecting water into a syringe of ammonia, the plunger is pushed in by atmospheric pressure since the pressure inside falls as the ammonia dissolves. The classic 'fountain experiment' makes a suitably spectacular follow-up demonstration. (This can be found as no. 79 in *Classic Chemistry Demonstrations*, Royal Society of Chemistry, 1998.)

Mass and weight of the gas state

The gas state presents many challenges, not least the idea that a sample of gas has a mass and therefore a weight due to gravity. Understanding that gases are substances ought to help with the idea of having mass. The particles of a substance have mass and this does not change with a change of state. However, understanding how a sample of gas exerts a weight is not at all straightforward. Consider an absolutely empty container (strong enough to resist atmospheric pressure) standing on a balance. If air is let into the container, the balance reading increases to show the weight of the air. However, since most of the air particles are not in contact with the bottom of the container, how can this be? The explanation lies with the concentration and hence pressure gradient caused by gravity acting on the sample. The pressure acting downwards on the bottom is greater than the pressure acting upwards on the top, which gives an overall downwards force. This so happens to equal the weight of all particles. Such arguments must wait for later (perhaps in physics). We raise the issue to point out that it is not at all obvious why a sample of gas has a weight. The increase in weight when particles from the gas state end up combined in the solid state is less problematic (such as oxygen in magnesium oxide).

Diffusion involving the gas state

■ Ammonia and hydrogen chloride

Perhaps the best known diffusion experiment is that shown in Figure 2.9.

Figure 2.9 Diffusion of ammonia and hydrogen chloride in air

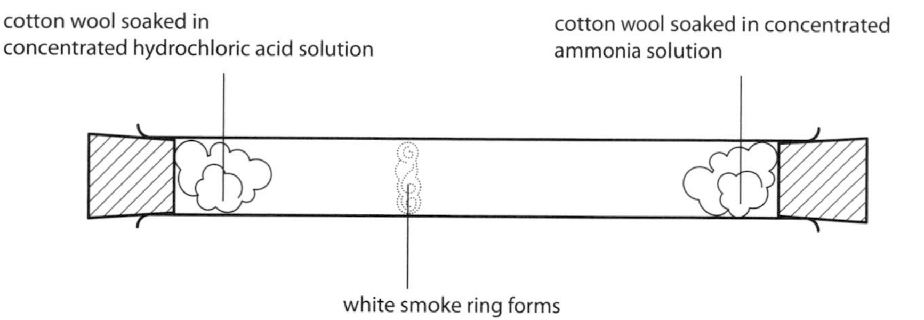

Once solutions of gaseous solutes have been addressed, students will be in the position to appreciate that some ammonia particles escape from concentrated ammonia solution and likewise for hydrogen chloride. The appearance of the white rings shows that the respective substance particles have met and have therefore moved by themselves (making their way in among air particles). The faster progress of the ammonia particles is also evident and this could just be left as an observation. Explaining the faster movement will draw on ideas of kinetic energy ($\frac{1}{2}mv^2$). Assuming the same average kinetic energy, ammonia particles have a higher speed to compensate their smaller mass.

■ Perfume

Pouring perfume or a smelly substance like propanone into a dish involves evaporation and mixing into the air. Some students think 'smell' is something else infusing the substance so it is important to establish smell as an interaction between particles and nose receptors which leads to a stimulation in the brain (theories suggest either through shape or vibrational frequency). Interestingly, some animals can smell carbon dioxide which enables them to catch prey in the dark. Also be aware that this demonstration gives a false impression of the rate of diffusion. Most of the spreading out is due to particles being carried by air currents rather than their individual random motion.

■ Samples of substances in the gas state

The simplest way to demonstrate diffusion in the gas state is to use substances in the gas state. Traditionally, samples are positioned in alternate vertical arrangements (Figure 2.10a). However, this adds the complication of weight and different densities. Arranging the tubes horizontally avoids the issue (Figure 2.10b).

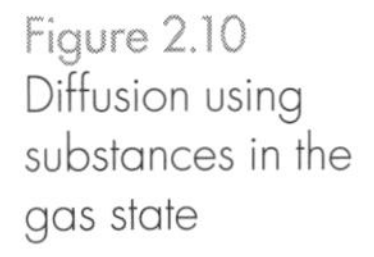

Figure 2.10 Diffusion using substances in the gas state

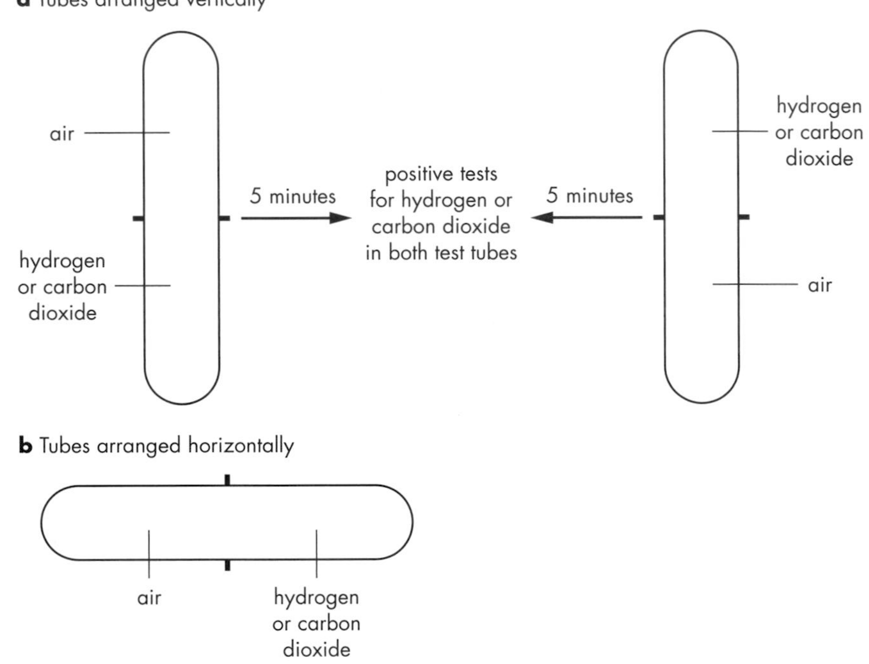

2.7 What, no 'solids, liquids and gases'?

Most likely, you will have been expecting to read about 'solids', 'liquids' and 'gases' in a chapter on introducing particle theory. This is the conventional approach. In fact, we have avoided all reference to 'solids', 'liquids' and 'gases' because fundamentally there are no such things. For the chemist, there are only substances and their states. The room temperature state is quite arbitrary and has no fundamental significance (Figure 2.11).

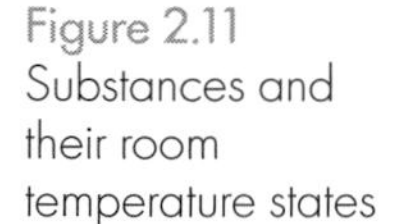
Figure 2.11 Substances and their room temperature states

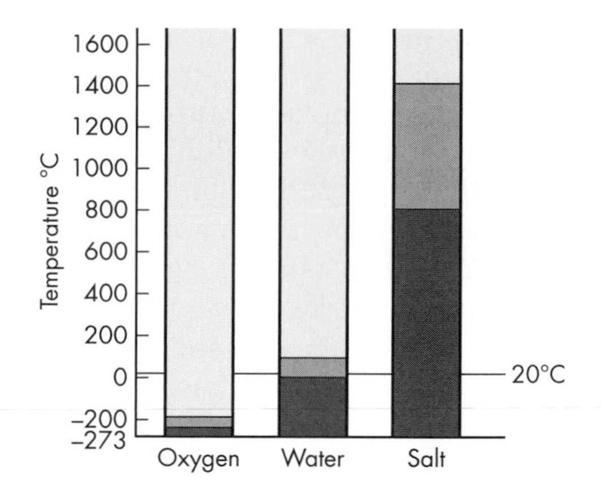

Three types of matter?

At first sight, basing our teaching on 'solids, liquids and gases' seems appropriate since this reflects direct experience and is more user friendly. Classifying materials into the three categories makes a nice activity. However, this carries the danger of promoting the idea that 'solids', 'liquids' and 'gases' are three separate kinds of matter, akin to different species. Students could come out thinking something is either 'a solid', 'a liquid' or 'a gas': many do. Changes of state are then viewed as the anomalous and somewhat confusing behaviour of some substances (for example, is wax 'a solid' or 'a liquid'?). Thinking of 'gases' as a third kind of matter preserves their mystery and inhibits the idea that gases are substances just as much as anything else.

The concept of a substance is important

By focusing on the states generically, the 'solids, liquids and gases' approach also ignores the concept of a substance. States do not have an identity and therefore the particles lack identity – the talk is in terms of 'solid particles', 'liquid particles' and 'gas particles'. This suggests different types of particle, which reinforces the idea of three different kinds of matter and that particles carry the macroscopic properties – the very common misconception noted earlier.

Ignoring the concept of a substance precludes the important distinction between pure samples of substances and mixtures of substances. All 'solids' are treated as one and similarly 'liquids' and 'gases'. In classification activities, this can lead to the very unsatisfactory situation where difficulties associated with classifying gels and pastes are seen as limitations of the framework rather than their nature as mixtures. Strictly speaking, the idea of a state only applies to pure samples of substances and in these cases deciding which state is relatively unproblematic. Students should not be asked to do the impossible as if it were possible. The first thing a chemist wants to know about a sample of stuff is whether it is a pure sample of a substance or not. Separation techniques are so important because pure samples are wanted (well, pure enough).

Distinguishing between mixtures and substances helps explain the difference between melting and dissolving. Otherwise, if sugar solution is just 'a liquid', why has the sugar not melted? Similarly, there would be no differentiation between boiling and evaporation below boiling point. In the 'solids, liquids and gases' approach, both are treated as a change of state. However, as noted earlier, it makes no sense to say that water changes to the gas state under such different conditions. Evaporation into the air below boiling point is better viewed as a mixing phenomenon akin to dissolving.

Another limitation of the 'solids, liquids and gases' approach is the way it ties the strengths of forces to the state. Thus 'solids' have strong forces, 'liquids' medium forces and 'gases' very weak forces. Of course, this pertains if comparing different substances at, say, room temperature. However, it does not deal well with changes of state. On melting there is not a significant weakening of forces since the average distance changes little. Furthermore, oxygen in the solid state does not have stronger forces than iron in the liquid state. More insidious is encouragement of the idea that force strength is a consequence of the state (because X is a solid etc.) rather than a factor in determining the state along with particle energy. Again, ideas of three types of matter are boosted.

Introducing a 'basic' particle model through substances covers everything the 'solids, liquids and gases' approach seeks to do but avoids the problems. Scientifically, it is more accurate and coherent. As noted earlier, the concept of substance is not obvious and needs to be taught alongside particle theory in mutual support. Identifying basic particles with the substance also lays a secure foundation for ideas of atoms. For example, when there is talk of 'oxygen', students must be able to distinguish between oxygen atoms and the substance oxygen (O_2) – they are quite different. (This was covered in greater depth in Chapter 1.)

Other resources

Stuff and Substance is a multimedia package developed by the Gatsby Science Enhancement Programme (SEP) which provides teaching materials to support a substance-based approach to introducing the particle theory. *Stuff and Substance: Ten Key Practicals in Chemistry* is a booklet which complements the multimedia package. It looks at how practical work can be used to support the development of key ideas. The *Stuff and Substance* materials are available online in the National STEM Centre elibrary (www.nationalstemcentre.org.uk/). Registered users of the National STEM Centre can also download the individual videos and animations.

Some articles describing the research background are:

Johnson (1998). Progression in children's understanding of a 'basic' particle theory: A longitudinal study. *International Journal of Science Education*, **20** (4), 393–412.

Johnson & Papageorgiou (2010). Rethinking the introduction of particle theory: a substance-based framework. *Journal of Research in Science Teaching*, **47** (2), 130–150.

Johnson & Tymms (2011). The emergence of a learning progression in middle school chemistry relating to the concept of a substance. *Journal of Research in Science Teaching*, **48** (8), 849–984.

3 Introducing chemical change

Keith S. Taber

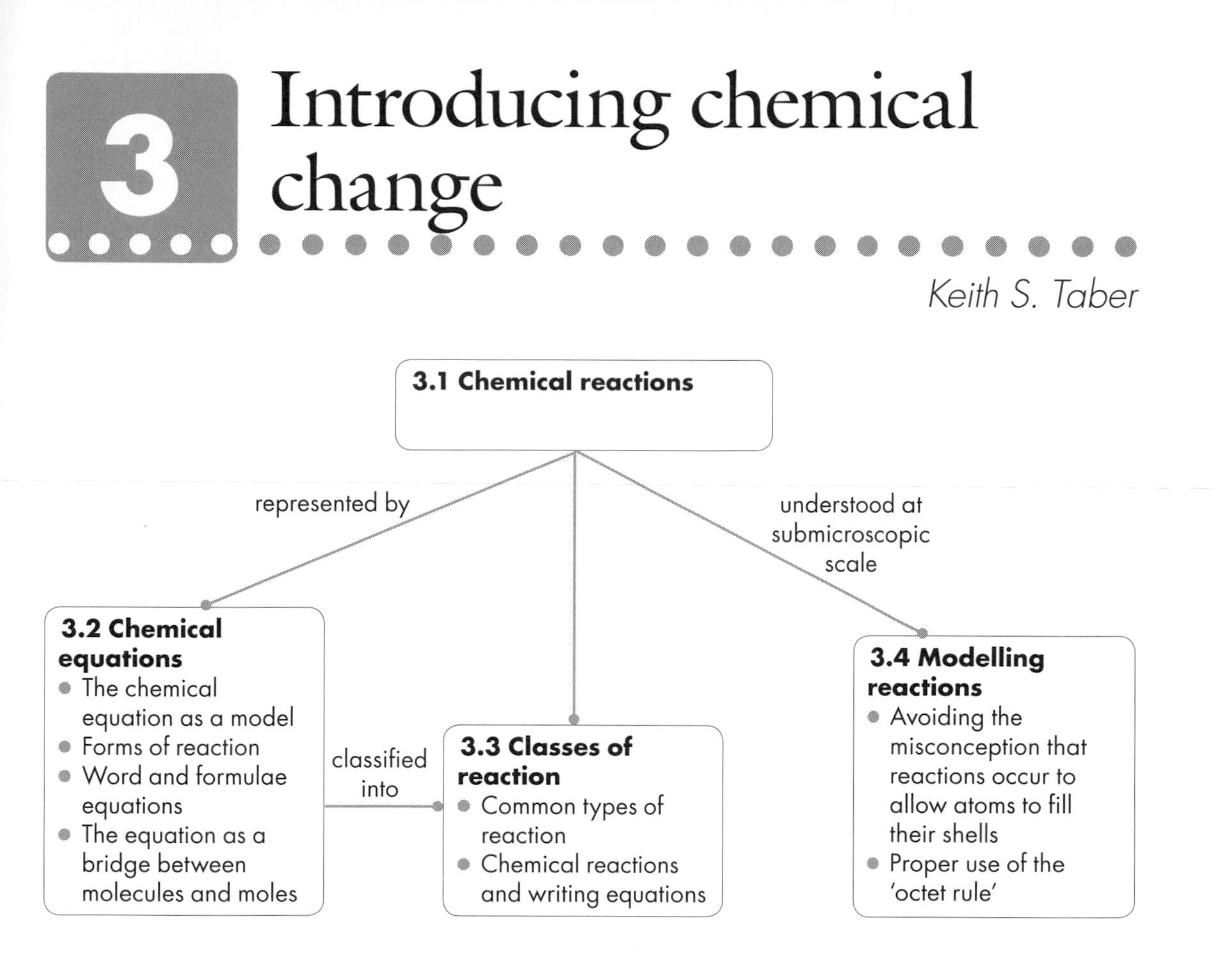

Previous knowledge and experience

Students will come to secondary school with experience of chemical reactions such as combustion (gas cookers, bonfires, etc.) and aware that cooking brings about changes in foodstuffs (although these are rather complex examples), as well as knowing that some household 'chemicals' such as bleach are used to 'kill germs'. Unfortunately most experience in everyday life does not provide a clear basis for characterising chemical changes, with many household 'chemicals' (which are usually mixtures of different substances, a topic covered in more depth in Chapter 1), such as washing powders, largely working without chemical change (but rather facilitating separation techniques). Moreover, not only are chemical changes not a clear distinct category of phenomena (Chapter 1), but students often tend to think of changes in terms of an active agent (such as the bleach) which acts upon a substrate, which is not helpful for thinking about the reaction between two substances.

As chemical change is so central to chemistry, it should be introduced early in secondary education, and then reinforced through the secondary years. A key concept in understanding chemical models of what is going on during chemical change is that of chemical bonding (the basis for the 'holding power' between atoms and molecules as discussed in Chapter 2). This is an abstract and challenging topic that depends upon understanding models of molecular and atomic structure, and so will only be explored in the more senior years (during upper secondary chemistry). This chapter discusses the introduction of teaching about chemical change, and the following chapter considers the related but more advanced topic of chemical bonding. The key difference between these two topics is that although both rely on models of matter at the submicroscopic scale, chemical changes can be directly observed, whereas chemical bonding is not an observable phenomenon but a theoretical construct used in explaining chemistry at the 'quanticle' (see Chapter 1) level.

Choosing a route

The comments made in Chapter 1 about teaching chemistry in an iterative fashion apply here as well (and indeed there is some inevitable overlap in developing basic chemical ideas across Chapters 1–3). The topics included in this chapter have been arranged in an apparently logical order to build up an understanding of the basic ideas relating to the key concept of chemical reaction. However, the teaching needs to be iterative and to follow a spiral curriculum: learning about each of the key concepts of this chapter will be supported by, and will support, learning about the others.

Moreover, even if students appear to have a good grasp of these ideas when they are first introduced, research suggests that abstract concepts such as this will become muddled over time unless there is careful regular reinforcement. Luckily, the ideas met in this chapter (as with the other chapters up to Chapter 5) have applications across the teaching of chemistry. It is important, therefore, to refer to these ideas explicitly in the teaching of other topics, in order to review understanding of the key ideas regularly. Indeed, this is necessary to ensure that students are making the links that chemists and science teachers take for granted. Again, research suggests that what seems obvious to teachers (because we have stressed previously how it always applies in chemistry; because the technical language we use should imply it clearly) is readily missed by many students. So when teaching about topics such as acids, geochemistry, chemical analysis and so forth, it is useful to ensure that students are thinking about the new teaching in terms of the desired understanding of the basic concepts of the subject.

The route set out here reflects a very common feature of teaching and learning chemistry – the shift between (i) macroscopic (bench) phenomena; (ii) their formal representation in conventional symbolic language (formulae, chemical equations); and (iii) the explanatory models that chemists use that are based on theoretical ideas about the structure of matter at a submicroscopic scale (molecules, ions, etc.). The flow chart at the start of this chapter has been set out to illustrate the main relationships between these three 'levels' of communicating about chemistry. Whatever route is taken through this material, it is important to signpost the shifts between these three levels (the three 'columns' in the flow chart), to help students begin to appreciate how chemistry explains the material world by working at these three distinct levels.

3.1 Chemical reactions

The distinction between physical and chemical change was discussed in Chapter 1. There it was suggested that although this is not a clear-cut distinction, it can often be a useful simplification. In chemical terms, the distinction concerns whether there is a change of substance: in a physical change the same substance changes its form or state; in a chemical change, different substances are present after the change.

In chemistry we refer to chemical changes as 'reactions' or 'chemical reactions'. Like many terms used in a technical sense in the sciences, 'reaction' can have unhelpful associations for students. The term react can imply a response to something, and research suggests that for many students a chemical reaction is understood as one chemical in some sense provoking a reaction in another. That is, one chemical is seen as being the active substance, bringing about change, while the other is more a victim of chemical intimidation! For example, when acids react with other substances, students may assume that it is the acid that is actively bringing about the reaction in the other substance. Students may also see the chemical reaction as being a 'reaction' to heating or stirring, or even as a 'reaction' to the chemist adding a reactant. It may be useful to adopt a class motto along the lines 'reacting with, not reacting to'.

Colleagues teaching physics also use the term reaction in relation to forces. Whenever there is an interaction between two bodies (a person standing on the Earth, for example), there are forces of equal magnitude acting on both bodies. These forces have traditionally been called 'action' and 'reaction', but any implication that one body initiates the interaction and the other responds is completely wrong: the interaction is always mutual, without any time lag in the symmetry.

The tendency of students to see chemical reactions in a similar way, initiated by one partner, is perhaps not so problematic. After all, chemists often think and talk in terms of 'attacking' species in reaction mechanisms. However, there are some chemical reactions where this does cause a problem. Consider the following change, which occurs when copper carbonate powder is strongly heated:

copper carbonate → copper oxide + carbon dioxide

This is an example of a decomposition reaction (considered further later in the chapter). This is a chemical change, as the substance present at the start (copper carbonate, a green solid) is no longer present after the change. Instead two new substances have been produced: black copper oxide powder and invisible carbon dioxide gas. Copper carbonate is reacting, but it is not 'reacting to' another chemical substance.

Thermal decomposition reflects how material tends to become less aggregated at higher temperatures (covered in greater depth in Chapter 2): condensed matter (substances in the solid or liquid state) will vaporise if heated sufficiently; complex molecules will break down to simpler ones or fragments. If we carried on heating, we could atomise materials and ionise the atoms; in theory, sufficient heating would decompose nuclei and even the nucleons themselves. Chemists do not have to worry about these latter stages (as they only occur under extreme conditions, such as in the very early Universe), but it is important to recognise that when we talk about the 'stability' of a substance or a chemical species, we should always bear in mind the conditions being referred to. At room temperature copper carbonate seems perfectly stable, but at a higher temperature (as reached by a Bunsen burner flame) it spontaneously decomposes.

Some students will, therefore, have difficulty considering thermal decomposition as a 'proper' chemical reaction, as the copper carbonate does not react with any other substance. Other students will not have this difficulty, because they think of the powder 'reacting to' the heat. The thing to watch out for here is whether students are confusing substances and energy – that is, if they think that heat can be considered as a substance. It is important that we emphasise that energy and substances are quite distinct, even though changes in the latter tend to involve transfers of the former from one energy store to another.

As a teacher, it is useful to emphasise continually that the key criterion for a chemical reaction is change of substance, regardless of how many substances are involved before or after the change. So the thermal decomposition of copper carbonate is a chemical reaction because the products present after the change, copper oxide and carbon dioxide, are different substances to the copper carbonate that has reacted and no longer exists.

A further note on 'stability'

When using the term 'stable' we should also bear in mind that something may be unstable – that is, it will spontaneously change – but only show very slow changes. In Chapter 1 we met the example of diamonds, which are not stable at the conditions at the Earth's surface. Yet they are very inert (meaning very slow to change) so it is not surprising that people think of them as stable. Only when something is both unstable and labile (meaning it will change quickly) can we readily observe that it is not stable. In everyday life, these two pairs of technical terms (stable and inert; unstable and labile) tend to be used interchangeably, but in chemistry they strictly relate to two different themes (Table 3.1). These are discussed further in Chapter 5.

Table 3.1 Terms associated with stability

Term	Meaning	Relates to
Stable	Does not spontaneously change	Energetics
Unstable	Spontaneously changes	Energetics
Inert	Slow to change	Kinetics (rates of reaction)
Labile	Readily changes	Kinetics (rates of reaction)

3.2 Chemical equations

Chemical equations are ubiquitous in chemistry because they are such useful tools for communicating between scientists. A chemical equation summarises a particular chemical reaction and such reactions are the central phenomena in chemistry. An equation has the general form:

$$\text{reactant(s)} \rightarrow \text{product(s)}$$

where the substances present before the reaction, the reactants, are signified (by the $\rightarrow$ sign) to change into the substances present after the reaction, the products.

Key teaching points

■ What exactly is equal in the equation?

The term 'equation' might lead students to expect us to use an equals sign, something like:

reactant(s) = product(s)

but the normal practice is to use an arrow to show the direction of the change. The term 'equation' reminds us that there is a sense in which the two sides are equal (but *not* that there are the same amounts of products as reactants during the reaction). The total mass of products at the end of a completed reaction must be the same as the total mass of reactants present at the start (as we saw in Chapter 1). If we think in term of the chemists' models of the structure of matter at the submicroscopic scale, then we can understand that there is a conservation of the most basic entities: there are the same number of electrons, protons and neutrons after the reaction as before (which in terms of this quanticle model can be understood as the reason that mass must be conserved). It is useful to reinforce this point for students: conservation of the 'quanticles' present, each having a fixed mass, implies conservation of mass in the reaction. The type of analysis presented in section 1.6 of Chapter 1 may be worth revisiting at various times in the context of different examples of chemical reactions.

■ What about energy transfers?

As well as mass being conserved, so is energy. This is not something we tend to emphasise strongly in chemistry, but we should be aware that when our students study physics topics they are taught that energy is never created or destroyed in any process. So although chemical reactions involve energy transfers, sometimes very obvious and significant ones, the total energy of the system does not change. Chemical systems are considered to have different amounts of chemical potential energy. During reactions there may be a transfer of energy from the chemical store to the thermal energy store (or sometimes vice versa) of the surroundings. This is usually only discussed in chemistry lessons near the end of the secondary years, but it is important that when we teach younger students we do not seem to be telling them something inconsistent with what they are learning in other science lessons. So when we talk about energy changes involved in chemical reactions, we should be careful to refer to the energy transferred and not to suggest that chemical reactions can 'make' or generate (or use up) energy.

■ Directionality of a reaction

We write chemical reactions as going from reactants to products and this is a useful way of thinking – a good model – that fits most reactions we come across. However as students advance through the school, they will meet reactions that do not go 'to completion', such as the key reaction in the industrially important manufacture of ammonia:

hydrogen + nitrogen $\leftrightarrows$ ammonia

The decision to write the reaction as above, rather than as

ammonia $\leftrightarrows$ hydrogen + nitrogen

is somewhat arbitrary in terms of chemical principles, but reflects our use of this reaction to make ammonia from nitrogen and hydrogen. We manipulate the conditions to produce ammonia, which is useful for making, for example, fertilisers, from the more readily available hydrogen and nitrogen.

There is a key teaching point here: all reactions are to some extent reversible. Even in a reaction such as exploding an oxygen and hydrogen mixture, it is possible theoretically to calculate some extent of reverse reaction (even if it would be of no practical importance at all in such an extreme case). More advanced students will study equilibria (covered later in Chapter 5), from which perspective we consider that for any possible reaction, the balance between reactants and products found 'after' a reaction will depend upon the conditions of temperature, pressure, etc. The word 'after' is placed in inverted commas here, as when reactions reach an equilibrium there is still plenty going on at the submicroscopic level, but just no further net change (since forward and reverse reactions proceed at the same rate and cancel each other out).

It should also be noted that theoretically the approach to equilibrium involves an ever-decreasing rate of change, although in many reactions there is soon a point when there is no further observable change. We can understand this in terms of a simple feedback cycle (Figure 3.1, overleaf). If we consider the driver for change to be the extent to which the reaction mixture is 'out of equilibrium', then the further from equilibrium the mixture, the greater the net rate of reaction (shown by a '+' symbol), which of course moves the reaction closer to equilibrium. However, that reduces the driver, and the rate of reaction slows, and the rate at which the reaction approaches equilibrium also slows.

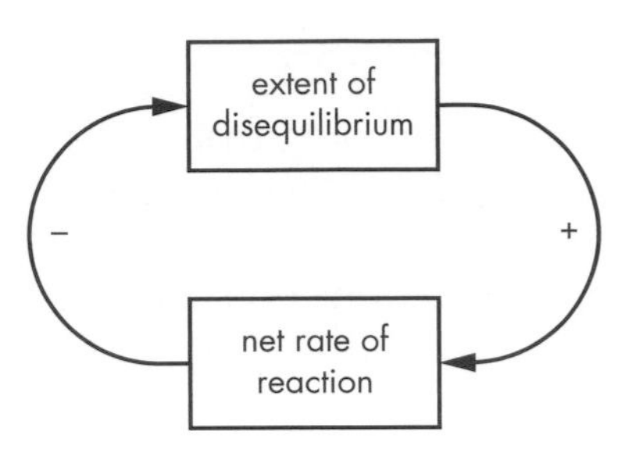

Figure 3.1 The approach to equilibrium is subject to a negative feedback cycle

Commonly in chemistry, we do not teach about feedback cycles in this way, but they can be a helpful way to think about an abstract explanation with older students. The feedback cycle offers a useful tool for thinking about systems that link across the sciences. Homeostasis in the body depends upon such cycles. In physics there are a number of phenomena that can be described with the same very simple feedback structure as in Figure 3.1 (cooling of a hot object, radioactive decay, capacitor discharge, for example). Feedback cycles are also very useful in thinking about aspects of Earth and environmental sciences (for example the possibility of a positive feedback cycle when atmospheric warming leads to warmer seas in which the greenhouse gas carbon dioxide is less soluble).

In introducing ideas about reactions to younger students, we do not want to overburden them with such considerations as equilibria, as there are many examples of reactions where it is clearly sensible to designate reactants (what we start with) and products (what we end up with), and where for all practical purposes the reaction goes to completion and can be readily observed to do so. However, in following the principle that we should avoid students developing ideas that will make progression in learning difficult, it is probably useful to make sure they are aware that some reactions never go to completion, and also that some reactions only occur very slowly under common conditions. More detailed advice on teaching these areas can be found in Chapter 5.

The chemical equation as a model

The points discussed above lead to the suggestion that the best way of thinking about the chemical equation

reactant(s) $\rightarrow$ product(s)

is as a model. Chemists and other scientists commonly use models which they know simplify or over-generalise where they are useful for thinking about aspects of nature. This equation is a very useful model because we can use it to describe any reaction. However, like all models it has its limitations. So for some reactions, those which are more obviously equilibria and where there will be some 'reactant' present no matter how long students observe the reaction mixture, the model has to be amended to reflect these equilibria. For the majority of reactions, although they are technically equilibria, the model does a 'good enough' job of describing the reaction for most purposes.

Presenting the idea of the chemical equation as a very useful model that can help us summarise what is going on in reactions, and which does a good enough job for most reactions, will both enable students to see the value of the formalism and stop them making unfortunate generalisations about all reactions going to completion (which could then later act as a misconception when they meet reactions that do not fit the model).

Forms of reaction

Various forms of reaction equation are possible, depending upon the number of different reactant and product substances, including:

$$\text{reactant}_1 + \text{reactant}_2 \rightarrow \text{product}$$

$$\text{reactant}_1 + \text{reactant}_2 \rightarrow \text{product}_1 + \text{product}_2$$

$$\text{reactant} \rightarrow \text{product}_1 + \text{product}_2$$

$$\text{reactant}_1 + \text{reactant}_2 \rightarrow \text{product}_1 + \text{product}_2 + \text{product}_3$$

Some classes will accept these different possibilities readily and will not need to spend much class time discussing this. For other students, this may seem quite complicated. A useful sorting and classifying activity for these learners would be to present them with a set of word equations for a range of reactions that they will come across during their course, and have them work in pairs or small groups to see which of these reactions fit into the different patterns. This will help students become familiar with the idea of using equations to represent chemical reactions. If you have a class with a wide range of abilities, you can set this task for some students and ask other students (for whom this would be a trivial activity) to identify examples of reactions not on your list in the different categories using a textbook (or the internet if readily accessible in your teaching room).

A good extension question to ask students is whether we could get chemical reactions of the form:

$$\text{reactant} \rightarrow \text{product}$$

We can suggest changes that would fit this general pattern:

$$\text{ice} \rightarrow \text{water}$$

$$\text{diamond} \rightarrow \text{graphite}$$

In these examples one material changes into another. However ice and water are different states of the same substance and this example is a physical change (melting, covered in more depth in Chapter 2). Diamond and graphite are very different materials but are both forms of carbon – we say they are different allotropes. Again, this is not usually considered a chemical change (although there is a change in the bonding within the structure when the carbon changes form), but this example does highlight how our basic chemical concepts run into difficulties in some situations. (Chapter 1 discusses this concept further.) This reminds us that the distinction between chemical and physical change should be seen as a useful model for helping us make sense of chemistry, rather than a fundamental division found in nature.

Word and formulae equations

Two most common types of chemical equation are those written in words using the names of the substances and those representing the formulae of the substances involved. (Later, students will meet representations involving structural formulae. These are covered in Chapter 10.) Consider the following examples:

1 hydrogen + oxygen → water
2 hydrogen + oxygen → water + energy
3 $H_2 + O_2 \rightarrow H_2O$
4 $2H_2 + O_2 \rightarrow 2H_2O$
5 $2H_{2(g)} + O_{2(g)} \rightarrow 2H_2O_{(g)}$
6 $H_2 + \frac{1}{2}O_2 \rightarrow H_2O$

Each of these equations describes the reaction between hydrogen and oxygen to give water (although example 3 would normally be considered inadequate). Examples 1 and 2 are word equations, whereas examples 3–6 are formulae equations.

■ Word or formulae equations?

It is sometimes thought that it is better to first introduce word equations with younger students and then later progress to formulae equations. This can seem sensible, as formulae equations look more abstract and technical. However, formulae equations may have some advantages, even with younger students. For one thing, some common names of substances (water and ammonia, for example) do not offer much of a clue to whether a substance is an element or a compound – and, if a compound, which elements it is a compound of. Formulae equations are much more explicit here, even if looking more technical and unfamiliar to younger learners.

Research has shown that when students are asked to complete simple word equations (describing types of reactions familiar from school science and with just one word missing), many struggle to know how to proceed. Many of the correct answers are lucky guesses or are based on chemically very dubious logic. If you want to test this out with your classes, a diagnostic task ('Completing word equations') can be downloaded from the Royal Society of Chemistry website (see the 'Other resources' section at the end of this chapter).

Formulae equations are much more explicit about the elements represented in the reaction and allow students to check readily if everything on one side is present on the other. (As pointed out in section 1.3 of Chapter 1, we need to be careful about how we talk about elements 'present in' compounds.) The assumption that word equations are easier is probably misguided, at least for some students.

The recommendation I would make here is to introduce formulae equations as early as possible and use them to complement word equations, so that younger students get used to seeing the same reaction represented in both forms. This looks like 'more' information for students to handle, but actually once they are used to this approach, the two formats are mutually reinforcing, which will prove helpful to most students. This also reflects research that suggests student learning is supported when learners are able to use multiple forms of representations for the same information.

Example 2, opposite, refers to energy in the equation. Some textbooks do this to emphasise how some reactions transfer (note, not create) large quantities of energy. Alternatively the term 'heat' might be used. (Heat is the term used to refer to energy being transferred due to a difference in temperature. However, many students incorrectly use the word heat to refer to the thermal (or internal) energy of a hot object, and this should be avoided. The Teaching Secondary Physics handbook in this series includes a very useful chapter discussing how best to talk about energy in school science.) We know that many students at lower secondary level do not distinguish well between materials and energy. For example, a common conception is that heat may be thought of as a substance – as a kind of fluid (which is how it is sometimes modelled). Therefore equations including energy like this should be avoided, unless a clear formalism is used that students know indicates that energy is not a chemical substance (for example, always writing the energy term in the same contrasting colour). However, technically it does not make sense to include an energy term like this, as we should consider the energy separately, as the following example shows.

hydrogen	+	oxygen	→	water		
potential energy associated with hydrogen	+	potential energy associated with oxygen	→	potential energy associated with water	+	heat

these two energy terms refer to the energy in the chemical store before the reaction

this energy term refers to the lesser amount of energy in the chemical store after the reaction

this is the energy transferred from the chemical store along a heating pathway

When we use a form of representation such as this, it is clearer that the heat transferred is the difference between the energy in the chemical store before and after the reaction: energy is being transferred by heating, but is not being created.

Some (mainly older) students may have heard that mass can be considered a form of energy for some purposes, and that changes in mass and energy are intimately related through Einstein's famous equation $E = mc^2$. It is technically the case that in any chemical reaction where there is an energy change, there will also be a calculable (if incredibly minute) change in the mass of the materials present (just as it is technically the case that warming a beaker of water changes its mass). However these effects are much too small to be of any significance at all in the chemistry laboratory, and for all practical purposes, mass and energy can be considered to be – separately – conserved in chemical changes. It is best to avoid introducing this complication, but any precocious student who raises the idea might be asked to do some research into the magnitude of the 'mass defect' that can be calculated for a typical reaction involving molar quantities (for example, 2 mol of hydrogen reacting with 1 mol of oxygen).

Example 3 on page 84 is an unbalanced formulae equation. This equation shows the formulae of the substances involved in the reaction, but has not been 'balanced' to make sure the amount of each element is the same on both sides. That needs to be done to work out the ratio of the reacting masses of the reactants (Chapter 1). Although there may well be points in lessons when such (in) equations may be considered while developing ideas, it is strongly suggested that teachers should always then move on to the balanced equation (here, example 4) when representing reactions with formulae in this way.

The difference between examples 4 and 5 is the level of detail provided. Example 5 offers additional information: the physical state in which substances are involved in the reaction. This is useful information when thinking about reactions we will observe in the laboratory, but often with younger students we omit this detail when using equations to represent reactions, to keep the focus on the substances.

The difference between examples 4 and 6 is trivial for a chemist, as reaction equations show the mole ratios of substances involved (Chapter 1) and multiplying throughout does not change this. So, for example, consider the following equation:

$$10H_2 + 5O_2 \rightarrow 10H_2O$$

This is not substantially different from equations 4 and 6 on page 84 – all represent the same reaction, as do an infinite number of other possible versions! We would not usually use the equation above, as it is not the simplest way of writing the equation. Equation 4 is more common as it is the simplest way of writing the equation that uses only integers (compare this with the $\frac{1}{2}$ in equation 6). However, sometimes equation 6 is used (as two of the three substances involved are present as one unit).

If some students find these ideas confusing, it might be worth providing them with variations of equations with different multipliers for several reactions they will meet in their course, and asking them to work in groups to identify (a) which examples represent the same reaction and (b) for each reaction, which representation is the simplest version. This should be a simple task for most learners and can readily be tweaked to provide differentiated versions to challenge the most able students in a class. For example, distractors which have the wrong ratios could be included for higher-attaining students, whereas only valid equations should be included for less confident learners. The task becomes somewhat more difficult if it includes versions of similar but distinct reaction equations, for example:

$$2Fe + O_2 \rightarrow 2FeO$$

$$4Fe + 3O_2 \rightarrow 2Fe_2O_3$$

$$8Fe + 6O_2 \rightarrow 4Fe_2O_3$$

$$6Fe + 3O_2 \rightarrow 6FeO$$

$$12Fe + 9O_2 \rightarrow 6Fe_2O_3$$

$$8Fe + 4O_2 \rightarrow 8FeO$$

$$12Fe + 6O_2 \rightarrow 12FeO$$

The equation as a bridge between molecules and moles

One of the most important features of modern chemistry is that it is a science that provides explanations of the reactions seen in the world at observable scales (at the molar or macroscopic level) in terms of theoretical submicroscopic entities such as molecules and ions and electrons. As chemistry teachers we are constantly shifting back and forth between the macroscopic descriptions and submicroscopic models: this shifting takes place in our thinking and is reflected in our classroom talk. To follow our arguments and explanations, our students need to be able to follow these shifts.

The symbolic language of chemistry acts as a bridge for us (Figure 3.2).

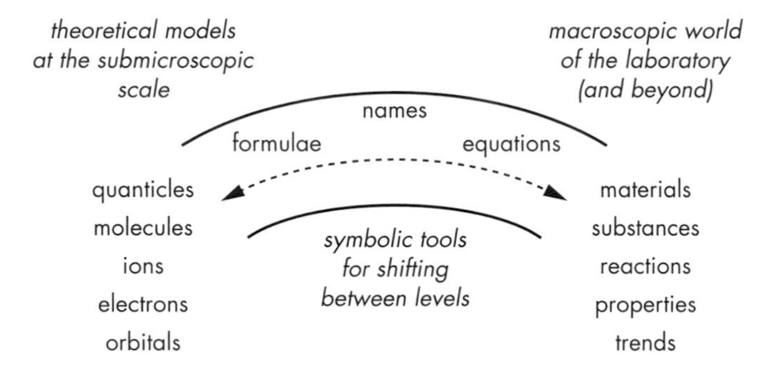

Figure 3.2 Chemistry teachers use symbolic tools to shift student thinking between what they can see and the explanatory models used to understand observations.

When we say words like 'hydrogen' and 'copper', or when we write on a board 'H_2O' or '$CuSO_4$' or 'hydrogen + oxygen → water', we are of course using symbols. However, a special property of these symbols is that they are usefully ambiguous. They are ambiguous symbols because we could be referring to actual samples of substances at a scale we can manipulate in the classroom laboratory or we could be referring to individual quanticles. 'Hydrogen' could mean the element in abstract, an actual sample of gas we have collected, or a single molecule. In the sentence 'hydrogen has only one proton' we would be referring to none of these, but rather a single atom of hydrogen.

This is incredibly useful because it enables us as teachers to make shifts between the observations students can make in the laboratory and the explanations chemists develop in terms of theoretical models at the submicroscopic level. However, we may get so used to talking in this way that we can easily forget that students may not always readily follow these shifts in thinking, and what we say may seem like some magician's sleight of hand. So we might be talking about an actual reaction students have seen and then we summarise this by writing (or projecting) a chemical equation, where the terms represent the substances the students have observed in the reaction. Then we might start talking about how, for example, the lone pairs on one species are attracted to the positive region on another; and now the same equation is referring to something rather different.

This tool of having labels that apply to both substances and molecules, and equations that refer to both bench-scale reactions and reaction mechanisms at the level of individual molecules and ions, is incredibly powerful – but it also has great potential to confuse students if they do not spot when we have crossed the symbolic bridge (Figure 3.2) to shift between levels. As teachers, therefore, we need to be very explicit about using this tool, so that students not only realise when we make these shifts, but come to see the power of this tool to help their own thinking. I strongly suggest modelling the use of this bridging process in teacher talk, as in the following example:

'... so we have seen that the hydrogen combusts with a squeaky pop and we can write this as

$$2H_2 + O_2 \rightarrow 2H_2O$$

which describes the chemical change. In the reaction, the substances hydrogen and oxygen are changed into a new substance, water. The large numbers tell us we need twice as much hydrogen as oxygen in this reaction, but we must remember that this is measured in moles.'

'The equation can also tell us why this is, because it can also summarise ...'

'... what is happening at the level of molecules. We see that two molecules of hydrogen are needed for each molecule of oxygen, to produce two molecules of water ...'

3.3 Classes of reaction

Chemistry as a science makes use of a range of classification schemes. The periodic table (discussed in Chapter 1) offers an excellent example of this, assigning elements to 'blocks', groups and periods. Reagents are characterised into types: acids (covered in Chapter 6), oxidising agents (Chapter 7), etc., and reactions can also be classified in similar ways. Given that there are an almost unlimited number of substances that can be formed in chemistry, and most undergo a range of reactions with various other substances, the sheer number of possible reactions is immense.

However, the ability to classify elements and compounds into groups according to similarity in properties (reaction behaviour) allows us also to classify reactions themselves into useful groups. There are many specific named reactions in more advanced chemistry (such as the Wittig and Diels–Alder reactions), but even in introductory chemistry it is useful to set out some common types of reaction that students will meet in their studies. Every particular reaction (that is, every different combination of reacting substances) will have some distinct, unique features, but a classification scheme provides a useful starting point for dealing with what would otherwise be an overload of information for the chemist (let alone the school student).

Common types of reaction

The reaction types suggested here will link to many of the reactions met in school science or chemistry courses, and in each case we can write a general reaction equation that shows the types of substance involved in that type of reaction. In specific cases, the general labels for types of substance (such as metal, acid) are substituted by different examples. It should be noted, however, that in some general reactions there are common specific products. When an acid reacts with a metal, for example, the substance hydrogen is usually produced. The other product formed varies depending on the acid and the metal reacting (although it will be a substance in the general class of salts).

Binary synthesis

Binary synthesis is the formation of a compound from two elements:

general form	**element**	+	**element**	→	**compound**
for example	hydrogen	+	oxygen	→	water
or	hydrogen	+	nitrogen	⇆	ammonia

■ Decomposition

This is the breaking down of a compound into simpler products (sometimes elements, not always) on heating:

general form	**compound**	→	**element/ compound**	+	**element/ compound**
for example	copper carbonate	→	copper oxide	+	carbon dioxide
or	ammonia	⇆	hydrogen	+	nitrogen

■ Neutralisation (acid–alkali)

The term neutralisation is used to refer to acid–base reactions, and so can include reactions between acids and metal oxides and carbonates (which are basic), but in introductory chemistry usually refers to acid–alkali reactions. (Chapter 6 gives a more detailed discussion of acids, alkalis and bases.)

general form	**acid**	+	**alkali**	→	**salt**	+	**water**
for example	hydrochloric acid	+	sodium hydroxide	→	sodium chloride	+	water
or	nitric acid	+	potassium hydroxide	→	potassium nitrate	+	water

■ Acid–metal

general form	**acid**	+	**metal**	→	**salt**	+	**hydrogen**
for example	hydrochloric acid	+	zinc	→	zinc chloride	+	hydrogen
or	nitric acid	+	magnesium	→	magnesium nitrate	+	hydrogen

■ Acid–carbonate

general form	**acid**	+	**metal carbonate**	→	**salt**	+	**water**	+	**carbon dioxide**
for example	hydrochloric acid	+	zinc carbonate	→	zinc chloride	+	water	+	carbon dioxide
or	nitric acid	+	magnesium carbonate	→	magnesium nitrate	+	water	+	carbon dioxide

■ Acid–metal oxide

general form	**acid**	+	**metal oxide**	→	**salt**	+	**water**
for example	hydrochloric acid	+	zinc oxide	→	zinc chloride	+	water
or	nitric acid	+	magnesium oxide	→	magnesium nitrate	+	water

■ Displacement/competition

In displacement reactions, a more reactive element 'displaces' a less reactive element from one of its compounds. The metaphor of 'competition' suggests that the more reactive element outcompetes the less reactive element for the other element or radical (such as a nitrate ion).

general form	**element 1**	+	**compound of element 2**	→	**compound of element 1**	+	**element 2**
for example	magnesium	+	zinc nitrate	→	magnesium nitrate	+	zinc
or	chlorine	+	sodium bromide	→	sodium chloride	+	bromine

These reactions normally occur in solution, so the actual reaction concerns a redox process (Chapter 7) involving two elements. In the examples above this would be (with the oxidation states – o.s. – shown in brackets):

Mg (o.s.: 0)	+	Zn^{++} (o.s.: +2)	→	Mg^{++} (o.s.: +2)	+	Zn (o.s.: 0)
Cl_2 (o.s.: 0)	+	$2Br^-$ (o.s.: –1)	→	$2Cl^-$ (o.s.: –1)	+	Br_2 (o.s.: 0)

Background note for the teacher

The term oxidation state refers to the charge on a simple ion (e.g. +2 on Zn^{2+}), but also to a nominal assignment of charge to the individual elements in a compound, based on a thought experiment that asks what ions would be formed *if* we were to break up the compound into simple ions. So, for example, in the compound ammonia, NH_3, we can imagine that if a molecule were broken up into ions it would give N^{3-} and H^+ ions. So we say that in ammonia, nitrogen is in the oxidation state –3 and hydrogen in the oxidation state +1. These are formal assignments that can be made even when compounds are very unlikely to break up into ions, but which allow us to extend the useful idea of oxidation and reduction to a wide range of chemical reactions.

■ Precipitation

general form	**ionic compound containing ions 1 and 2**	+	**ionic compound containing ions 3 and 4**	→	**ionic compound containing ions 1 and 4**	+	**ionic compound containing ions 2 and 3**
for example	silver nitrate	+	sodium chloride	→	silver chloride	+	sodium nitrate
or	potassium carbonate	+	magnesium nitrate	→	potassium nitrate	+	magnesium carbonate

These are called precipitation reactions because they take place in water ('aqueous' or 'aq') solution and one compound is precipitated out of the solution (and can then be filtered, washed, recrystallised etc.). As with displacement reactions, some of the species present do not actually take any active part in the process (and so are called 'spectator' ions). So in the first example, sodium chloride and silver nitrate are both soluble, so what is actually being mixed is a solution containing hydrated sodium ions and hydrated chloride ions, with one containing hydrated silver ions and hydrated nitrate ions. The compound silver chloride has very low solubility (the bonding between the ions is not readily broken down to allow hydration of the ions), so is precipitated from the solution:

$$Ag^{+}_{(aq)} + Cl^{-}_{(aq)} \rightarrow AgCl_{(s)}$$

(It is worth noting that this equation makes it clear that ionic bonds form simply because of the attraction between ions: there is no need for a process of 'electron transfer' between the metal and non-metal. This is followed up in Chapter 4.)

Precipitation reactions are sometimes referred to as 'double decomposition' reactions. This term is confusing and best avoided, but the teacher needs to be aware that students may come across such synonyms in their reading.

Chemical reactions and writing equations

It is important to make it clear to students that being able to write an equation for a reaction does not mean the reaction will happen (under observable conditions). So, for example, as gold is a metal we can write an equation for its reaction with hydrochloric acid as hydrochloric acid + gold → gold chloride + hydrogen. However, gold is an unreactive ('noble', rather than 'base') metal and, should your school be in a position to keep a gold sample in the chemical stores, it would be safe from reacting with bench acid (normally $2\,mol\,dm^{-3}$) at room temperature. Similarly, the equations for the displacement and precipitation reactions can be written in reverse, but the reactions 'go' in one direction, determined by the energetics of the reaction at the conditions it is carried out (Chapter 5).

■ Learning activities

Some students will pick up the idea of general equations for types of reactions very readily and will see the general forms as providing templates into which specific examples can be fitted. It will also seem obvious to many students how the general form of the equation (once learnt) allows us to see what reactions might be possible. So, for example, the reaction between sulfuric acid and potassium hydroxide is an example of the general type of reaction that produces a salt and water, and in this case the salt must be potassium (from the particular alkali) sulfate (from the particular acid). However, research suggests that, for many students, completing word equations is a mixture of guesswork, half-remembered ideas and recollected patterns (some of which are inappropriate generalisations, or simply completely wrong).

This suggests that it is important to spend time helping students appreciate the general patterns and practise relating these general equations to a range of specific examples. As with many ideas in chemistry, it is also important to revisit whenever the opportunity arises. There are likely to be many potential opportunities for applying these ideas in different contexts, providing a useful basis for building up student proficiency and confidence. The important thing to remember is that once familiar, these ideas seem very obvious, but it takes most students a good deal of engagement with using the ideas before they become familiar enough for this to be the case. Therefore, it is important for teachers to be explicit in using these ideas – to model the logic clearly when opportunities arise, until students are able to see the steps themselves.

It makes sense to introduce one general equation, and work through examples, before then gradually introducing other types of reactions, so that students can effectively learn and discriminate between the different patterns. Scaffolding activities where students are given most of the information, but asked to complete examples and explain their reasoning, can be differentiated for students with different levels of prior attainment, and made incrementally more difficult as students master the ideas. This can be the basis of group work, followed by a plenary where the teacher asks groups to explain the reasoning behind their answers, giving students opportunities to practise using the logic of relating general and specific equations.

3.4 Modelling reactions

Chemical changes involve reorganisations of matter to produce different substances – and so involve both conservation and change. It was suggested earlier that it is important that students learn to appreciate what is conserved in a reaction and what is modified.

Particle models, such as those discussed in the previous chapter, are important here, for chemists use structural models at the submicroscopic level to develop explanations of chemical processes. These explanations often include consideration of:

- the structure of reactants and products at the 'molecular' level
- the relative stability/energy states of different species present before and after reactions
- the reaction mechanism – a model of how the interactions between molecules and ions lead to change taking place.

At an introductory level, we might model reactions by using diagrams of molecules and ions, for example using space-filling type representations, as in the figure below:

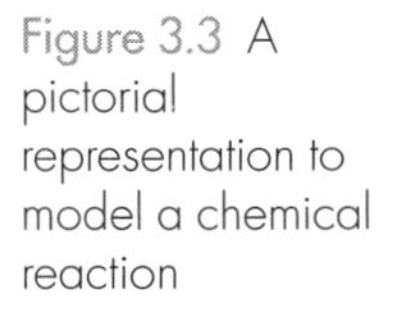

Figure 3.3 A pictorial representation to model a chemical reaction

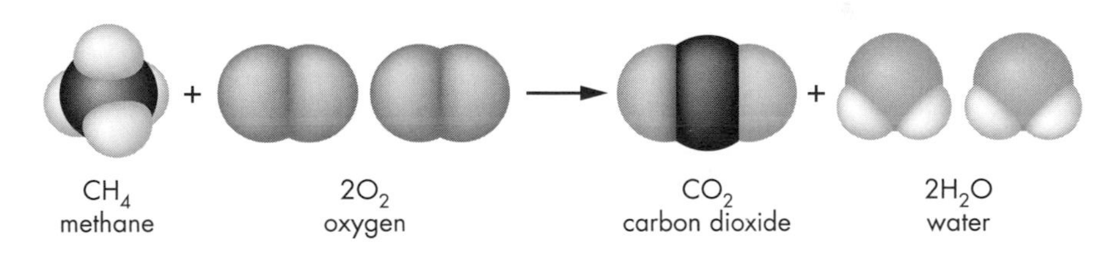

This type of representation can be used, to complement the word and formulae equations discussed earlier in the chapter, as in the example here of combustion of methane. The representation shows the different types of molecules (or ions) involved in a reaction, and the stoichiometric relationships – so here two molecules of oxygen are needed to interact with one molecule of methane (and so also two moles of oxygen are needed to react with each mole of methane). If different colours (or different hatching or shading) is used, as in the example here, this type of representation also allows students to check on the conservation of atomic cores: so here one carbon, four hydrogen, and four oxygen 'atoms' are represented both before and after the reaction. In this simple model atoms are rearranged (but not created or destroyed) in chemical reactions. This is a simplification, as intact atoms are not present in molecules, and the apparent boundaries between different atoms 'in' molecules shown in these representations may be misleading. However this is a useful model, as long as students do appreciate it is a model, and do not think it is meant to be a realistic picture of how molecules 'really' are.

More advanced explanations often focus on the electrical structure of species – the way charge is distributed in a molecule, that is whether a species has a region of relatively low electron density where the positive atomic core can attract (and be attracted by) electrons on another species, and whether the existing charge configurations in species are effective at binding it together.

A useful way of thinking about chemical change at the submicroscopic level is to consider species such as ions, molecules and lattices as comprising atomic cores (a nucleus, usually surrounded by one or more 'inner' electron shells) and associated 'valence' electrons (which can be thought of as those from the outer or valence shells of the corresponding neutral atoms). Consider, as an example, the reaction between hydrogen and fluorine:

hydrogen + fluorine → hydrogen fluoride

$$H_2 + F_2 \rightarrow 2HF$$

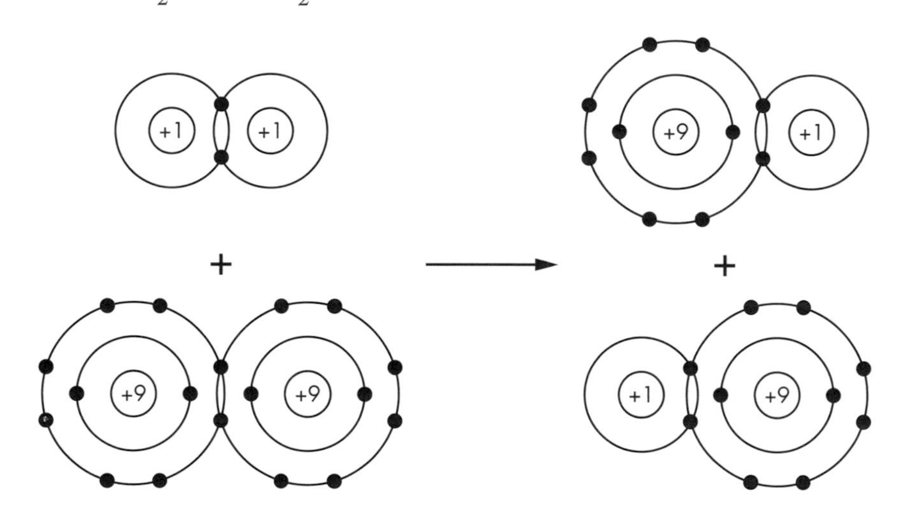

Figure 3.4 Reactant and product species when hydrogen reacts with fluorine

In this reaction the molecules of hydrogen and fluorine are broken up and new molecules of hydrogen fluoride are formed. Matter is conserved at the level of the 'quanticles': the valence shell electrons that were around the hydrogen nuclei in the hydrogen molecules and around the fluorine atomic cores in the fluorine molecules are reconfigured around the new arrangement of the same atomic cores in the new molecules. As hydrogen is in the first period, so a hydrogen atom has only one occupied electron shell, the atomic core in this case is just the nucleus. For elements in period 2 and above, the valence electrons surround an atomic core of nucleus plus inner shell(s) of electrons.

We can imagine that when we teach students about chemical reactions in secondary school, they might well wonder about why reactions occur. Yet this is not a question with a simple answer. A chemist looking to explain why this reaction occurs – at least under conditions where there is sufficient energy to break up some of the molecules – would need to find reasons why the molecules in the product species are more stable than those in the reactants. Another way of saying this is asking why are the hydrogen nucleus and fluorine core bound together more strongly in HF molecules than these species bind together in the molecules of the reactants? A chemist might also want to suggest a reaction mechanism to explain how the reconfiguration proceeds – usually conceptualised in terms of the electrical interactions between the species present (and sometimes described in terms of the overlap of electron orbitals on different species). This level of thinking is clearly not accessible by most secondary students still familiarising themselves with our basic models of matter at the level of molecules and ions, but becomes important at college ('sixth-form') level.

Avoiding the misconception that reactions occur to allow atoms to fill their shells

Considering the reasons for reactions is quite advanced and is usually only met by those who choose to continue with chemistry beyond school. It is certainly not usually included in introductory chemistry classes. However, it is worth noting that by the time students do get taught about these ideas, they have already had years of observing and studying chemical reactions. Not surprisingly, they have usually developed their own ideas about what is going on among all these reacting molecules. Very commonly, students who have successfully completed secondary chemistry in school will explain reactions in terms of atoms 'needing' to acquire full shells. They will argue (in relation to our example above) that a hydrogen atom only has one electron in its outer shell and 'needs' two to be stable, and that a fluorine atom has seven electrons in its outer shell and 'needs' eight; and that by the hydrogen atom and fluorine atom sharing electrons they are both able to be 'happy'. Students will often still give this type of explanation even after learning about the explanations that chemists develop in terms of energetics (Chapter 5). If you teach students at this level, and want to check if they think this way, you can download a diagnostic task ('Why do hydrogen and fluorine react?') from the Royal Society of Chemistry website. Full details are given in the 'Other resources' section at the end of this chapter.

Yet, of course, such an answer is not tenable, as it does nothing to explain why hydrogen molecules and fluorine molecules (which already provide the 'full shell' patterns students focus on) should interact and rearrange to allow the reaction to occur.

It seems important, then, that although we usually consider some chemical ideas too complicated to teach to younger students, teachers should avoid encouraging the development of misconceptions that can later get in the way of progression in learning. The misconception that chemical reactions occur to allow atoms to fill their shells is very pervasive, and very tenacious, and teachers should avoid making comments that might encourage such ideas.

Proper use of the 'octet rule'

Even the idea that full shells are associated with chemical stability needs some careful presentation. The electronic structures of the species present in most stable substances tend to have particular patterns – and usually this means that there are eight valence electrons around each atomic core (apart from hydrogen and helium where there are two valence electrons). This is a very useful rule of thumb, although there are many exceptions – stable substances where this pattern is not found.

The 'octet rule' is very useful in highlighting which species are likely to be more stable when comparing like with like. So if a student is not sure whether nitrogen (electronic configuration, e.c., 2.5: that is two electrons in its first shell, five in its second shell) forms a hydride that has the formula NH_2, NH_3 or NH_4, then the octet rule will correctly suggest that the trihydride would be more stable. This is represented in Figure 3.5 which shows that NH_3 (but not NH_2 or NH_4) provides an octet of electrons around the central nitrogen atomic core (which, comprising a nucleus with charge +7 surrounded by a shell of two electrons, is shown as having a +5 charge).

Similarly, if a student is not sure whether the common ion of magnesium is Mg^+, Mg^{2+} or Mg^{3+}, then the octet rule tells us that Mg (e.c. 2.8.2) will form Mg^{2+} (e.c. 2.8) rather than another ion. In these situations the octet rule is very useful. This is represented in Figure 3.6. (Note how the Mg^+ ion is shown as having a larger, but less charged, atomic core, as it has an electron in the third shell, unlike the Mg^{2+} and Mg^{3+} ions.)

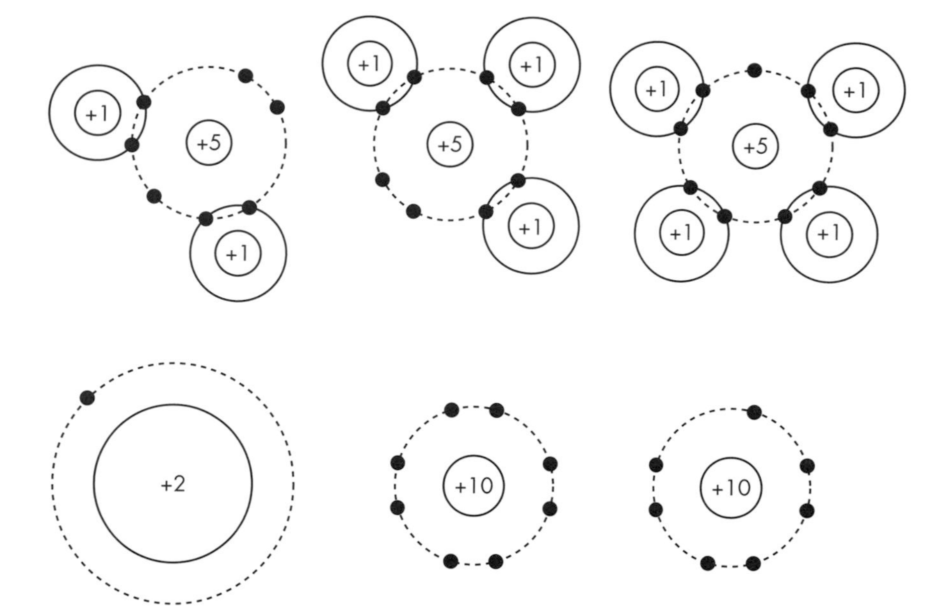

Figure 3.5 Using the octet rule to identify the stable molecule

Figure 3.6 Using the octet rule to identify the stable ion for magnesium

However, the octet rule is only a heuristic or rule of thumb. There are plenty of exceptions, including compounds such as carbon monoxide (CO) and borane (BH_3). Transition metals commonly form a range of ions (iron, for example, commonly exists in compounds as both Fe^{2+} and Fe^{3+}). A key point is that because most substances stable enough to be found in our normal surroundings, or indeed in chemistry laboratories, have electronic structures that already 'obey' the octet rule (as in the hydrogen and fluorine example shown in Figure 3.4), it is of no help in explaining why they do or do not react.

Moreover, the stability of ions is always relative to the chemical environment. Sodium is a reactive metal that readily forms compounds containing the sodium ion, Na^+, so we think of Na^+ as a stable species. This is a fair judgement, as the ion is stable in normal chemical contexts: in the metal lattice, in salts, in aqueous solution.

However, what students often do not realise is that the (isolated) neutral atom is more stable than the (isolated) ion with its outer electron removed. Chemists actually do strip the electrons from atomised sodium (and other elements) to measure the ionisation energy, and energy is needed because the atom is more stable, despite not having a full shell. The removed electron will be attracted back if no other chemical species are around.

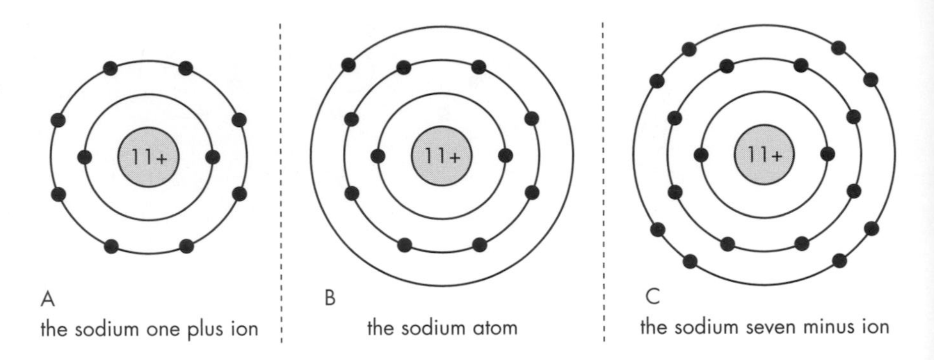

Figure 3.7 Which of these sodium species do students consider stable?

Ionisation of atomised metal may seem an extreme case, so perhaps it is not important if students think of the Na^+ ion as more stable than the atom. However students have a strong tendency to see any species with an octet of electrons as stable, and research shows that by the end of secondary education, students will commonly rate a whole range of dubious ions as more stable than atoms because they have full shells or octets of electrons. So not only do students tend to think Na^+ is a stable ion, they make the same judgement about the chemically quite ridiculous species Na^{7-} shown in Figure 3.7. It is important, therefore, that teachers make sure that students do not over-generalise the octet rule from a very useful rule of thumb for identifying the most likely formulae for molecules and ions, and adopt it as an absolute principle to judge stability and explain why reactions occur.

You may believe that none of the students you teach are likely to think that Na^{7-} and other chemically unlikely species, such as Cl^{11-}, will be stable, but you may be surprised. You can download a diagnostic task ('Chemical stability') from the Royal Society of Chemistry website to test this (see below).

Other resources

Websites

The Royal Society of Chemistry (RSC) publishes a range of resources to support chemistry teaching in schools. *Chemical Misconceptions – Prevention, Diagnosis and Cure* (Taber, K.S., 2002; Volume 1: Theoretical background; Volume 2: Classroom resources) includes resources for probing student thinking and finding out whether students have acquired common misconceptions on a range of topics relevant to this chapter. The classroom resources may be downloaded from the RSC site: www.rsc.org

Examples of students' alternative ideas and ways of thinking about chemical reactions are given on the ECLIPSE (Exploring Conceptual Learning, Integration and Progression in Science Education) project website:
www.educ.cam.ac.uk/research/projects/eclipse/

Incorporating effective group discussion tasks in secondary science and mathematics was a focus of the epiSTEMe (Effecting Principled Improvement in STEM Education) project. Information about this project is available at the project website:
http://www.educ.cam.ac.uk/research/projects/episteme/

Books

Information about analytical techniques may be found in Faust, B. (1997) *Modern Chemical Techniques*, London: Royal Society of Chemistry.

4 Developing models of chemical bonding

Keith S. Taber

4.1 From holding power to chemical bonding
- Properties of substances
- Structure of materials
- Explaining holding power

4.2 Principles underpinning the bonding concept
- The nature of the chemical bond
- Quantum theory

4.3 Modelling varieties of chemical bonding
- Metallic lattices
- Ionic lattices
- Covalent bonding
- Bond polarity
- Other forms of bonding

4.4 Relating type of bonding to changes in chemistry
- Melting and boiling
- Dissolving
- Bonds and chemical reactions

Chemical bonding is a theoretical idea that is a key part of the way matter is modelled by chemists at submicroscopic scales (as composed of extremely tiny 'quanticles', as described in Chapter 1) and so is not likely to be familiar to students from their experiences outside the classroom. However, if particle models are introduced in the manner recommended in this book (Chapter 2), students will have been introduced to the idea that the particles of which substances are made have 'holding power' of varying strengths, which tends to cause them to clump together unless they have sufficient movement to overcome this.

Choosing a route

The ideas presented in this chapter, then, build upon prior learning that chemists think of matter as composed of myriad tiny little parts (usually called particles but, as suggested in earlier chapters, this can be a misleading term), which have a holding power that leads to them tending to stick together. The topic of chemical bonding should therefore be seen as part of a learning progression from introductory particle theory.

One route into thinking about chemical bonding is in terms of the states of matter (the (submicroscopic) structures that different solid substances take) and the properties of different substances. As Chapter 2 points out, different substances undergo phase changes (solid to liquid; liquid to gas) at very different temperatures. We also find that substances have different behaviours if we try to dissolve them in solvents, or pass electricity through them, for example. These differences in behaviour present the chemist with phenomena to be explained, and provide the chemistry teacher with one basis for exploring the nature of the 'holding power' students learn about early in their secondary chemistry.

A second route to approach the idea of chemical bonding is through modelling chemical reactions. In the previous chapter, ideas about modelling chemical reactions in teaching were discussed, and it was suggested that understanding the way chemists conceptualise chemical change requires students to think about what is happening during reactions at the level of the 'quanticles' involved: the molecules and ions. A simplified model presenting chemical reactions as involving the rearrangement of 'atoms' can offer a starting point for thinking about chemical change at the scale of atoms, but raises the question of why atoms seem to come 'unstuck' from each other during reactions, and then stick together in new combinations. This provides a second motivation for developing theories of chemical bonding.

Chemical bonding is a challenging topic, because it relates to theoretical models about entities much too small to be seen. Despite this, because of its importance in making sense of chemical phenomena, it has long been recognised as a core concept area in chemistry. The abstract nature of the topic no doubt contributes to its challenge as part of a secondary curriculum, and students commonly demonstrate serious misconceptions. The teaching sequence suggested here is informed by research into student learning difficulties. It is especially important that we emphasise something of the electrical nature of bonding interactions before students can develop false ideas about the octet rule being a 'cause' for bond formation (see the previous chapter).

Whilst the subject of this chapter is one of core theoretical ideas of chemistry, there is room for discussion of how much detail different groups of students should be taught. Many secondary students will find this material dull and difficult, and it may be that the curriculum for some groups will not prescribe much discussion of the material covered here. Conversely, some groups of students will find this material fascinating (see Chapter 12), and a full exposition at secondary level provides a very useful background for those who may go on to study post-secondary chemistry courses. Where students are required to study chemical bonding in any depth, I would strongly recommend that the approaches offered here should be adopted. In particular, understanding bonding as primarily the outcome of the same physical forces studied in secondary physics is educationally and scientifically much more valid than allowing students to develop ideas of chemical bonding as related to mystical notions of atoms actively seeking to fill their electron shells or to acquire mystical numbers of electrons.

4.1 From holding power to chemical bonding

As chemical bonding is an abstract, theoretical, idea, it is important that students are motivated to learn about the topic. After all, models of different kinds of chemical bonding must seem obscure and largely irrelevant to most learners. This first section of the chapter, then, builds upon earlier chapters by setting out what it is that the chemical bonding concept can help us understand and explain. Providing some background of this type is important for all students, but some groups will need more persuading than others.

Properties of substances

It is useful to remind students that samples of substances can exist as a solid, liquid or gas depending upon the conditions (see Chapter 2) and in particular temperature. This could be a useful context for setting up a data handling exercise. For example, Table 4.1 overleaf presents the melting and boiling temperatures of the group 2 (alkaline earth) and group 7 (halogen) elements in periods 2–5 of the periodic table (see Chapter 1).

Table 4.1 Melting and boiling temperatures for some elements

Period	Group 2 element	Melting temperature/K	Boiling temperature/K	Group 7 element	Melting temperature/K	Boiling temperature/K
2	Beryllium	1556	2750	Fluorine	53	85
3	Magnesium	923	1390	Chlorine	172	239
4	Calcium	1123	1765	Bromine	266	331
5	Strontium	1043	1640	Iodine	387	456

There are various ways this data could be displayed, including the very useful form of diagrams for comparing the temperatures of phase changes of different substances introduced in Chapter 2 (see Figure 2.6, page 59), but drawn to a specific scale. For some groups of students, asking them to produce a set of these diagrams to compare between elements will be a suitably challenging task. Depending upon the time available and the speed at which the students work, they could be asked to produce a set of diagrams for one or other group, or a pair to compare the two elements in the same period. A good way to check understanding is to ask students which state these different elements are found in at room temperature, and how readily they can be frozen/melted etc. A more capable group of students can be asked to find alternative ways to display the information shown in Table 4.1 (e.g. line graphs showing trends, bar charts, etc.), and to offer a commentary on the relative strengths and weaknesses of different ways of presenting the data.

Having reminded students about the different states in which substances are found, and how this depends upon their melting and boiling temperatures, another useful activity focuses on solids. Students can be presented with a sample of different solids in sealed test tubes, and asked to suggest how these different solid samples of substances may have different properties. A suitable selection might include:

- sulfur powder
- graphite rod
- copper turnings
- hydrocarbon wax shavings
- iron filings
- magnesium ribbon
- zinc sheet
- copper sulfate crystals
- a piece of glass rod
- salt crystals
- polythene beads.

It is best to avoid composite materials (such as wood), and to just use samples that can be considered to be particular substances. This activity can be set up as group discussion work, with students exploring and pooling their ideas in small groups. Ask the students not only to base suggestions on their observations, but also to use their background knowledge of these (and other) substances. After providing a period for discussion, ask each group to share suggestions with the class, and list these on the board. All sensible suggestions should be considered acceptable.

After compiling suggestions, highlight to the students those features of particular interest to chemists. One important teaching point is that samples of the same substance may take different forms. So although graphite was presented as a rod, it is also available as a powder. Similarly, iron filings have a relatively small particle size, but single discrete blocks of iron of considerable size are possible. So some material properties are not inherent in a substance, but are contingent on how a particular sample is prepared.

Some relevant ideas that students might present (and which you may wish to point out otherwise) are:

- Some substances are translucent or transparent, where others are opaque.
- Some substances have (various) colours.
- Some substances are shiny, whilst others are dull.
- Substances have different densities at the same temperature.
- Some substances are obviously crystalline (regular-shaped solids with flat faces).
- Substances will dissolve in water to different extents.
- Substances will dissolve in organic solvents (e.g. a vegetable oil) to different extents.
- Some, but not all, substances often feel cold to the touch.
- Some substances are good conductors of electricity, whereas others are insulators.
- Some substances are attracted to magnets, but most are not.
- Some substances are hard, and some soft.
- Some substances are elastic (they deform and then return to their original shape) and others plastic (they deform, and keep the new shape) or brittle (they readily snap).
- As pointed out earlier, some solids will melt at much lower temperatures than others.

There may be scope here for including some class practical work, or some demonstration practicals, to illustrate some of these properties, if time allows. These types of properties are often referred to as 'physical properties' to contrast them with 'chemical properties' (the profile of chemical reactions different substances undergo).

Structure of materials

As a science, chemistry seeks to find explanations for chemical phenomena. So chemists are interested in the reasons why different substances have such different properties. This leads chemists to try and find out about the structure of different substances at the submicroscopic scale: how the component quanticles are arranged into structures. However, the scale of this structure is far too small to be investigated with ordinary light microscopes, so chemists have adopted a wide range of special techniques to investigate structure at this level.

As these techniques are generally too advanced to be explained in a secondary course, it is probably sensible to present them largely as 'black box' tools that students are not expected to understand – similar to the approach taken in Chapter 1 to the chemist's special 'elemental analyser' (a mass spectrometer). A variety of modern chemical analytical techniques have been used to explore structures, including X-ray diffraction and various forms of spectroscopy. These techniques depend upon physics (of waves) that some students will study at upper secondary level, and finding out about these techniques might be a suitable extension or homework activity for some (but probably not most) students. These techniques provide information that tells chemists about the special arrangements of the atomic centres in different substances – the geometry of the structures, and how close together the atomic cores are.

Crystal structures have very regular patterns of atomic centres. Although students are not normally expected to know details of crystal structures at secondary level, many students will be interested to see models of common structures. A model of NaCl structure will show students that the sodium and chlorine atomic cores have a very regular arrangement, which builds up into a cubic pattern. A model of diamond structure will show an extensive regular arrangement of the same (carbon) unit with a tetrahedral geometry. Common metallic structures are based around each atomic core having 8 or 12 near neighbours.

A very important finding from structural studies is that sometimes a substance in the solid state has an extensive arrangement of atomic cores, all held closely to all their neighbours, whereas sometimes there is a different type of structure with small numbers of atomic cores closely bound together – but with these small units being more loosely bound with others like themselves. This suggests that the idea of the 'holding power' (see Chapter 2), which holds particles together into structures, is not straightforward.

■ Ideas and evidence about chemical structures

Not only are analytical techniques that reveal the structures of substances at the submicroscopic scale complex, but the data they produce often needs considerable interpretation. This may be another good reason to consider these techniques as 'black boxes' – as tools chemists use, but which we do not need to understand at secondary level.

However, with some groups of students (see Chapter 12), it might be appropriate to look at some examples of the output of some of these techniques: such as infrared (IR), (ultraviolet) UV-visible, NMR (nuclear magnetic resonance) spectra, and X-ray diffraction patterns. Many examples can be found on the internet. What all of these techniques have in common is an output that is quite *different* from the structures being elucidated, offering indirect information. That is, there is a major task of interpreting features of different diffraction patterns or spectra to identify the information needed to specify the structures themselves (the film 'Life Story' offers an excellent dramatisation of this – see the list of 'Other resources' at the end of this chapter). This makes the important teaching point that often there is no simple path from the data obtained in scientific experiments to an obvious and clear conclusion, and that often the development of new techniques not only involves building the machines – the physical apparatus – but also developing (constructing, testing, confirming) a mathematical and analytical apparatus that allows us to draw inferences from the data.

Whilst this offers students a confusing and complex image of the nature of chemistry, it is an image that is more realistic than is likely to be gained from many school chemistry practicals that have been set up to give readily interpreted results. This is important, as a key aspect of scientific literacy is appreciating why there is often intense debate between scientists about the interpretation of evidence (for example the level of risk in using nuclear power; the extent to which there is human-caused global warming), and why accepted scientific ideas are sometimes found to be wrong. One of the core aims of science education, including chemistry education, should be to allow students to come to appreciate science as a reliable source of knowledge based on logical analysis of evidence, whilst also recognising that scientific knowledge is never absolute and beyond revision.

Explaining holding power

Chapter 2 highlighted the way in which a sample of a substance could be a solid, liquid or gas depending upon the 'holding power' of the 'particles' and on how fast they are moving: the higher the temperature of a sample, the faster the movement of its component particles, and the less effective the holding power at holding them together.

'Bonding' is the term used in chemistry as a formal label for the holding power. In the previous section it was suggested that structural studies report that this holding power or bonding does not always act in the same way – so sometimes it seems to produce organised arrangements that keep particles very close together, and sometimes it leads to more haphazard, more weakly bound, clumps. Chemists therefore seek an explanation not only for the origin of the holding power, but also for these different types of effects. It transpires that chemical bonding is a very complex topic, but that much can be explained by the simple principle that electrical charges attract and repel each other (Table 4.2).

Table 4.2 Attraction and repulsion of electrical charges

	positive charges (such as on atomic cores, cations, nuclei)	negative charges (such as on electrons, anions)
positive charges (such as on atomic cores, cations, nuclei)	repel	attract
negative charges (such as on electrons, anions)	attract	repel

We also need the idea of chemical bonding to produce more detailed models of what is happening (at submicroscopic scales) when chemical reactions occur, as we have seen this involves some previously bound combinations of quanticles breaking apart, and new arrangements forming (see Chapter 3). Again this is a complex topic, but a useful rule of thumb is that reactions allow rearrangements which lead to the component quanticles holding together more tightly – or being more strongly bound (or bonded). When we consider this idea in relation to bonding being an electrical effect, we can think of chemical reactions occurring when they allow the charges present (the atomic cores and the outer shell electrons) to rearrange into a more stable pattern. As we will see later in this chapter, that is not the whole story, but it provides a useful starting point for an explanation of what is going on that, if not complete, is at least scientifically sound and can be understood by upper secondary level students. To summarise, this is a topic that may not be studied in any detail by some secondary students (see Chapter 12), but is important enough that, when it is part of the curriculum, we should teach it using scientifically sound models in ways that help students see the nature of chemistry as a science that develops theoretical models that help us explain phenomena.

4.2 Principles underpinning the bonding concept

In considering chemical reactions, or indeed processes such as changes of state and dissolving, in terms of submicroscopic particle models, it is clear that the quanticles (ions, molecules, etc.) tend to clump together and that the changes we observe at a macroscopic scale may be explained in terms of reconfigurations of those theoretical particles. We might think of these changes in terms of two types of competitive considerations:

1 The effect of the attraction between quanticles working against the inherent movement of the quanticles. (Higher temperatures mean greater movement. This is covered in greater depth in Chapter 2.)
2 The effect of competitions between different possible arrangements – with chemical reactions occurring when there is enough energy available to disrupt one arrangement and allow a reconfiguration into a more stable arrangement. This idea of 'activation' energy is discussed further in Chapter 5. A key concept here is that electrical binding holds together different quanticles in various configurations.

The nature of the chemical bond

At the level at which we talk about bonding in secondary school chemistry, we can think of it as an electrical phenomenon. The basic components of matter are electrically charged (positive nuclei, containing positive protons, and negative electrons) and in chemistry it is often most useful to think of chemical change in terms of reconfigurations of positive atomic cores surrounded by negative electrons. Similar charges repel (atomic cores repel each other; electrons repel each other) but opposite charges attract (cores attract and are attracted by electrons) and the attractive forces pull the cores and electrons together into arrangements until similarly charged components are close enough for the repulsion to balance the attraction. This forms an equilibrium arrangement, which is stable because any small disturbance from this position will be resisted by the electrical forces. This applies to atoms, ions, molecules, lattices, etc.

As often in school science, here we are presenting a model. It is a model that has limitations and advanced students will go on to meet more sophisticated ideas. This need not be problematic, as long as (a) we are explicit that we are teaching models and (b) our students understand that developing such models is a core part of the work of sciences such as chemistry. Provided that students appreciate the role and nature of models in chemistry, then we can teach them about models which chemists find useful as thinking tools even though they are necessarily limited descriptions of nature. This will avoid students thinking these models are either absolute accounts of the molecular world (making later progression in thinking difficult) or pointless things they are asked to learn for examinations, but will then be told are wrong if they later go on to study the subject at a higher level.

In my conversations with advanced students it is clear that a great deal of anguish is caused by those teachers who seem to take pride in rubbishing simplifications met at earlier levels of study, offering the promise that in this course the student will find out ways things really are. It is surely much better that as teachers of chemistry we both reflect the nature of science by acknowledging that all our models and theories are partial, limited accounts of nature, and also enact good educational practice by only teaching ideas of some intellectual merit. We must continue to value students' prior achievements by acknowledging that more basic learning has an important role in developing understanding of what can be a very abstract subject.

When students first meet the idea of chemical bonding, they have limited experience on which to construct a model in terms of electrical forces. Lower secondary students usually have limited appreciation of forces between charges and they commonly think of bonds as not just being physical (a force) but actually material. They may think in terms of glue, sticks, springs and various other physical connectors – and some of these may well derive from the physical models we use in teaching (where plastic spheres are connected by such material means). It is difficult for students to think in terms of bonding that is not due to some material link, even if (and this is not always the case) they appreciate that when considering electrons and ions and molecules, it is not possible for there to be smaller material links holding them together. Handling this as a teacher requires some subtlety. When teaching abstract ideas, it is important to find connections with students' existing experience and ideas, however the notion that the chemical bonds are material links is clearly a serious misconception. As teachers we need to welcome students making comparisons with what is familiar – the bond acts *like* glue or *like* an elastic band joining two things together – while trying to ensure that this analogy does not become seen as an identity (in other words, not that the bond *is* a spring between two atoms).

Some students may think of the bonding in the atom as being like that due to magnets. This is not a perfect comparison (as magnets have two poles, unlike charged particles), but students will be familiar with magnetism acting as a physical force that can occur without material connection – that is through a field – and in this sense this may be a more appropriate analogy. However it remains important to emphasise that the bonding is electrical. (At higher levels of study those students who continue with chemistry will find that actually there is a magnetic component to atomic structure and bonding, but at the introductory level this is an unhelpful complication. Indeed, ultimately, magnetism is an electrical phenomenon, but again that is an unhelpful complication to introduce here.)

Quantum theory

The electrical model is not the whole story, as it does not explain why the stable arrangements produced by the electrical forces so often result in species with particular electronic configurations (two electrons in the first shell and often eight in the outermost shell of a species). This is not normally tackled until more advanced levels of study, as it is usually considered to be a difficult topic. In principle, however, this relates to how at the tiniest scale, everything is quantised: such quantities as energy and angular momentum are found to exist in minimally sized packets. This is very similar to the way in which matter and charge are quantised, giving us basic units such as electrons that cannot be divided into anything smaller. That there is a fundamentally minimum packet or quantum of something like charge often seems a very counterintuitive idea, but one that students usually come to accept given plenty of opportunity to explore and use our particle models of matter.

A strong clue that the electrical model cannot be the whole story is how in atomic and molecular structures electrons are often considered to be found in pairs – pairs of electrons form covalent bonds between atomic cores, and non-bonding pairs of electrons on some atoms form hydrogen bonds with hydrogen atoms bonded to other atoms. If electrical forces were the only important factor, then electrons would not be expected to act as if paired up. Yet students rarely seem to raise this as an issue, something that indicates how difficult it is to persuade them to think primarily in terms of electrical interactions. A teacher who finds students raising this objection can probably consider they are doing better than most in helping students think about matter at the submicroscopic scale in the way chemists do. (The electrons do repel, but the existence of quantum mechanical spin can reduce this effect for pairs of electrons. One model of this is to think of electrons as tiny magnets which can be arranged anti-parallel (N–S and S–N) so that they have a magnetic attraction to counter the electrical repulsion.)

4.3 Modelling varieties of chemical bonding

At a basic level, the different types of bonds discussed in chemistry can all be understood as due to the electrical interactions between different species, attracting each other together until the repulsions between similar charges balance the attractions between opposite charges. There is often a net attraction between neutral species such as molecules, for example, until they get very close when this is balanced by repulsions. This is an essential prerequisite for matter as we know it to exist (Figure 4.1) – otherwise the Universe would probably either have no condensed matter or would be one large neutron star!

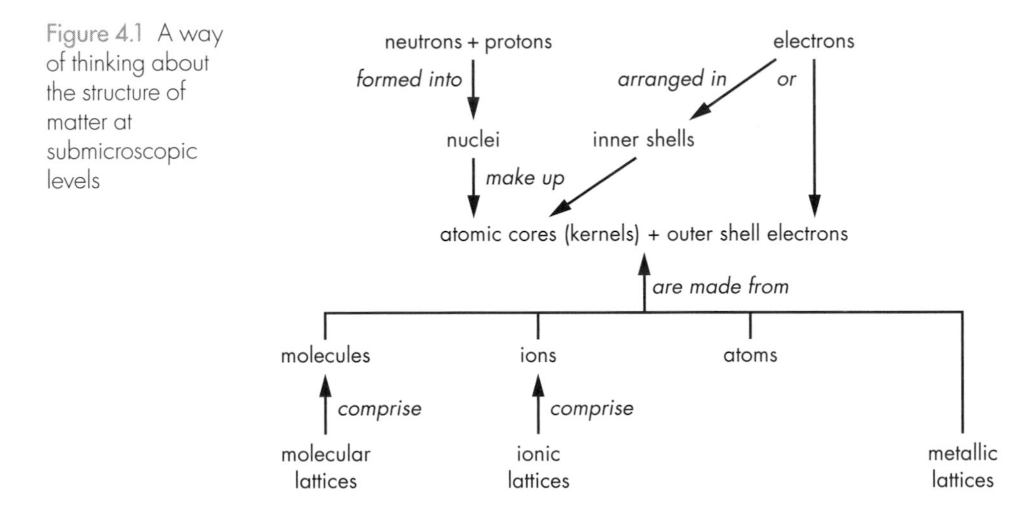

Figure 4.1 A way of thinking about the structure of matter at submicroscopic levels

One language issue to be aware of is the use of the terms bond and bonding. Chemical bonding applies to the general phenomena, whereas the term bond is applied to a specific interaction. Covalent bonding, which is usually localised (between two atomic centres) and so can be considered 'directional' (along the axis between the two atomic centres), is usually considered to comprise discrete bonds. Similarly, hydrogen bonds can also usually be understood as single interactions that are often represented by lines joining specific atomic centres. In ionic structures, there is less agreement about whether it is appropriate to think of each interaction between adjacent oppositely charged ions as a discrete ionic bond – although this is taken as reasonable in this chapter. However, metallic bonding involves interactions primarily between atomic cores (cations) and delocalised electrons moving about the structure, when the overall effect is to bind the atomic cores into a tightly bound regular lattice. Some chemists would argue that it does not make sense to talk of metallic bonds between the atomic cores (rather than metallic bonding holding the whole lattice together), but this is a semantic issue, as it is only problematic if it leads to students misunderstanding our models.

Ultimately language is used for communication, and what is important is that as a teacher you are clear that you are referring to electrical forces between charges that hold structures together. The type of bonding that holds separate molecules together is usually known as van der Waals' forces, and here we do not usually talk of discrete bonds. There is also some ambiguity among chemists and in different textbooks over whether the term 'chemical bonding' should even be applied to all or just some of these interactions (Table 4.3).

Table 4.3 Different types of bonding interactions

Usually agreed to be chemical bonding	Not always seen as chemical bonding
Covalent (including dative) Hydrogen Ionic Polar Metallic	Intermolecular interactions (such as van der Waals' forces) Solvent–solute interactions

Such a distinction is usually based on considering those interactions generally labelled as bonds as being stronger than the others. However, that is not a clear distinction, as we see when something like NaCl (with strong chemical bonding) readily dissolves in water – where the strength of the interactions between the ions and the polar water molecules are a major factor in its solubility.

Even the term 'intermolecular' bonding needs to be applied carefully. Hydrogen bonding (usually considered to be a chemical bond, although sometimes that status has been questioned) can be intramolecular (being very important in determining the shape of proteins and nucleic acids, for example) or intermolecular.

Dative bonding could also be considered intermolecular, so if gaseous ammonia (NH_3) and hydrogen chloride (HCl) are mixed, they form a solid (NH_4Cl). This may be considered to be an 'adduct', the name given when two already stable molecules are able to join into a new larger molecule. (In other words, NH_4Cl might also be written as $NH_3.HCl$.) If the product here, ammonium chloride, is gently warmed then it thermally decomposes back to hydrogen chloride and ammonia:

$$HCl + NH_3 \leftrightarrows NH_3.HCl \text{ (or } NH_4Cl\text{)}$$

The formation of the ammonium chloride is easily demonstrated in the lab using a long glass tube with cotton wool at either end (see Figure 2.9, page 69). As little energy input is needed to bring about the decomposition, there is a tendency to see the bonds formed as intermolecular, and consider that the adduct is not a fully stable molecule. Clearly such complexities are unhelpful when introducing the topic of chemical bonding to students.

It is more sensible to consider all of the electrically based interactions that hold together molecules, lattices and complexes such as adducts and solvated species as forms of chemical bonding, and make it clear that some bonds are much stronger than others, and so much more energy has to be transferred to disrupt the stronger bonds.

Research suggests that students sometimes have quite idiosyncratic ideas about how and when to use terms such as bond, bonding and attraction. A diagnostic task ('Interactions') to explore student ideas around this topic can be downloaded from the Royal Society of Chemistry website (see the 'Other resources' section at the end of this chapter). This task might be most appropriate as a revision task that could be worked on in pairs or small groups after the topic of chemical bonding has been taught. This would be most suitable for classes that would benefit from being challenged and where students have learnt good ground rules for productive discussion work.

A recommended teaching sequence

Students are prone to see all materials as made up of molecules, and in particular to see ionic compounds as containing discrete molecule-like entities. This tendency is encouraged when introducing covalent bonding first. It is recommended that you take metallic bonding as a starting point for thinking about bonding, moving on to ionic and then covalent bonding. It is important to be explicit that what you are teaching are models that chemists use to explain properties of different substances. These 'first-order' models offer a good deal of explanatory power, but do not explain everything and do not fit all cases.

Metallic lattices

Metals are elements and their relatively simple structure provides a good place to start modelling bonding. At the simplest level, the structure of metals can be modelled as a regular arrangement of atomic cores, that is metallic cations, which – despite all having a positive charge – are held in their lattice positions because of a large number of electrons able to move around and between them, acting as a kind of 'electrical glue'. This is, of course, a metaphor and should be used carefully. You can download a diagnostic task ('Iron') to identify common alternative conceptions students may hold about metals from the Royal Society of Chemistry website (see the 'Other resources' section at the end of this chapter).

A very common metaphor used to describe this arrangement is that the electrons form a 'sea' in which the cations are immersed. However, students using the 'sea of electrons' idea sometimes understand it to mean that there is a vast excess of electrons, whereas the stability of the metallic lattice depends upon its overall neutrality. In teaching this model it is important to stress that the number of electrons per cation is the same as the magnitude of the core charge, as the overall neutrality of the lattice is an important factor in its stability. So in the example of magnesium (Figure 4.2), the electronic configuration of the atom is 2.8.2, which can be modelled as a positively charged atomic core (2+), plus two valence electrons.

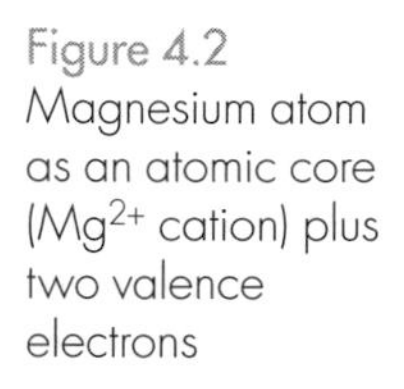

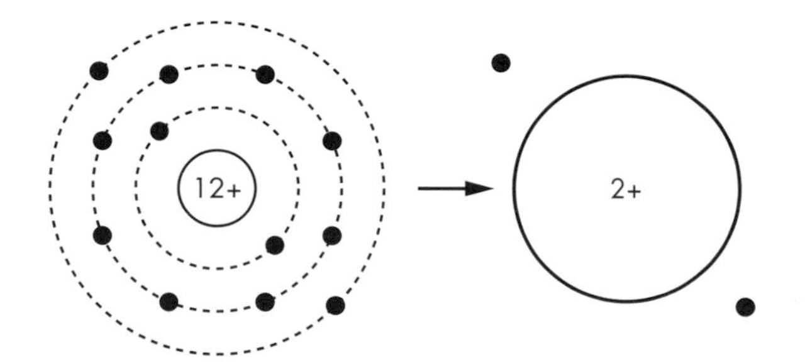

Figure 4.2 Magnesium atom as an atomic core (Mg^{2+} cation) plus two valence electrons

The metallic lattice consists of a vast array of these cores, Mg^{2+} cations, with the associated electrons. We can think of the lattice forming when a great number of magnesium atoms come close enough together for their outer shells to overlap and merge so the electrons in them can move throughout the array. This is shown in Figure 4.3 overleaf, where a sectional slice is shown – with the electron positions shifting from one moment to the next. It is also helpful to use three-dimensional models to emphasise that the structure is not just ordered in two dimensions.

Figure 4.3 Metallic bonding holds together the lattice in metals such as magnesium. The bond comprises the mutual attraction between the metallic atomic cores and the delocalised electrons able to move around the structure.

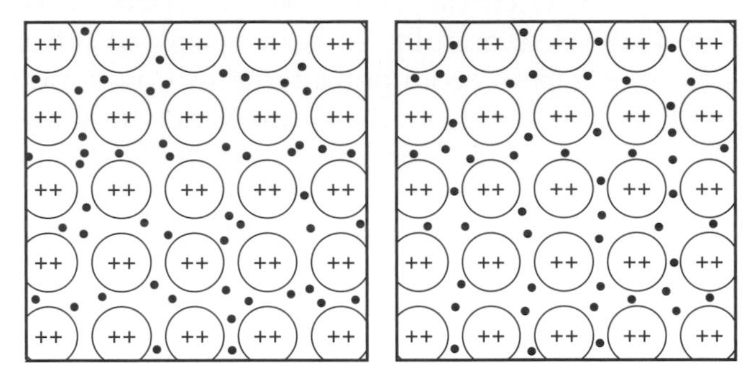

A simple way to reinforce the electrical neutrality of the metallic lattice would be to present students with a series of images (similar to those in Figure 4.3) showing core charges of +1, +2 and +3, some showing a balance of charge and some with substantially too few or too many electrons. Students could be asked to work in groups to identify which figures represent plausible structures and to justify their decisions. This activity might be especially suitable for less highly achieving students.

The simple arrangement in Figure 4.3 reflects a 'cubic' arrangement of cations. This is just a model. Although some metallic structures are cubic, most are based on each ion in a slice of the lattice being surrounded by six others in a hexagonal arrangement (Figure 4.4), as this actually gives better packing ('close' packing) when the next layer is offset.

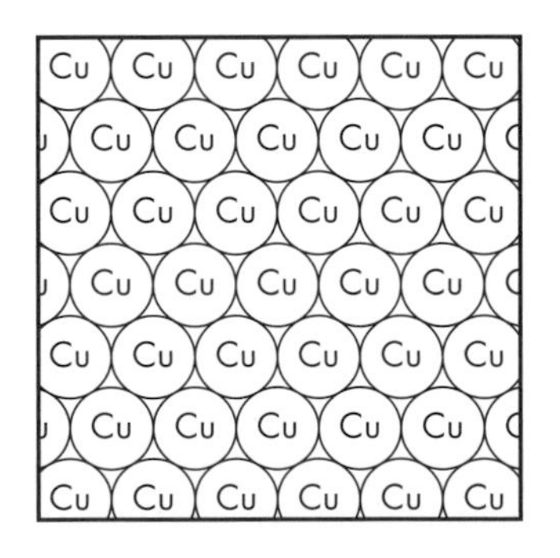

Figure 4.4 A representation of close packing in copper

Modelling this with spheres (such as marbles or expanded polystyrene balls) shows that there are actually two regular ways of building up such layers, depending upon whether the third layer sits directly above the first ('ABAB') or is offset from both the first two layers ('ABCABC'). This can provide the basis of a simple practical modelling activity for students to build models of the two different arrangements using suitable spheres (such as expanded polystyrene balls of one size). Although secondary students are not usually expected to know about these specific structures, the modelling activity provides a suitable group practical that can reinforce teaching about the regular nature of the metallic lattice. With some students, close instructions for building the model will be appropriate, whereas for others they could simply be issued the challenge of producing two non-identical structures with all the spheres close packed.

The apparent ambiguity between representations such as Figure 4.3, which shows space between the atomic cores for the electrons, and physical models showing the close packing provides an important teaching point about the difficulty of modelling quanticles – atoms and ions do not actually have clearly defined surfaces or boundaries, but rather become more tenuous further away from the nucleus.

A more advanced way of thinking, usually only met at college level, explains metallic structure in terms of a more complex orbital model, where the metallic bonding is formed by the overlap of atomic orbitals and the outcome is an enormous number of 'molecular' orbitals that will have a complex pattern of geometries. However, they will also form a virtual continuum of energy levels (the 'conduction band'), so although each particular molecular orbital may put restrictions upon occupying electrons, the available thermal energy is sufficient for electrons to readily move between orbitals in the band.

Although this is a more complex picture, and will only be met by advanced learners, it is important to ensure introductory teaching will not act as an impediment to later progression for those students who do continue with the subject. So, if teaching in terms of overlap of shells, we should be careful to stress that this is a model, and somewhat simplified, so that students who may study chemistry at higher levels do not become too committed to that particular picture. Given this proviso, an overlapping shells model can act as a much simpler (introductory) 'version' of the molecular orbital/conduction band model to explain the delocalisation of the valence electrons – the 'conduction' electrons so important to the properties of metals.

The key points when introducing metallic bonding are that the cations form a regular pattern and are bound by the attraction between the positive cations and the negative electrons. As always, our diagrams need to be presented to students as representations designed to emphasise certain points rather than as realistic images of how metals actually are.

Ionic lattices

Somewhat more complicated than the metallic case is that of ionic compounds. Research shows us that students very commonly misunderstand ionic bonding. You can download a diagnostic task to identify where students hold common alternative conceptions about ionic bonding from the Royal Society of Chemistry website (see the 'Other resources' section at the end of this chapter).

Ionic bonding is often taught through a convention of considering atoms of an electropositive metal and an electronegative non-metal – often sodium and chlorine is used as the example – and then considering how they might interact to form ions, Na^+ and Cl^-, which would then bond together (Figure 4.5). This does *not* reflect a likely chemical process, but is just a kind of 'thought experiment' in how ions might be formed from atoms, in the unlikely event that such labile species as individual atoms should be around to happen to interact. However, students often think that such a scheme is how ionic bonds are actually formed and – worse – may think that the 'electron transfer' depicted in such schemes 'is' the bond.

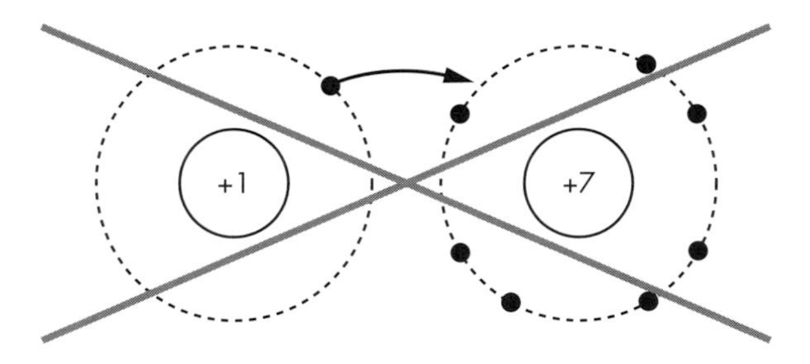

Figure 4.5 A common student misconception of the ionic bond

However, we do not have to consider how ions come about in order to explain ionic bonding. Indeed, presenting such schemes only encourages students to think of all chemical processes as starting with atoms, rather than more feasible reactants. In the natural world there are many materials that already contain ions like Na^+ and Cl^-, but we never find atomic sodium or chlorine under natural conditions. (There are very few materials that contain discrete atoms: samples of the noble gases being the obvious exceptions.) In the case of metallic bonding it was useful to start thinking about atoms, but thinking in terms of atoms in explaining ionic bonding is an unhelpful mindset.

When introducing ionic bonding to students, it is more useful to think in terms of a more feasible chemical context, such as in terms of reactions they are expected to be familiar with. One example might be the neutralisation of an acid and an alkali, such as:

$$\text{hydrochloric acid} + \begin{matrix}\text{sodium}\\ \text{hydroxide}\end{matrix} \rightarrow \begin{matrix}\text{sodium}\\ \text{chloride}\end{matrix} + \text{water}$$

In this case the reactants are solutions containing ions and the actual chemical reaction is between hydrogen ions and hydroxide ions to form water. This leaves sodium and chloride ions in solution: $Na^+_{(aq)}$ and $Cl^-_{(aq)}$. Solid sodium chloride does not form because the ions are too strongly hydrated (they are each bonded to a sheath of solvent molecules, which forms an ad hoc complex in the solution). However, if the solvent, the water, is allowed to evaporate (see Chapter 1), then this leaves the ions, which organise into a regular array of cations and anions because of the mutual attraction between oppositely charged ions (Figure 4.6).

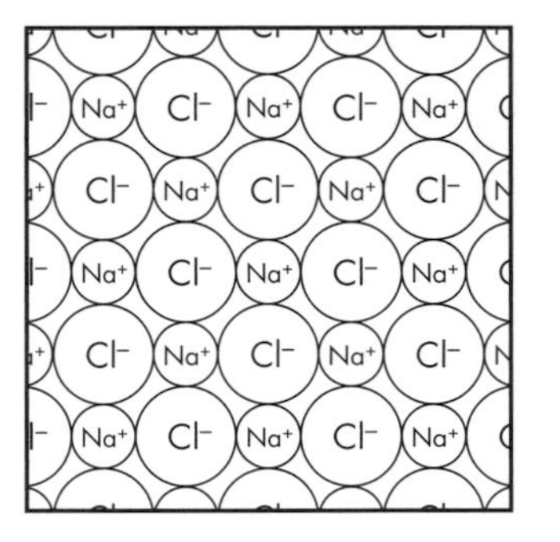

Figure 4.6 Representation of the ionic sodium chloride lattice in two dimensions

An alternative context for introducing ionic bonding would be a precipitation reaction (see Chapter 3), because students can see the formation of the ionic product immediately. So if silver nitrate solution is added to sodium chloride solution, the formation of the precipitate, silver chloride, is immediate.

$$\text{silver nitrate} + \begin{matrix}\text{sodium}\\\text{chloride}\end{matrix} \rightarrow \text{sodium nitrate} + \text{silver chloride}$$

$$AgNO_{3(aq)} + NaCl_{(aq)} \rightarrow NaNO_{3(aq)} + AgCl_{(s)}$$

It is important to stress to students that although we call the reactant solutions 'silver nitrate' and 'sodium chloride' (as they are solutions of these compounds), the solutes do not exist as bonded compounds in the solution, but rather as ions which are mixed into the solvent. However, when the two solutions are mixed together so that the resulting mixture contains silver, sodium, chloride and nitrate ions, the attraction between silver and chloride ions is strong enough that when they collide in the mixture they bind together, eventually forming large clumps that settle from the solution to give a solid silver chloride 'precipitate' (Figure 4.7).

In this process an ionic lattice is formed because of the attraction between the oppositely charged ions. In essence, that is the bonding. There is no need to explain how the ions came to be –

they exist in materials available to chemists, in substances we call salts (NaCl, NaBr, KCl, K_2SO_4, $Ca(NO_3)_2$, etc.).

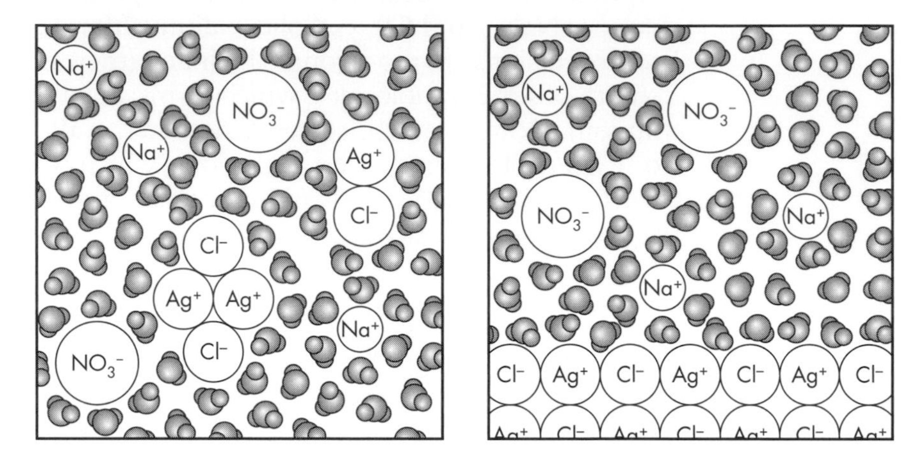

Figure 4.7 When solutions of silver nitrate and sodium chloride are mixed, the silver and chloride ions bond together forming a solid with an ionic lattice.

Most of the material on Earth originates from the output of nuclear processes that took place in stars, where the temperature is much too high for individual atoms to exist (and so matter is in the form of plasma, a kind of gas comprising separate nuclei and electrons). After this material was exploded into space, at the end of the star's 'life' cycle, it cooled and formed into more stable combinations of nuclei and electrons: sometimes atoms, but often ions or molecules. Very little of the matter that formed the Earth is in the form of discrete atoms. The formation of the ionic lattice due to ionic bonding does not involve any mysterious 'electron transfer', and if ionic bonding is taught as suggested here, then there is no reason for students to think in those terms.

Of course the formation of sodium chloride by binary synthesis (see Chapter 3) does involve the formation of chloride ions (arguably the sodium ions are already present in the metallic lattice), but from molecules, not individual atoms. Students may see this reaction demonstrated, but they are unlikely to carry it out (on safety grounds); and neither metallic sodium nor chlorine gas are common laboratory reagents. While the binary synthesis route certainly offers an exciting demonstration of a vigorous reaction, forming sodium chloride by neutralisation followed by evaporation is a much more practical way of producing sodium chloride.

This is important because research tells us that students often think that in the NaCl lattice, for example, there are NaCl molecules, or at least discrete ion pairs which are bound because they have a history of having transferred electrons. So despite the symmetry of Figure 4.6, students often interpret such figures as a collection of NaCl molecules which have ionic bonds within them and are then attracted to each other just by forces. This is unfortunate, as such misconceptions (a) cannot help students understand why NaCl is hard and has a high melting temperature and (b) often lead them to expect NaCl molecules to be the solvated species when a solution is prepared. These misconceptions all seem to derive from teaching the ionic bond through fictitious electron transfer events between isolated atoms, which actually are quite irrelevant to the chemistry. This idea appeals to students so much that they will sometimes explain precipitation reactions, such as our AgCl example above, in terms of:

- the silver ion getting its electron back from the nitrate
- the chloride ion giving its electron back to the sodium atom, so that silver and chlorine are back to being atoms
- allowing the silver atom to then give an electron to the chlorine atom to reform the silver and chloride ions (which were already present, of course) with an ionic bond between them.

This scheme is actually much more complicated than the scientific model of silver ions sticking to chloride ions because of their opposite charges (Figure 4.7) and illustrates just how tenacious some misconceptions can be once they have a hold of a student's imagination. You can download a diagnostic task to identify common alternative conceptions students may hold about how bonds form in a precipitation reaction ('Reaction to form silver chloride') from the Royal Society of Chemistry website (see the 'Other resources' section at the end of this chapter).

An activity which asks students to work in groups to evaluate two models for thinking about ionic bonding is included in a publication available from SEP (the Science Enhancement Programme; again see the 'Other resources' section). The second of two group-work tasks in the activity 'Judging models in science' asks students to consider whether the scientific model, or an alternative model deriving from common student conceptions, better explains the observed properties of NaCl. This activity should help students appreciate the superiority of the scientific model.

Covalent bonding

Covalent bonding tends to occur between non-metallic elements and is often described using the metaphor of 'sharing' electrons. A covalent bond is understood to occur when the valence shells of two atoms overlap so that one (or more) pair(s) of electrons falls within the valence shells of both atoms. At more advanced levels this is described in terms of the interaction of atomic orbitals on different atoms forming molecular orbitals (and the pair of electrons occupying the lower-energy 'bonding' molecular orbital), but at an introductory level, atoms are often simply represented as having overlapping outer electron shells.

We can represent covalent bonds in a variety of ways, which can be confusing for students. Experienced chemists and teachers see past these differences in representational formalism, but the reasons for different ways of drawing the same thing may seem arbitrary to learners. Indeed, when looking at a range of student textbooks there is often no obvious reason for the preferred forms of diagrams used. These images represent 'quanticles' – entities that are fuzzy and often better thought of as clouds of charge than as tiny 'billiard balls' with definite surfaces. This may not be what our common forms of representation suggest (for example, drawing electron shells may give the impression that they are solid structural elements of atoms), so it is up to the teacher to emphasise the limitations of images which are often a compromise between what we can easily draw and the specific points we wish to represent in particular images.

Some examples of how covalently bound molecules may be represented are illustrated here. Figure 4.8 shows a molecule of fluorine where the covalent bond is represented as the pair of electrons where the outer shells overlap.

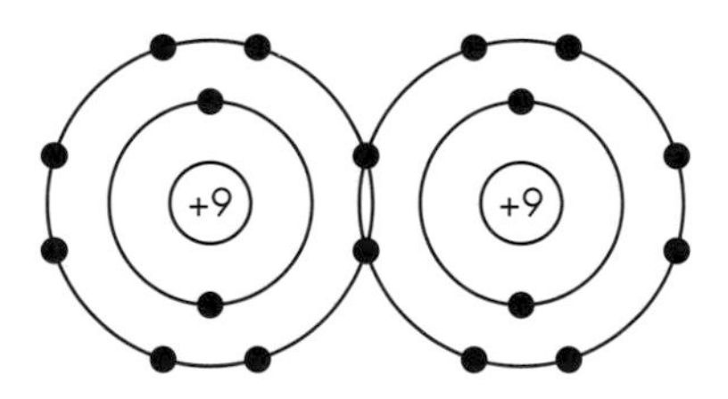

Figure 4.8 One possible representation of the fluorine (F_2) molecule

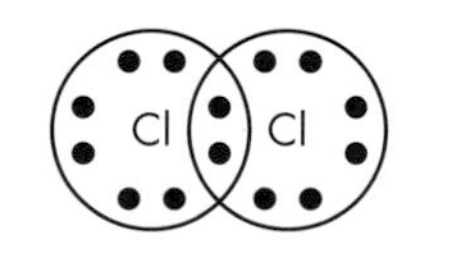

Figure 4.9 One possible representation of the chlorine (Cl_2) molecule

In Figure 4.9, a chlorine molecule is represented in an alternative way, with only the valence shell electrons shown, and the bonding pair of electrons shown inside the overlapping atoms. It is not sensible to ask which of these pictures is a more accurate representation of a molecule (which is actually physically too small to be visible, and three dimensional, with electrons in motion) itself, but rather it is important to explain to students which features are being foregrounded in different forms of representation. In terms of the covalent bond itself, the essential feature is that the pair of electrons is electrically attracted to, and by, the positive nuclei and so acts as a bond.

Figure 4.10 shows some alternative ways of representing the methane (CH_4) molecule. The first of these uses a 'dot-and-cross' style of showing valence shell electrons. This can be helpful for students in 'keeping account' of electrons by showing the electrons from different atoms in a different style. However, in the molecule, there is no difference between the electrons, and the interactions between the electrons and the positive nuclei are completely independent of where the electrons derive from. This should be stressed, as some students assume that each electron in the bond is more strongly attracted to its 'own' atom and that on bond fission, the electrons will always go back to the atoms they came from (which will interfere with later learning about heterolytic bond fission). Students should be nudged from thinking in terms of the 'ownership' and 'history' of electrons, to instead thinking simply in terms of the electrical forces acting between the different charges present.

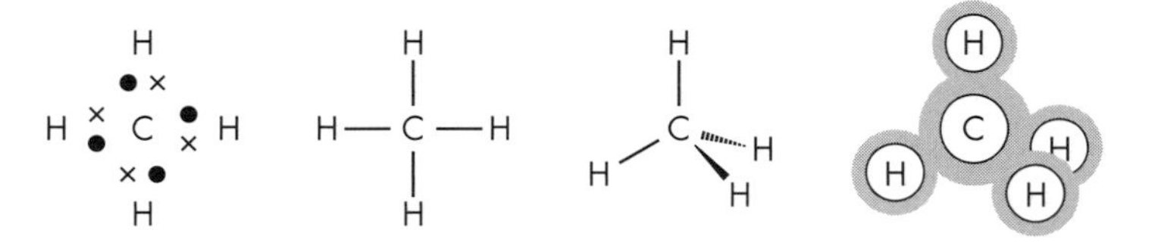

Figure 4.10 Several representations of the methane (CH_4) molecule

The second representation in Figure 4.10 simply shows the bonds as lines connecting the parts of the molecule together. This is an easy representation to draw (which can be important with more complicated molecules), but is probably best only used once students are happy with the idea that the bond is the electrical interaction due to the electrons found between the different atomic nuclei. The third representation shows how the molecular structure is not flat, but takes up a three-dimensional arrangement (tetrahedral in this case, due to the mutual repulsion of the electron pairs in the bonds). The final image replaces the representation of discrete bonds with a representation of the 'clouds' of electron density around the atomic cores – in other words, how we might imagine a time-averaged image of where the electrons are to be found. (From a quantum mechanical perspective, the structure of molecules is better described in terms of the probability of finding electrons in particular positions, but it is useful to think of this as the electrons moving about and so smearing out their charge density.) A more sophisticated version could show the variations in electron density – more like a contour map.

As I have suggested throughout the chapter, it is not helpful to talk about which of the various possible representations is 'best'. Students should be made aware that molecules and other quanticles are not easily drawn and that scientists will model them through representations that stress particular relevant features. There is a key issue here for teachers, as it usually makes good sense to adopt particular conventions in teaching and then to use them consistently to limit learning demand for students. Yet the range of figures students see in books and online will be diverse, and reliance on one form of representation can lead to students treating that form as a realistic image of how molecules actually are. A sensible compromise would seem to be in order: that is initially using a preferred form to represent molecules but later (once students are used to seeing and drawing images of molecules) introducing variation where context makes other forms useful to make particular teaching points (for example about molecular shape or the presence of double bonds).

Bond polarity

At an introductory level, students tend to get the impression that bonding in compounds is covalent or ionic, as if this is a dichotomy:

Covalent bonding	Ionic bonding
in non-metallic elements and compounds of non-metals	in compounds of metals with non-metals

Students often come to see these two forms of bonds as fundamentally very different, making it difficult for them to appreciate later how few bonds are 'pure' covalent and indeed no bonds are 'pure' ionic.

The ionic bond as represented in introductory chemistry texts is an ideal and most compounds thought of as ionic are actually some way from having fully ionic bonding. (At post-secondary levels, advanced students will learn how tables can be used to estimate the percentage of ionic and covalent character, depending upon the electronegativity difference between the elements.) Fully covalent bonds usually only exist between atoms of the same elements, and strictly then only where those atoms are not themselves bonded to very different atoms. So the C–C bond in ethanol is not pure covalent, as one of the carbon atoms is bonded to an electronegative oxygen atom, which will influence the carbon–carbon bond through an 'inductive' effect, distorting the geometry of the electron density in the carbon–carbon bond (Figure 4.11).

Figure 4.11 The C–C bond in ethanol is not a 'pure' covalent bond, because the oxygen atom (core charge +6) attracts electron density and distorts the bond indirectly.

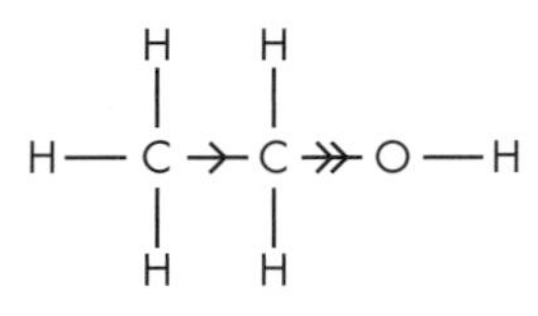

These issues are ignored in introductory treatments, but some options for representing molecules offer a better starting point for later progression in students' thinking. So Figure 4.12 is a representation of a tetrafluoromethane (CF_4) molecule, showing the core charge and valence electrons. If students are taught about bonding as an electrical interaction, it will seem clear that the electron pairs in the bond will – all other things being equal – be pulled closer to the +7 charge (fluorine atomic core) than the +4 charge (carbon atomic core). So the bond here will be polar rather than purely covalent.

Figure 4.12 A simple representation of the tetrafluoromethane (CF_4) molecule

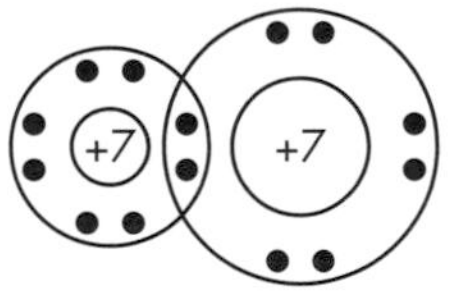

Figure 4.13 A representation of the Cl–F molecule

Figure 4.13 shows a representation of an interhalogen compound, ClF, using the same format. Here both halogen atoms have the same core charge, +7, but because of the difference in size of the two cores (for fluorine, nucleus +9 and one shell of two electrons; for chlorine, nucleus +17 and two inner shells, 2.8), the equilibrium position for the bonding pair of electrons will be nearer the fluorine nucleus. So again, this is a polar bond.

At post-secondary levels there are various ways of showing bond polarity (such as electron position or the use of ∂+ and ∂– symbols to indicate 'partial' charges on atomic centres). While such detail is not needed in introductory treatments, it is important to teach bonding as primarily an electrical interaction, to help students later appreciate how ionic and covalent bonds can be understood as extremes on a continuum, and not a simple dichotomy where all bonds in compounds fit one or other category easily.

The extent to which a bond will be polar then depends on the difference in electronegativity of the elements involved:

No electronegativity difference .... Large electronegativity difference

Covalent .... Polar .... Ionic

Other forms of bonding

In secondary chemistry it is common to limit explicit discussion of bonding to the metallic, ionic and covalent cases. However, students will have come across the idea that substances in the solid state are held together by some form of holding power or bonding, when learning about the basic particle model of matter (Chapter 2), and so are likely to assume that this bonding will be ionic in ionic compounds, metallic in metals and covalent in material with covalent bonds. The latter assumption would be correct in carbon, silicon and other substances with giant covalent lattices. However, there are many materials with discrete covalent molecules, which exist in the solid state at room temperature because of the weaker interactions *between* molecules. For example, wax and polyethylene (polythene) both contain molecules that have covalent intramolecular bonding, but are attracted to each other by a weaker form of (intermolecular) bonding.

Figure 4.14 shows the molecules in sulfur (S_8), which is in the solid state at room temperature and can be melted by heating in a test tube over a Bunsen flame. (Take care if this is demonstrated – some of the sulfur may burn, leading to noxious vapour being released.)

Sulfur comprises molecules in the form of rings of eight atomic centres, with each atomic core bound to two neighbours by a covalent bond. However, sulfur exists in the solid state at room temperature because the molecules are attracted to each other (and actually fit together to give a crystalline structure). This is despite there being no electron 'sharing' between molecules, nor any ions present. This is normally explained in terms of the electron movements in adjacent molecules becoming synchronised to give 'fluctuating transient dipoles' that allow temporary areas of higher electron density (overall negative charge) on one molecule to attract and be attracted by temporary areas of lower electron density (overall positive charge) on an adjacent molecule.

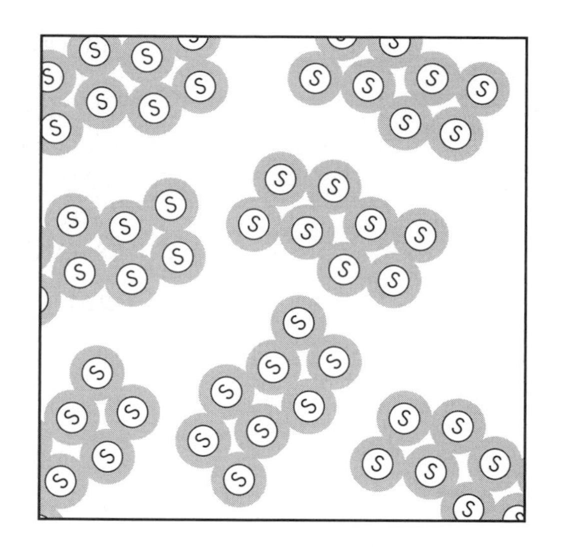

Figure 4.14
Molecules in sulfur

This explanation is difficult to visualise and is not usually discussed with students in introductory chemistry. However it is important that students appreciate that neutral molecules will attract together (Chapter 2) because of the charges present, even if no details are offered. This will help to avoid the common misconceptions that so-called 'molecular solids' have covalent bonds throughout, and the corollary that, as many of these substances melt readily when in the solid state, covalent bonds are often quite weak.

Students need to appreciate that bonding effects generally are explained in terms of interactions between charges (and it may be worth pointing out that the forces attracting molecules of sulfur together cannot be explained in terms of forming octets or full shells). Figure 4.15 develops this principle (summarised earlier in Figure 4.1) by showing how different chemical structures are built up of different configurations of atomic cores and valence electrons. Although it is certainly not sensible to present such a scheme to be learnt by students first meeting bonding ideas, it can usefully inform teaching. Teaching that is consistent with this way of thinking is more likely to help students appreciate the principles common to different forms of bonding, and will better support progression for those who go on to more advanced study.

The types of interactions found in substances such as wax, polythene and sulfur, often called van der Waals' forces, are not the only important types of bonding beyond metallic, ionic and covalent bonds. Solvent–solute interactions may be due to transient dipoles, but can also often involve permanent (rather than just transient) polarity on molecules. This is why water is a good solvent for ionic materials.

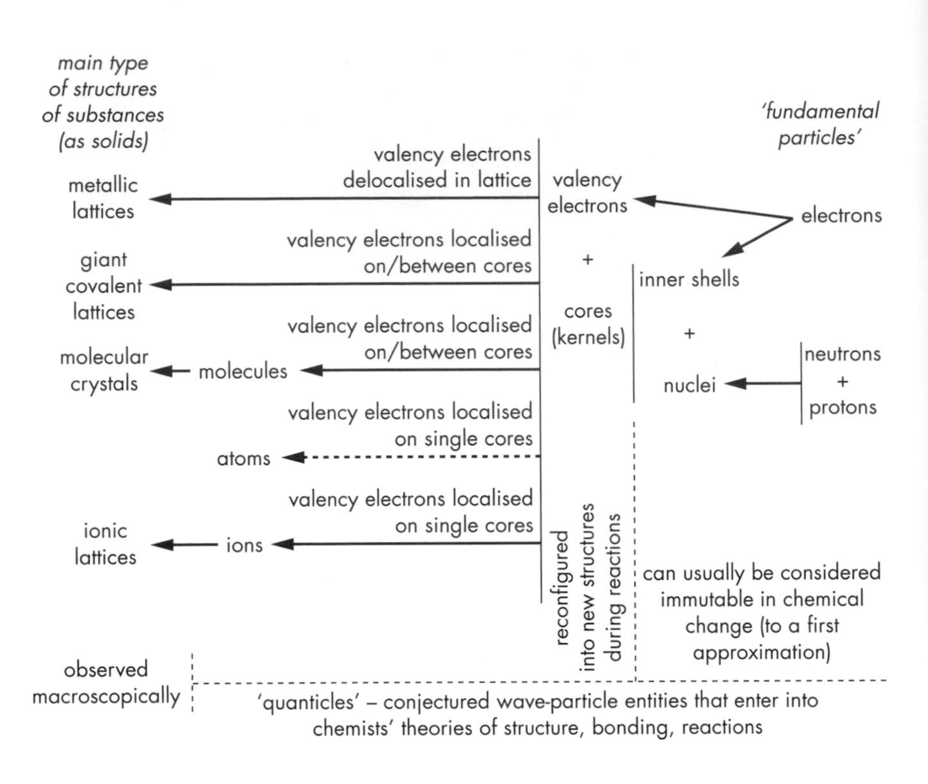

Figure 4.15 Macroscopic structures reflect different ways of binding atomic cores together with valence electrons.

An important example of an interaction due to bond polarity is the hydrogen bond, which is very important for the properties of water, proteins and nucleic acids. Unfortunately, students sometimes come across hydrogen bonding discussed in biology lessons (where the nature of the bond may not be explained) before it has been introduced in chemistry, leaving them to infer what is being referred to. (A common guess in this situation seems to be that it is just a covalent bond involving a hydrogen atom, so for example, the bonds in methane would then be called hydrogen bonds.) Hydrogen bonds form between electronegative atoms (usually O, F, N, sometimes S, Cl; all of which have one or more pairs of non-bonding outer shell electrons, so called 'lone pairs') and hydrogen atoms that have polar bonds to other electronegative atoms. Figure 4.16 shows hydrogen bonding between two molecules (a dimer) of ethanoic acid (as found in 'glacial', i.e. solid, ethanoic acid).

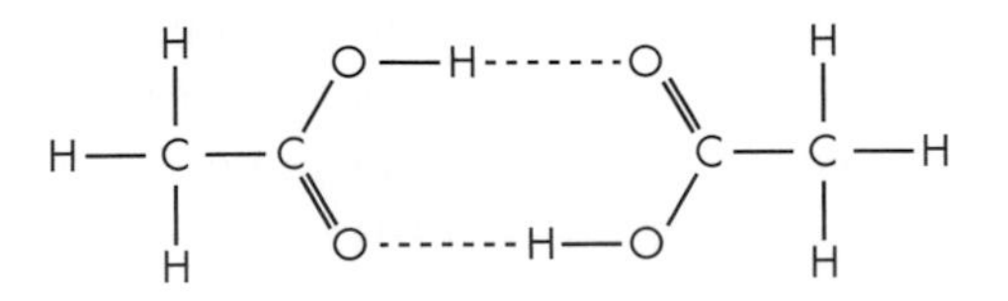

Figure 4.16 Hydrogen bonding in ethanoic acid

Hydrogen bonds can be quite strong for intermolecular bonds. (A hydrogen atom bonded to an electronegative atom is a positively charged proton which has very limited electron density covering its 'rear' and so is readily attracted by/to 'lone pairs' of electrons on other atoms.) Researching into the nature of the hydrogen bond could be a useful extension activity for more able ('gifted') students in a mixed-ability class. For example, hydrogen bonds have a specific geometry (the hydrogen bond being approximately opposite the polar bond), whereas most intermolecular bonds are non-directional. This could be followed up by a gifted student by exploring orbital models of bonding. A small group of more able students could be set the task of building a model of ice structure (or even part of a nucleic acid molecule) showing the importance of the intermolecular bonds in the crystal structure.

4.4 Relating type of bonding to changes in chemistry

Sometimes the weaker forms of interaction, such as van der Waals' forces and solvent–solute interactions, and even hydrogen bonding, are considered not to really count as chemical bonds (see Table 4.3, page 115). However, as the strength of most types of interactions varies considerably, it makes more sense to think of there being a range of different types of bond, some of which are usually stronger and some of which tend to be weaker. So sulfur, held together by van der Waals' forces, has a higher melting temperature than mercury, although the latter has metallic bonding.

Melting and boiling

A simple change of state, such as melting, has different consequences for substances depending upon the type(s) of bonding present. See Table 4.4, overleaf.

Table 4.4 Relating type of bonding to melting and boiling

Metals	Metallic bonds are not strongly disrupted by melting. (The metallic bond is not dependent upon a particular geometry, so the metal is still held together by metallic bonding in the liquid state and, for example, continues to conduct electricity.) Some metals have quite modest melting temperatures (such as sodium, 98 °C), but boiling requires overcoming the metallic bonding completely (so sodium has a boiling temperature of 890 °C). Group 2 metals have higher melting temperatures than group 1 metals (as would be expected from having greater core charges and more delocalised electrons). Transition metals tend to have higher melting temperatures than main group metals and their bonding is said to include some 'covalent character'. (Here our model that bonding involves only the outermost shell of electrons is found to have limitations.) Boiling a metal is basically a form of atomisation – although metal vapours may also include small clumps of atoms.
Ionic solids	Ionic solids tend to have high melting temperatures, as the ionic bonding depends upon the ions being arranged in the lattice so cations are next to anions and not other cations (and similarly for anions). Vaporisation of ionic materials usually leads to vapours containing discrete ions and some clumps (such as ion pairs).
A substance with a lattice of covalent bonds, such as diamond (C or C_{∞}) or silica (SiO_2)	As covalent bonds are directional and strong, these substances tend to have high melting temperatures. Melting the material requires breaking (not just weakening) of the bonds.
Molecular solids	Molecular solids tend to have relatively low melting and boiling temperatures (for example, nitrogen, methane, carbon dioxide, ammonia are in the gaseous state at room temperature), as the bonds between molecules are weak and the bonds within molecules do not need to be broken for the change of state.
Hydrogen-bonded solids	Solids with hydrogen bonding tend to have higher melting and boiling temperatures than other materials with similar size molecules. In ice, water molecules form a lattice with each molecule hydrogen-bonded to four others. This is disrupted on melting, although there is a constant flux of hydrogen bonds being formed and broken in the liquid.

Dissolving

Dissolving of a substance in the solid state involves breaking of bonds in the solid, disruption of bonds in the solvent and formation of new interactions between solvent and solute. See Table 4.5.

Table 4.5 Relating type of bonding to dissolving

Example	Involves	Feasibility
Alloying	Disruption of original metallic lattices and formation of new lattice – however retains delocalised electrons between positive cores	Often feasible – some mixtures allow a better 'fitting' lattice than in pure metals.
Salt in oil	Would require breaking strong bonds between ions, but the ions would not bond strongly to non-polar molecules.	No significant dissolving occurs.
Salt in water	Requires breaking strong bonds between ions, and disrupting hydrogen bonding in liquid water, but ions often become strongly hydrated as polar water molecules are attracted to the ions.	Some, but not all, ionic solids are very soluble in water.
Wax in water	The forces between wax molecules tend to be modest, but hydrogen bonding in water is quite significant, and only weak interactions are formed between water molecules and wax molecules.	No significant dissolving occurs.
Wax in oil	The interactions formed between non-polar solvent and non-polar solute molecules are similar in kind and strength.	Mechanical agitation of the solid by collisions from solvent molecules is sufficient to allow a solution to form.
Glass in oil or water	Strong covalent bonds in materials such as glass are difficult to break and would not be compensated for by solvation interactions.	No significant dissolving occurs.

Bonds and chemical reactions

Chemical reactions seldom occur between substances in an atomic form. Although some reactions of the noble gases have been achieved, these are exceptions. Generally a chemical reaction, such as the various ones described in the chapters of this book, involves both the breaking of bonds in the reactants and the formation of bonds in the products. Consider the following examples.

1 Covalent bonds broken and formed:

hydrogen	+	oxygen	→	water
covalent bonds in hydrogen molecules broken	+	covalent bonds in oxygen molecules broken	→	covalent (polar) bonds formed to give water molecules; on condensing hydrogen bonds form between molecules

2 Metallic and covalent bonding broken:

sodium	+	water	→	sodium hydroxide solution
metallic bonding in sodium broken	+	covalent (polar) bonds between oxygen and hydrogen broken in some molecules	→	solvent–solute interactions formed; sodium ions and hydroxyl ions hydrated by polar water molecules

3 Ionic bonding formed:

silver nitrate solution	+	sodium chloride solution	→	sodium nitrate solution	+	silver chloride
solvent–solute interactions between water molecules and silver ions are broken	+	solvent–solute interactions between water molecules and chloride ions are broken	→	(solvated sodium and nitrate ions are 'spectators' that are unchanged during the reaction)	+	ionic lattice formed due to mutual attraction between silver cations and chloride anions

4 Metallic and covalent bonding broken, ionic bonding formed:

sodium	+	chlorine	→	sodium chloride
metallic bonding in sodium broken	+	covalent bonds in chlorine molecules broken	→	ionic bonding formed between sodium ions and chlorine molecules

Once students have been taught about bond types, it is useful when discussing different reactions in various topics to ask students about the types of bonding broken and formed. This will reinforce learning, help to shift students from thinking of bonding and reactions in 'octet' terms to electrical interactions, and encourage them to try to visualise what is occurring at the submicroscopic level.

Other resources

Websites

The Royal Society of Chemistry (RSC) publishes a range of resources to support chemistry teaching in schools. *Chemical Misconceptions – Prevention, Diagnosis and Cure* (Taber, K.S., 2002; Volume 1: Theoretical background; Volume 2: Classroom resources) includes resources for probing student thinking and finding out whether students have acquired common misconceptions on a range of topics relevant to this chapter. The classroom resources may be downloaded from the RSC site:
www.rsc.org

The group-work activity 'Judging models in science' is included in *Enriching School Science for the Gifted Learner*. Taber, K.S. (2007). London: Gatsby Science Enhancement Programme. Available from 'Mindsets Online':
www.mindsetsonline.co.uk

A range of simulations can be downloaded from the 'Chemistry Experiment Simulations and Conceptual Computer Animations' page from the Chemical Education Research Group at Iowa State University. These include a simple simulation of a particle model of NaCl dissolving in water and a simulation showing hydrogen bonding between water molecules in the liquid phase.
www.chem.iastate.edu/group/Greenbowe/sections/projectfolder/simDownload/index4.html

The simulations available from The Concord Consortium are mostly suitable for more advanced learners, but a number of them may be useful for supporting learning with some secondary students. For example, there is a simulation showing how the electron density around two hydrogen atoms is distorted as one is moved towards and then away from the other, and a model showing that the distinction between ionic and covalent bonds reflects the extremes of a bond between two atoms (their electronegativities may be varied by 'sliders').
www.concord.org/activities/subject/chemistry

Examples of students' alternative ideas and ways of thinking about chemical bonding and chemical reactions are given on the ECLIPSE (Exploring Conceptual Learning, Integration and Progression in Science Education) project website:
www.educ.cam.ac.uk/research/projects/eclipse/

Further resources

The BBC co-produced feature film 'Life Story' (featuring the actors Jeff Goldblum, Tim Pigott-Smith, Alan Howard and Juliet Stevenson among others) offers excellent insight into aspects of the nature of science in telling the story of the elucidation of the structure of DNA: a story of scientific cooperation and competition; slow and careful work collecting and analysing data; model-building; dead-ends; lucky breaks; building upon the work of other scientists; and a great deal more.

5 Extent, rates and energetics of chemical change

Vanessa Kind

Choosing a route

This chapter offers insights into concepts and suggestions for answering three important questions relating to chemical reactions: 'how far', 'how fast' and 'why' do chemical reactions go? Helping students to learn qualitative responses to these questions at GCSE will aid formation of a secure platform for further development of chemical knowledge, as these issues are central to understanding how chemists measure and control chemical reactions. These topics are presented here in a logical order.

Previous knowledge and experience

Students will bring to this topic their previous experience of chemical reactions in science lessons, as well as experience of many reactions outside of the classroom (although they may not recognise these changes as chemical reactions without some prompting). The material presented in this chapter draws upon the basic concepts and ideas explored in Chapters 1–4. Students often develop alternative conceptions about this topic so attention is drawn to these in each of the three sections of the chapter.

5.1 The extent of chemical change: how far do chemical reactions go?

Many chemical reactions go to completion. This means that the reaction continues until all component particles (whether ions, molecules or atoms) of one reagent have reacted, then stops. The reaction between hydrochloric acid and marble chips (calcium carbonate) is an example:

$$2HCl_{(aq)} + CaCO_{3(s)} \rightarrow CaCl_{2(aq)} + H_2O_{(l)} + CO_{2(g)}$$

The reaction may stop because all of the calcium carbonate, all of the acid or all of both reagents have reacted completely.

Fuel combustion is another example: as long as a car has fuel in its tank, the reaction will continue:

$$2C_8H_{18(l)} + 25O_{2(g)} \rightarrow 16CO_{2(g)} + 18H_2O_{(g)}$$

As soon as the car 'runs out of petrol' (or diesel, LPG, etc.), it will not move, because although the supply of oxygen is plentiful, no fuel remains to react with the oxygen.

Some chemical reactions do not go to completion. Instead, significant amounts of products react together to reform the original reactants. For example, when a solution of silver ions is added to a solution of iron(II) ions, the following reaction occurs:

$$Ag^+_{(aq)} + Fe^{2+}_{(aq)} \rightarrow Ag_{(s)} + Fe^{3+}_{(aq)} \text{ (Reaction 1)}$$

However, as soon as some products form, a few particles react to form the original reagents:

$$Ag_{(s)} + Fe^{3+}_{(aq)} \rightarrow Ag^+_{(aq)} + Fe^{2+}_{(aq)} \text{ (Reaction 2)}$$

Reactions 1 and 2 together form a 'reversible' reaction. Reaction 1 is called the forward reaction and Reaction 2 is called the reverse reaction. In Reaction 1, solid silver is seen as a shiny deposit. The two reactions continue indefinitely, as long as no other changes, such as temperature or pressure, occur. Eventually the rate at which the products of Reaction 1 form is the same as the rate at which they react in Reaction 2. The rates of the forward and reverse reactions are equal. No further change is observed to the amount of silver deposited. This position is called equilibrium. (Textbooks often call this 'chemical equilibrium'.) The reaction is said to be in *dynamic* equilibrium because the forward and reverse reactions continue at the same rate, so no external changes are seen.

$$Ag_{(s)} + Fe^{3+}_{(aq)} \leftrightarrows Ag^{+}_{(aq)} + Fe^{2+}_{(aq)}$$

A double-headed arrow ($\leftrightarrows$) is used to represent reversible reactions.
At equilibrium:

- The rates of the forward and reverse reactions are equal.
- The concentrations of products and reagents are constant.
- The concentrations of products and reagents are not necessarily equal.
- The reaction mixture is called the equilibrium mixture.
- No further observable change occurs.

The main condition keeping a reaction at dynamic equilibrium is that the whole set-up remains isolated, that is, the reaction occurs in a closed system without intervention from the atmosphere such as changes in pressure, or temperature, or any other reagent.

Students' misconceptions about equilibrium and dynamic equilibrium

Students find the concept of equilibrium and dynamic equilibrium difficult to understand. Research has revealed that students think equilibrium means:

- balance
- see-saw
- static
- products and reactants present in equal amounts.

The first three meanings imply incorrectly that no change occurs at equilibrium. Students need to understand that reactions are still occurring, but that their net effect is no observable change. The fourth, that products and reactants are present equally, arises logically from the 'balance' notion. A see-saw balances when equal forces act on each side. This is not true for chemical equilibrium. The forward or reverse reaction may dominate, so products and reactants are not necessarily in equal concentrations, but the overall equilibrium position is unchanged. Students often think reversible means 'moving backwards' or 'reversing'. They may interpret 'moving' as meaning 'equilibrium position moves'. This is problematic because the equilibrium position does not change.

Other faulty ideas students may have include the following:

- 'No further change' means 'no further reaction' or 'everything stops' when a system reaches equilibrium. Students need to know that particles react continuously, albeit unseen.
- 'Equal rates' means 'equal concentrations'. Students need to understand that this is not so.
- Forward and reverse reactions are separate and occur alternately. This is wrong, because the reactions are two parts of one, reversible reaction.

■ A useful analogy to explain dynamic equilibrium

One way of helping students understand dynamic equilibrium is to show someone staying in the same position on a moving escalator. To stay still, the person must be walking, either taking steps up or down, depending on the direction. We cannot see this movement, but know it must be occurring at the same rate the escalator is moving, or else the person's position would change. This is analogous to a dynamic equilibrium, in which invisible 'movement' is going on. Two reactions occur at the same rate, resulting in no overall change.

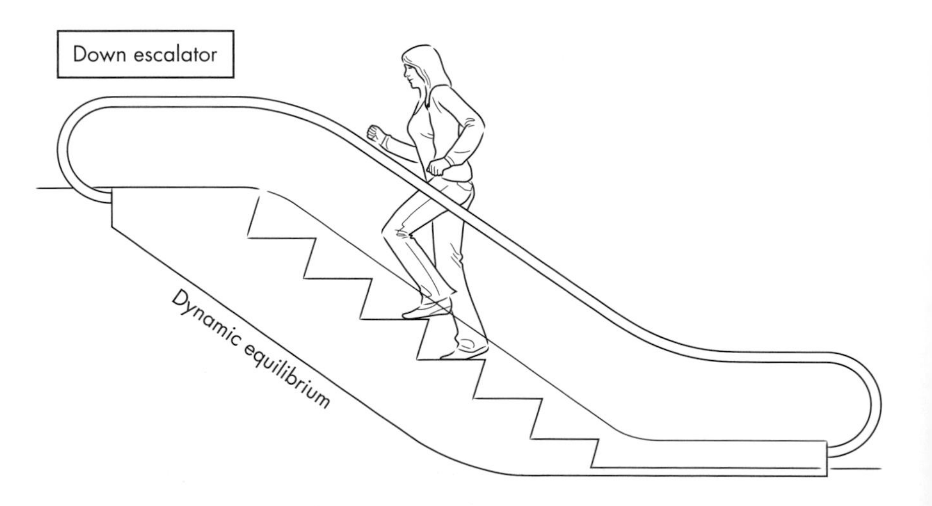

Figure 5.1 The idea of a person staying in the same position on a moving escalator can be used to help students understand dynamic equilibrium.

As noted above, 'dynamic equilibrium' only applies to reactions which are 'closed systems'. This means that the atmosphere is not involved. The reaction between silver ions and iron(II) ions (page 138) is an example. The two reactions described at the start of the chapter, between calcium carbonate and hydrochloric acid and between petrol and oxygen, are both 'open' systems. The calcium carbonate/hydrochloric acid reaction releases a gas into the atmosphere. The petrol/oxygen reaction involves a gas in the atmosphere in a reaction. 'Open' reactions go to completion. Closed systems reach an equilibrium position.

Two practical activities that may help students understand the principles of dynamic equilibrium are described next.

Why do fizzy drinks go 'flat'?

For this demonstration, you will need:

- a transparent plastic bottle full of a fizzy drink
- a cup or glass
- a 'fizz-keeper' device to pump air into a drinks bottle.

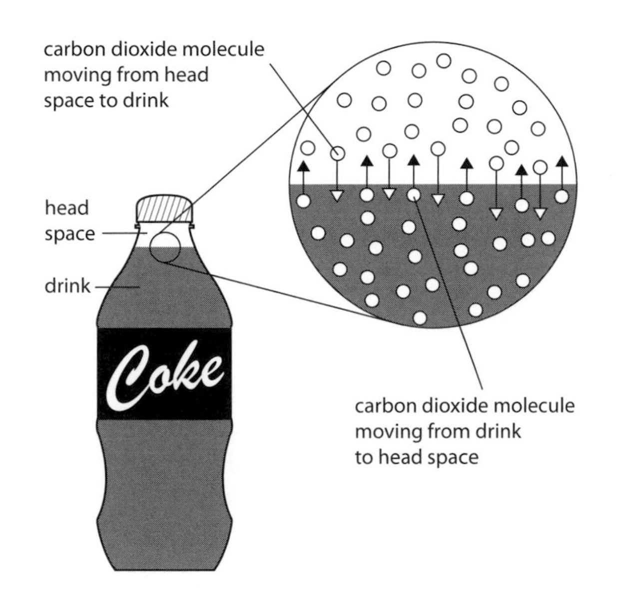

Figure 5.2 Dynamic equilibrium in a stoppered fizzy drink

A fizzy drink is a useful example to illustrate dynamic equilibrium. In making a fizzy drink, carbon dioxide gas is pumped into the liquid, at a pressure above atmospheric pressure. The gas dissolves:

$$CO_{2(g)} \rightarrow CO_{2(aq)}$$

The bottle is sealed. A gap, called the head space, remains between the liquid and bottle top. This allows a dynamic equilibrium to be set up:

$$CO_{2(g)} \leftrightarrows CO_{2(aq)}$$

The forward reaction is carbon dioxide gas molecules going into the drink. The reverse reaction is carbon dioxide molecules diffusing into the head space. At equilibrium, the rates of these reactions are equal. The reaction is at equilibrium as long as the temperature does not change. The bottle does not get bigger, shrink, get harder or softer to press, for example. If the conditions change, such as when the bottle gets hot, the pressure of carbon dioxide in the head space builds up, so not only does the bottle harden, but anyone opening it can get showered as the pressure releases. The amounts of aqueous and gaseous carbon dioxide are not equal. A full fizzy drink bottle contains far more aqueous carbon dioxide than gaseous carbon dioxide. The equilibrium position lies to the right, or favours the forward reaction.

If the bottle is shaken, the kinetic energy of the contents increases. This also disturbs the equilibrium, and helps to explain why a shaken fizzy drink bottle, especially on a hot day, can create significant 'fizz', and even 'explode' when opened.

When talking to students about the concept of equilibrium in a fizzy drinks bottle, it can be useful to ask the following questions about the full bottle to monitor their understanding:

- Why do we hear a 'fizz' when a full drink bottle is opened?
- Why should a full fizzy drink bottle be kept cool?
- What happens when a full fizzy drink bottle gets hot?

Ensure students explain what happens in terms of the reversible reaction between aqueous and gaseous carbon dioxide.

The next step is to think about what happens when the bottle is opened, a drink is poured out and the bottle reclosed. Opening the bottle means the system is no longer closed, so a change to the equilibrium position occurs. Usually, a 'fizz' is heard when the bottle is opened. The equilibrium position shifts towards the reverse reaction, releasing carbon dioxide pressure in the head space by diffusion from the drink.

As the drink is poured, gas bubbles form, creating the 'fizziness' we experience when drinking. When the top is replaced, the system is closed once again. The head space above the liquid is enlarged. To restore dynamic equilibrium, carbon dioxide diffuses from the liquid until there is no further change. The reverse reaction is favoured. When equilibrium is re-established, concentrations of reactant and product differ from the original, full bottle, situation. The fizz in the drink is lowered. Every time a drink is poured out and the top replaced, the equilibrium is re-established. Over time, less carbon dioxide remains in solution and more goes into the increasingly large head space above the drink. Eventually, when the top is released, the fizz does not happen and a flat, non-fizzy drink is poured out.

To monitor understanding, ask your students to think about what will stay the same and what will change when a drink is poured out. List the following and ask students to say whether or not it will change:

- the amount of carbon dioxide dissolved in the drink
- the volume of drink
- the amount of carbon dioxide above the drink
- the amount of carbon dioxide that diffuses (escapes) from the drink.

Students should hopefully understand that all will change except the volume of drink.

What will happen to the amounts of dissolved and gaseous carbon dioxide if some drink is poured out and the top replaced so the bottle is closed? Give students these possible answers and then discuss the merits of each:

- nothing
- more carbon dioxide dissolves into the drink, becoming aqueous
- more carbon dioxide diffuses from the drink, becoming gaseous
- the amounts of aqueous and gaseous carbon dioxide are equal straightaway
- the amounts of aqueous and gaseous carbon dioxide dissolving and escaping from the drink equalise gradually
- the amounts of aqueous and gaseous carbon dioxide establish a new equilibrium position.

The correct answers are that more carbon dioxide diffuses from the drink and that a new equilibrium position is established. See the previous page for explanations.

A 'fizz-keeper' can help preserve fizziness in a partially full bottle. The fizz-keeper is a device allowing air, mainly nitrogen and oxygen, to be pumped into the bottle. The extra gas molecules slow the rate at which carbon dioxide molecules diffuse from the liquid, delaying re-establishment of equilibrium. The fizz-keeper is not perfect – it will not prevent diffusion. The equilibrium position will be restored, but over a longer time period. In addition, the increased air pressure exerted by the fizz-keeper cannot compensate completely for the high pressure (greater than 1 atmosphere) used originally to pump the carbon dioxide into the bottle, so there is still plenty of room in the head space for more gas molecules. At best, fizziness will be preserved for a few hours. Regardless of advertising claims, a fizz-keeper will not work for days or weeks.

Ask students the following questions about using a fizz-keeper:

- What gases are pumped into the bottle by the fizz-keeper?
- How do these gases help 'keep the fizz'?
- Is the fizz-keeper perfect? Explain.

Students could carry out a further investigation into the length of time and volumes for which different styles of fizz-keeper work effectively. These could be compared with the fizziness retained by pouring the remaining drink into a smaller bottle to create a smaller head space.

The 'pink and blue' substance reaction

The equilibrium reaction between two cobaltion complexes provides a good way of introducing students to the concepts of chemical equilibrium and dynamic equilibrium by experiment. This is based on a strategy devised by Dutch researchers Van Driel, de Vos and Verloop in 1998. (Further details of their work are given in the 'Other resources' section at the end of the chapter.) The strategy takes a constructivist perspective. This means that rather than telling students up front, they are encouraged to work through the experiments. This allows correct scientific ideas to be built up gradually. Teaching by this method relies on asking questions to draw out students' thinking, discussing their ideas and introducing key terms (dynamic equilibrium, reversible reaction, etc.) when appropriate.

Non-chemists (or even chemists) should not be put off by apparently complicated formulae. The only essential knowledge is that one substance in the experiments is pink and the other is blue. The substances are formed when cobalt(II) ions (Co^{2+}) combine with chloride ions to make the blue substance and with water molecules to make the pink substance (Figure 5.3). The formal names for the substances are shown in the equation:

$$Co(H_2O)_6^{2+} \quad + \quad 4Cl^- \quad \leftrightharpoons \quad CoCl_4^{2-} \quad + \quad 6H_2O$$

PINK	**BLUE**
(cobalt(II) hexahydrate complex ion)	(cobalt(II) tetrachloro complex ion)

The colour changes when water or chloride ions are added or when there is a change in temperature.

Figure 5.3 The 'pink and blue' substance reaction

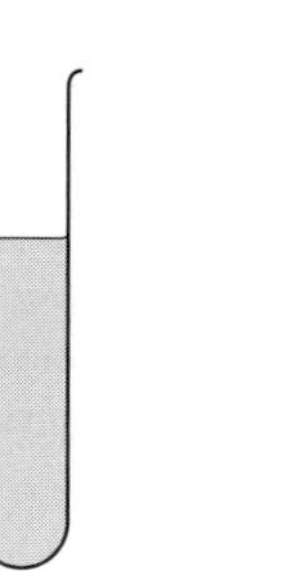

Initially, the principle of a reversible reaction is introduced. Ask students to carry out the first three experiments described. Keep the results from Experiment 1 so students can compare the colour – this is evidence that helps explain the results from Experiments 2 and 3.

PROCEDURE

For all three experiments, you will need:

- eye protection
- test tubes
- a test-tube rack
- pipettes – to permit drop-wise addition of liquid; one pipette per liquid per student group

Safety
All students should wear eye protection.
Teachers may wish to carry out this experiment as a demonstration.
Please note, teachers in Germany and Ireland will not be able to perform this experiment due to the hazard classification of cobalt compounds.

Experiment 1: Heating and cooling a pink substance

Materials

- one test tube containing 4 cm^3 of a solution of pink substance (pink substance solution: 1.0 g cobalt(II)chloride hexahydrate ($CoCl_2.6H_2O$) dissolved in 25 cm^3 propan-2-ol (isopropyl alcohol, IPA) by magnetic stirring (to produce a blue solution); add 5 cm^3 water to turn the solution pink)
- access to a water bath at about 80 °C and an ice bath at about 4 °C

Method
Tell students they are experimenting on the solution of a pink substance. They must repeatedly heat and cool the solution, note and discuss their observations in terms of the chemical reactions they think are occurring. (In hot water, the pink solution turns blue. In cold water, the blue solution turns pink.)

Experiment 2: Testing a blue substance

Materials

- one test tube containing 4 cm^3 of a solution of blue substance (blue substance solution: 1.0 g cobalt(II)chloride hexahydrate ($CoCl_2.6H_2O$) dissolved in 100 cm^3 propan-2-ol (isopropyl alcohol, IPA))
- two pipettes
- water

- access to a saturated solution of anhydrous calcium chloride (calcium chloride dissolved in propan-2-ol); this can be labelled 'chloride ions'

Method

Tell students to add water drop-wise to the solution of blue substance. They must stop when there is no further change (the solution will turn pink). Then, to the pink solution just made, they add about $1.5\,cm^3$ of chloride ions (saturated anhydrous calcium chloride solution). The solution turns blue.

They must note their observations and discuss what they see in terms of chemical reactions between the blue substance, water, chloride ions and pink substance.

Experiment 3: Testing a pink substance

Materials

- about $4\,cm^3$ solution of pink substance (as for Experiment 1)
- chloride ions (as for Experiment 2)
- water
- pipettes

Method

Tell students to add chloride ions drop-wise to the solution of pink substance, to stop when there is no further change (the solution will turn blue), then to add water drop-wise until there is no further change (the solution will turn pink).

Students should again discuss the events they observe in terms of chemical reactions between the pink and blue substances, chloride ions and water.

Students should be able to write down word equations representing the chemical reactions they observed, for example:

pink substance + chloride ions → blue substance

blue substance + water → pink substance

The next step is to introduce the term reversible reaction, suggesting that these reactions occur together, so can be represented as:

pink substance + chloride ions ⇆ blue substance + water

The point of the three experiments together is to show that the pink and blue substances can be formed by chemical reactions in the same test tube. The two reactions comprise a reversible reaction because careful addition of each reagent (water or chloride ions) creates the other coloured substance. From this, students should understand that:

- A chemical reaction can be reversible. This means that A + B turns into C + D, and vice versa.
- Chemical reactions do not all go to completion and stop when A or B is exhausted.

Questions to ask your students could include:

- Do you think the blue substance you started with in Experiment 2 is the same as the blue substance you made in Experiment 3? What is your evidence?

- Do you think the pink substance you started with in Experiment 3 is the same as the pink substance you made in Experiment 2? What is your evidence?

(Yes: students should use the results of Experiment 1 to help provide evidence for this.)

- What word equation represents how the blue substance forms?
- What word equation represents how the pink substance forms?
- Write one equation that combines both reactions.

Introduce the terms forward and reverse reactions to describe the combined equation. When the students have understood these points, the next step is to introduce the notion of dynamic equilibrium in a reversible reaction using the next experiment.

PROCEDURE

Introducing dynamic equilibrium: pink and blue substances in the same test tube

Materials

- one test tube containing 4 cm^3 of a solution of pink substance
- one test tube containing 4 cm^3 of a solution of blue substance
- a water bath at about 40 °C

Method

Tell students to put both test tubes in the warm water. They should note their observations and explain these using the chemical equation discussed on page 146. (Both solutions turn the same shade of purple.)

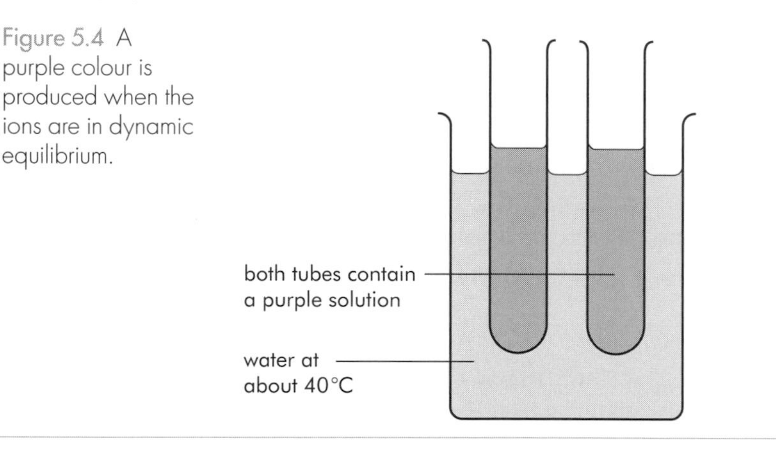

Figure 5.4 A purple colour is produced when the ions are in dynamic equilibrium.

Questions to ask your students:

- Do you think the forward and reverse reactions occur at the same time? What is your evidence?
- Suppose both reactions do take place – what can you say about their rates?

Students should recognise that both the pink and blue substances are present in the warm solution. They can be prompted to explain this using the equilibrium reaction shown previously. The answer must be that both reactions occur simultaneously, resulting in the purple colour. The purple solution represents the reaction at equilibrium.

At this point, the term dynamic equilibrium can be introduced. The purple solution is created when particles of pink substance react with chloride ions at the same time as particles of blue substance react with water molecules. The purple colour arises because both reactions occur at the same rate.

During discussions students may need help to develop the reversible reaction idea. When introducing dynamic equilibrium, they may think an equilibrium is static, see-saw or that a reaction stops entirely at equilibrium. The ongoing nature of the reactions may be difficult to understand. Ask students to explain their thinking or provide evidence for their thinking. For example, if they think the reaction is static, what is their evidence for this? Encourage students to imagine they can see the particles. What would they see?

5.2 The rate of chemical change: how fast do chemical reactions go?

Chemists want to know how quickly products are made in chemical reactions. Controlling reaction rate is important to the chemical industry, as this ensures products can be made in useful timescales. The 'rate' of reaction is the amount of product made in a unit of time (millisecond, second, minute, hour, etc.). Rates can be extremely fast, such as the reaction between hydrogen and oxygen gases:

$$2H_{2(g)} + O_{2(g)} \rightarrow 2H_2O_{(g)}$$

When the gases are mixed in a confined space and a spark applied, the product, water, is made so quickly the reaction is explosive. If no spark is applied, the reaction will still occur, but very slowly (and safely). Other reactions, such as fermentation of fruit juice to produce alcohol, take months or even years to complete. Reaction rate depends on a number of factors including:

- concentration of the reagents
- temperature
- surface area, or availability of the reagents
- whether or not a catalyst is present.

Reaction rate varies as a reaction progresses. The fastest rate is usually at the beginning, when reagents are first combined together. This is when availability of reagents is at a maximum, so products form easily. As time goes on, fewer unreacted reagent particles are available. These take time to meet in the reaction vessel, so the rate of product formation slows. Figure 5.5 represents the progression of the reaction between calcium carbonate and dilute hydrochloric acid:

$$2HCl_{(aq)} + CaCO_{3(s)} \rightarrow CaCl_{2(aq)} + H_2O_{(l)} + CO_{2(g)}$$

Carbon dioxide gas is a product. The gas can be collected in a gas syringe (Figure 5.6) and the volume produced over time recorded. The data recorded can be plotted.

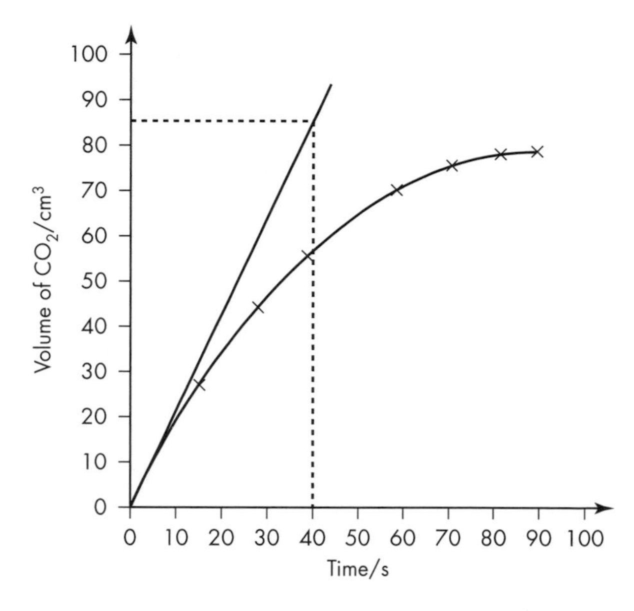

Figure 5.5 The rate curve for the reaction between dilute hydrochloric acid and calcium carbonate

Figure 5.5 is an example of a rate curve. This shows changes in the amount of product made with time. Changes occur in the steepness of the curve. The first part, from the origin, is steepest. The slope, or gradient, of this section gives the initial rate of reaction. Chemists often calculate initial rate values to compare rates of reactions, or to find out if they have been successful in speeding up or slowing down a reaction. The initial rate can be calculated by drawing a line (a tangent) against the steepest part of the curve, extrapolating it, then dividing the amount of product (volume of carbon dioxide) by the amount of time taken to produce it:

$$\text{rate} = \frac{\text{volume of carbon dioxide produced after 40 seconds}}{\text{time in seconds}} = \frac{85}{40} = 2.125\ \text{cm}^3/\text{sec}$$

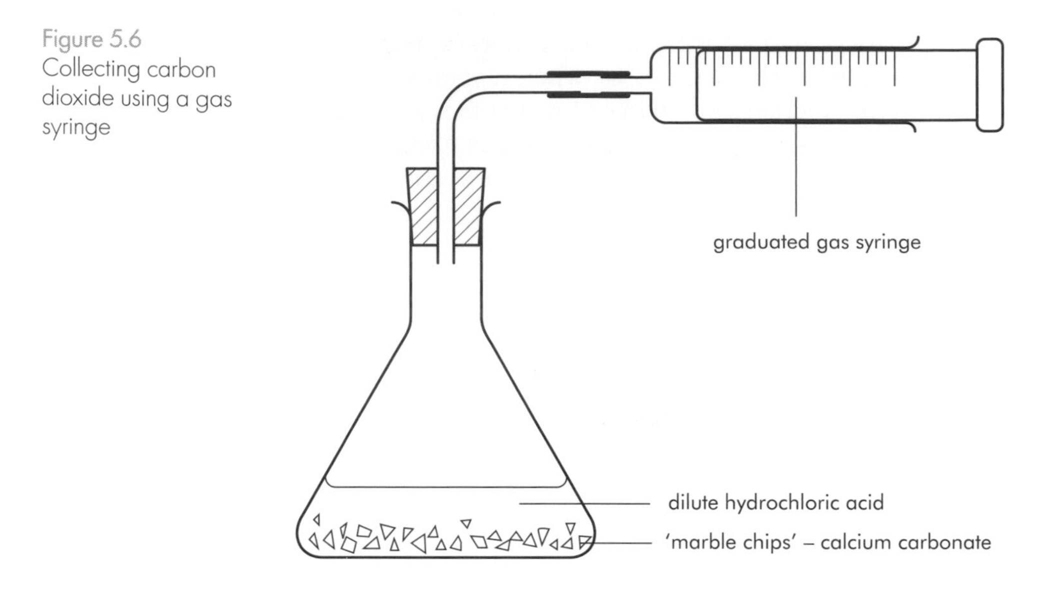

Figure 5.6 Collecting carbon dioxide using a gas syringe

Any points taken along the line will give the same result. Note the rate units combine the product measurement unit and time unit.

The latter parts of the rate curve are less steep, showing that product is produced more slowly. Eventually, the curve flattens to a horizontal line. In this section, no product is made and the reaction rate is zero. This indicates the reaction is complete, because the supply of one or other of the reagents is exhausted.

Students' difficulties with rates of reaction

Outlined here are four difficulties that students may find with rates of reaction. Each has implications for understanding more advanced aspects of the topic.

■ 1 Students think reactions are all 'fast'

Students may think all reactions are fast and that this is a definition for a chemical reaction itself. Events that go beyond time frames common in school may not be regarded as chemical reactions: there is a strong association with reactions taking place in chemistry lessons only and not outside a lab. Commonly, everyday chemical reactions take hours, months or years to complete. Explicit discussion is needed for students to recognise that the same rules apply to 'slow', non-laboratory-based reactions and those seen within the hour-long time-span common to many chemistry lessons. The first activity suggested in the next section (page 152) will help to address this. Other examples to discuss could include:

- the formation of rust – this may be regarded as a chemical reaction, but rate of production of iron(III) oxide, or rust, is probably not considered in the same way as carbon dioxide production in the reaction given earlier
- browning of a cut apple – the rate at which the brown chemical is produced occurs over minutes
- digesting food – this takes several hours, depending on the food (I once read that the average American male has about 500 g of undigested red meat in his intestines!)
- making yoghurt/bread – this requires hours to complete.

■ 2 Students think 'fast' reactions are always exothermic

Students may link fast rate with energy production. This may occur because exothermic reactions are often used to illustrate rate. The idea is problematic because students then find it hard to understand that exothermic reactions can be slow due to high activation energy (like the example of the hydrogen and oxygen reaction) and that some spontaneous, fast reactions are endothermic. There are links here to the next topic, energetics (section 5.3).

■ 3 Students think catalysts change the activation energy for a reaction

Use of the statement 'catalysts lower activation energy' should be discouraged. Catalysts do not lower the activation energy for the original reaction. This remains unchanged. A catalyst provides an alternative route, or reaction pathway, that has lower activation energy.

■ 4 Students think increased particle movement is responsible for rate increase

When temperature is increased by 10 °C, collision theory states that reaction rate doubles. Students often think the rate increases because particles move about much more than at lower temperatures, or have greater kinetic energy. This is only partially true. The increased temperature raises the average energy of the particles involved. This means more particles have the minimum energy needed to react. The rate increase is mainly because of the increased number of successful collisions between particles. This is the dominant factor. Note that full treatment of collision theory as an explanatory model is beyond the scope of the present chapter.

Probing prior knowledge about reactions

Before proceeding with experiments that investigate how different factors affect reaction rates, it is worth spending a few minutes finding out ideas students have about rates of different chemical reactions. Discussing these may help address any difficulties, such as those mentioned above, so will be time well spent. One suggestion is to provide a list of everyday and laboratory-based chemical reactions and ask students to indicate whether they think the rate of each is 'slow' or 'fast'. A sample list might include:

- burning petrol/fuel in a car engine (E)
- frying an egg (E)
- doing the 'pop' test for hydrogen gas
- a nail rusting (E)
- cleaning a sink with bleach (E)
- reacting sodium with water
- digesting chocolate (E)
- reacting marble chips and acid
- making yoghurt or letting milk go sour (E)
- burning gas in a Bunsen burner
- cleaning your teeth (E)
- reacting an acid and alkali together.

See whether students can separate out 'everyday' reactions (marked E) from laboratory ones. Find out what their criteria are for fast and slow reactions: do they associate rate with 'making product in a unit of time'; do they think fast reactions mean energy is produced; or is their idea of a fast reaction one that simply does not take long to do (for example cleaning teeth, perhaps)? Also, what do they think are the products of these reactions? What might change the rate at which the products could be made? Why is rate reaction important? Establishing that all chemical reactions have one or more products and a rate that can vary is a worthwhile exercise.

Factors affecting rate of reaction: concentration

Here are experiments that can be used to illustrate how concentration affects the rate of a reaction.

■ The iodine 'clock' reaction

Experienced chemistry teachers are likely to know this as a classic reaction, but it is very good, so no apologies are made for its inclusion. The iodine 'clock' reaction can be used as a demonstration to show that the time taken for a product to be made can change, or in a more advanced way as an investigation of precise effects of changing the concentration of a single reagent. The instructions provided allow for both.

The clock reaction involves the detection of iodine being produced in this reaction:

$$S_2O_8{}^{2-}{}_{(aq)} + 2I^-{}_{(aq)} \rightarrow 2SO_4{}^{2-}{}_{(aq)} + I_{2(aq)} \text{ (reaction 1)}$$

The $S_2O_8{}^{2-}$ ion is called peroxodisulfate(VI) or persulfate. Iodine is the only coloured substance in the reaction. Iodine forms a blue–black complex with starch, which is added as an indicator. A clock reaction measures the time required to produce a small, fixed amount of product, iodine in this example. To do this, a second reaction is added alongside the first one to 'hide' the product for a certain amount of time. The second reaction used here is:

$$2S_2O_3{}^{2-}{}_{(aq)} + I_{2(aq)} \rightarrow S_4O_6{}^{2-}{}_{(aq)} + 2I^-{}_{(aq)} \text{ (reaction 2)}$$

$S_2O_3{}^{2-}$ is called the thiosulfate ion and $S_4O_6{}^{2-}$ is called the tetrathionate ion. Reaction 2 has iodine as a reagent. As soon as iodine is formed in reaction 1, it reacts with thiosulfate ions to make colourless iodide ions again. Thiosulfate is called the limiting reagent because this limits when the iodine is seen: when all thiosulfate ions have reacted, no more are available to react with any iodine produced. Instead, iodine produced in reaction 1 will make the blue–black complex with starch and this colour will appear. The time taken for iodine to appear depends on the concentration of iodide ions. If this is high, iodine is produced in reaction 1 more rapidly, so thiosulfate ions in reaction 2 react and reach their 'limit' quickly. If the iodide ion concentration is low, the rate at which iodine is produced in reaction 1 is slow, so the thiosulfate ions also react more slowly.

PROCEDURE

Materials – class experiment, per student group

- eye protection for each student
- five boiling tubes
- one test tube
- stirring rod
- thermometer 0–100 °C
- burettes or graduated pipettes: 1, 2 and 5 cm^3
- stopwatch
- 1 $mol\,dm^{-3}$ potassium iodide solution, about 15 cm^3
- 0.04 $mol\,dm^{-3}$ potassium peroxodisulfate(VI) solution, about 10 cm^3
- 1 $mol\,dm^{-3}$ sodium thiosulfate solution, about 10 cm^3
- fresh starch solution, about 5 cm^3
- water

Safety
All students should wear eye protection.

Method
Each group can make five reaction mixtures (Table 5.1). The procedure for each reaction is:

1 Measure the water, potassium iodide solution, sodium thiosulfate solution and starch solution into a boiling tube.
2 Record the starting temperature of the reagents.
3 Measure the potassium peroxodisulfate solution into a separate test tube.
4 Set the stopwatch to zero.
5 Add the potassium peroxodisulfate solution to the other reagents in the boiling tube and start the stopwatch at the same time.
6 Record the time taken for a blue–black colour (iodine) to form in the reaction mixture.
7 Record the final temperature of the reaction mixture.

Table 5.1 Reaction mixtures for the iodine clock reaction

	Mixture				
	A	B	C	D	E
Potassium iodide solution / cm^3	5	4	3	2	1
Water / cm^3	0	1	2	3	4
Sodium thiosulfate solution / cm^3	2	2	2	2	2
Starch solution / cm^3	1	1	1	1	1
Potassium peroxodisulfate(VI) solution / cm^3	2	2	2	2	2
Total volume / cm^3	10	10	10	10	10
Concentration of iodide ions / $mol\,dm^{-3}$	0.5	0.4	0.3	0.2	0.1

Note that the total volume for each reaction mixture is $10\,cm^3$. Water is added to make up the volume. Only the volumes of potassium iodide and water change.

Handling the data
There are two ways in which the data can be handled. The first is relatively easy; the second is more difficult and may not be suitable for all students.

1 Easy: plot a graph showing the volume of potassium iodide solution (x axis) against the time taken for iodine to appear in each of the five reaction mixtures (y axis). This should be a straight line. The conclusion is that the rate of reaction is directly proportional to, or depends directly on, the concentration of iodide ions.
More difficult: convert the volume of potassium iodide solution to concentration of iodide ions, then plot these values (x axis) against the time taken for iodine to appear (y axis). The potassium iodide concentration was $1\,mol\,dm^{-3}$. In reaction mixture A, $5\,cm^3$ of this solution was diluted to $10\,cm^3$. Therefore the concentration of iodide ions is halved, so is $0.5\,mol\,dm^{-3}$. The volume was reduced from $5\,cm^3$ by $1\,cm^3$, successively, in each of reactions B–E, so the concentrations are 0.4, 0.3, 0.2 and $0.1\,mol\,dm^{-3}$, respectively. The graph should be a straight line. This shows that the rate of reaction is directly proportional to the concentration of iodide ions. Students should see that doubling the concentration doubles the rate of reaction.

The iodine clock reaction as a demonstration

The iodine clock reaction makes a good demonstration. To set this up, make the same reaction mixtures as above, but enlarge the volumes and carry out the reactions in conical flasks so the colour change can be seen easily. Set the flasks against white backgrounds (such as pieces of card or a white screen). Ask students to help – have a timer for each reaction and ensure that the potassium peroxodisulfate solution is added simultaneously to each flask.

To make a more dramatic impression, change reaction mixture E. Dilute the iodide ions further, for example 1/100 (0.1 cm^3 iodide ions, 4.9 cm^3 water). This will slow down the rate considerably compared to the other mixtures, so the student responsible for flask E may start to look puzzled when their reaction does not seem to 'go'. This can make discussion a little more interesting than a straightforward 'ticking' clock in which the iodine colour appears at regular intervals.

Alternatively, alter the reaction mixture compositions to change colour after specific time intervals, and perhaps set students specified work to complete before the next colour change occurs. This may provide some stimulation, because they will not know exactly when to expect the next change.

■ Get fruity: investigating how bleach reacts with red fruit juice

This is a simple experiment exploring how changing bleach concentration affects the reaction rate with red fruit juice. Rather than the colour being formed, the time taken for it to disappear is recorded. The experiment could make a good investigation by changing fruit juice and bleach concentration. The reaction is:

$$2NaOCl_{(aq)} + \text{red pigment} \rightarrow \begin{matrix}\text{colourless}\\ \text{pigment}\end{matrix} + 2NaCl_{(aq)} + O_{2(g)}$$

Tiny bubbles of gas may be seen. These are too small to collect.

PROCEDURE

Materials

- about 25 cm^3 bleach (such as high-strength hydrogen peroxide or sodium hypochlorite solution)
- about 5 cm^3 red fruit juice extract (such as blackcurrant squash)
- water
- about ten boiling tubes or 100 cm^3 beakers
- a rack for the tubes
- a stopwatch
- measuring cylinders – 50 cm^3, 25 cm^3, 10 cm^3
- pipettes
- eye protection

Safety
All students should wear eye protection. If using sodium hypochlorite a fume cupboard should be used.

Method
Add 0.5 cm^3 (about three drops) of fruit juice extract to 25 cm^3 water. Then add 5 cm^3 undiluted bleach. Start the stopwatch and time how long it takes for the colour to fade completely (usually around 5 minutes). An initial colour change to yellow is observed after about 2 minutes.

Students will need to establish how to judge the colour 'fading completely'. They could use bleach solution or an acceptably pale fruit juice colour as a control, for example.

How science works: investigation
Students can decide how to dilute the bleach to change reaction time. Halving the concentration doubles the time. Once this is established, students could test other chemicals to see how they react with the red juice extract – for example, hydrogen peroxide, other enzyme-based cleaners, etc. This would give the opportunity to investigate the best reaction conditions for optimum cleaning of a red juice spillage.

Factors affecting rate of reaction: temperature

Every reaction has an energy barrier that must be overcome if reactant molecules are to combine and become product molecules. Most reactions we show in school chemistry do not need much encouragement, as room temperature often provides sufficient energy for reactant molecules to cross the relevant energy barrier. The level of energy required is called the activation energy (Figure 5.7). The activation energy for any reaction is fixed – nothing can be done to raise or lower it.

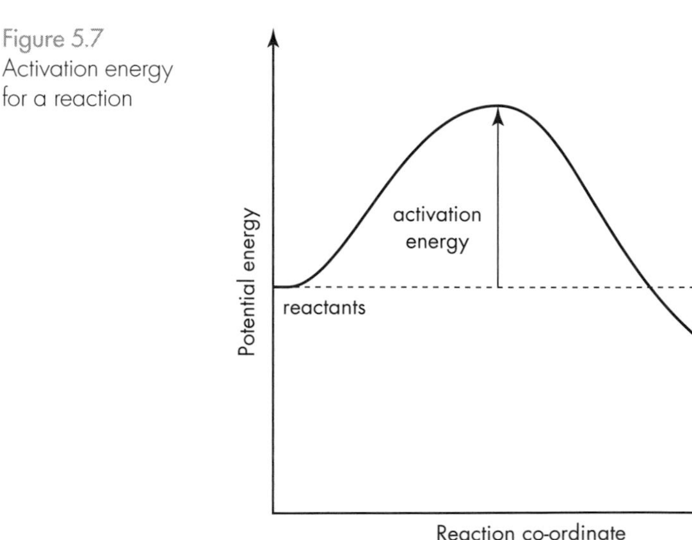

Figure 5.7 Activation energy for a reaction

However, a chemist can change the number of reactant particles that have the necessary energy level. Changing temperature is one way of achieving this. As a rule of thumb, reaction rate doubles for every 10 °C rise in temperature. The main factor influencing the rise in reaction rate is that more successful collisions occur between reactant particles. This is because increasing the energy level of reagents means more collisions occur. As the individual energy level of the particles is raised, more collisions involve reactant particles with the activation energy (or enthalpy) to react.

Investigating the effect of temperature on reaction rate: magnesium and water

The reaction between magnesium and water provides a simple illustration of the effect temperature has on reaction rate. The equation for the reaction is:

$$Mg_{(s)} + 2H_2O_{(l)} \rightarrow H_{2(g)} + Mg(OH)_{2(s)}$$

In water at room temperature, magnesium ribbon reacts slowly, producing few bubbles of hydrogen gas. At higher temperatures, significant numbers of gas bubbles can be seen.

PROCEDURE

Materials

- magnesium ribbon – about 10 cm
- emery paper
- a 250 cm^3 beaker
- water
- a Bunsen burner
- a heatproof mat
- a tripod
- a gauze
- a thermometer
- splints
- eye protection

Safety

All students should wear eye protection.

Stop heating when the temperature reaches about 80 °C.

Method

1. Clean the magnesium ribbon with the emery paper to remove the oxide layer.
2. Curl the magnesium ribbon around a pencil to help expose a larger surface area to the water.
3. Pour about 150 cm^3 water into the beaker.
4. Put the magnesium ribbon into the water.
5. Record the water temperature. Note any reaction between the magnesium and the water.
6. Set the beaker on the tripod and gauze.
7. Heat the beaker from underneath. Record the temperature and note qualitatively how reaction rate increases by observing the production of gas bubbles.

Discussion points
- Why are more gas bubbles produced at higher temperatures?
- What does this tell us about the rate of reaction?
- How might we measure accurately what happens to the rate?

Reaction of magnesium with steam – teacher demonstration

When steam, or water vapour, is passed over magnesium, a violent reaction occurs that creates a dramatic demonstration.

PROCEDURE

Materials
- a boiling tube with a hole in the closed end and a bung with a hole to fit the tubing from the steam generator
- a steam generator (a large conical flask containing plenty of water, with tubing, as shown in Figure 5.8)
- two Bunsen burners
- a tripod
- a gauze
- about 20 cm magnesium ribbon
- splints
- a clamp stand, boss, clamp
- emery paper
- eye protection
- screens

Safety
For teacher demonstration only. All students should wear eye protection and the teacher should wear safety goggles. The reaction should be carried out behind a safety screen. The boiling tube is weakened by the reaction and may break and shatter on the bench as the reaction cools down.

Method
1 Set up the apparatus as shown in Figure 5.8.

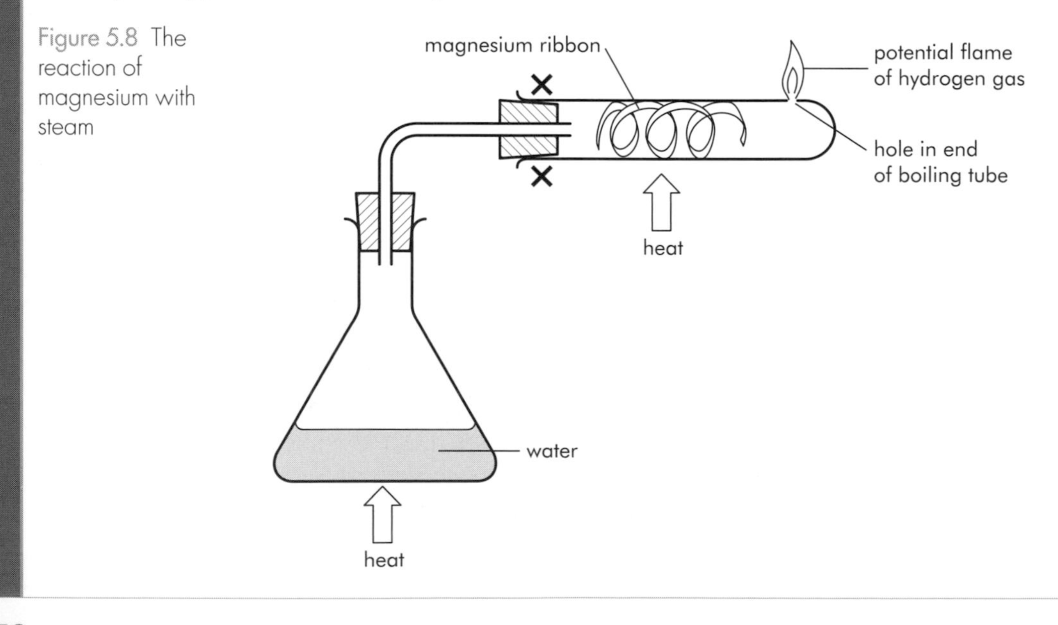

Figure 5.8 The reaction of magnesium with steam

2 Clean the magnesium ribbon thoroughly with the emery paper.
3 Curl the ribbon around a pencil, then place it inside the boiling tube, ensuring very good contact between the metal surface and glass (this requires careful preparation beforehand). The reaction will not 'go' if contact is poor. Note the boiling tube will be destroyed as the reaction fuses the magnesium and glass together.
4 Fit the boiling tube to the steam generator. Clamp the boiling tube at the neck, not the main body of the tube over the magnesium. Ensure the hole in the tube points upwards.
5 Set screens in front of the arrangement to protect the class from any breakages.
6 Dim lights so the bright glow from the reaction and hydrogen flame can be seen clearly.
7 Heat the steam generator from underneath with a strong blue Bunsen flame. Note that producing steam takes time, so start with water at about 60 °C, for example from a hot tap, or start the generator beforehand.
8 Simultaneously, heat the magnesium ribbon gently, holding the second Bunsen burner by hand to ensure even heating. When a steady jet of steam is passing over the magnesium, heat it more strongly.
9 The magnesium will react vigorously with the steam, glowing brightly. As soon as this occurs, stop heating the magnesium. Hydrogen gas will be emitted from the hole at the end of the tube. This can be lit with a lighted splint.

The drama of the experiment lies in showing that water, normally an innocuous substance, reacts vigorously with magnesium at high temperatures. The bright light and hydrogen flame produced create dramatic visible effects.

Have two boiling tubes ready in case the first one turns out to be a damp squib. Usually this is because the contact between glass and magnesium is poor, reducing heat transferred from the Bunsen flame. Also, sometimes only part of the magnesium ribbon reacts. The best examples occur when the magnesium ribbon reacts completely from one end to the other, showing a bright flare moving spirally down the boiling tube.

Factors affecting rate of reaction: the presence of a catalyst

A catalyst provides an alternative route to the products. The alternative route has a lower energy barrier than the original reaction. Students need to understand that the catalyst does not lower the activation energy. The activation energy for the original reaction remains unchanged (see Figure 5.9 overleaf).

The alternative routes vary depending on the catalyst type. A heterogeneous catalyst is in a different physical state to the reagents, such as a solid catalyst for gaseous or liquid reagents. Solid catalysts provide a surface on which reagent particles meet, changing the frequency of collisions. An example is the metal mesh, usually coated with catalysts containing platinum and palladium, present inside catalytic converters in car exhaust systems. These provide a surface on which carbon monoxide and hydrocarbons from car exhausts combine to produce water and carbon dioxide.

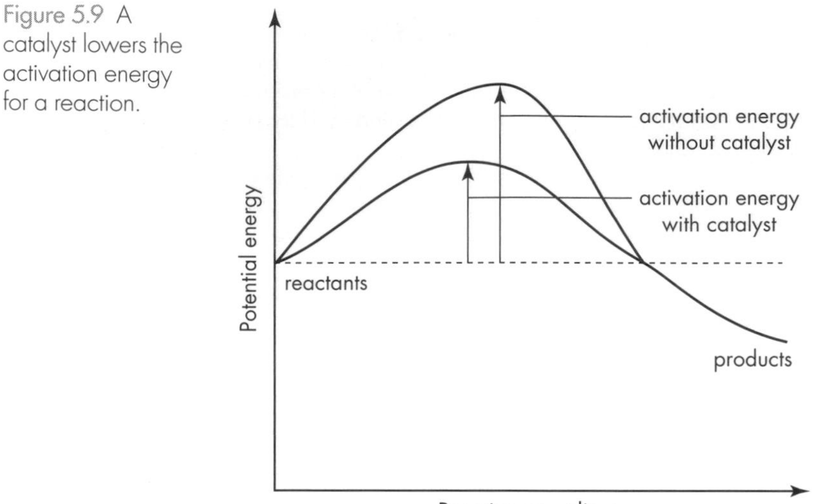

Figure 5.9 A catalyst lowers the activation energy for a reaction.

A homogeneous catalyst is in the same physical state as the reagents. This provides an alternative route by combining with one reagent, making an intermediate particle which goes on to react with another reagent. The formation of the intermediate and the next reaction step involve less energy than the original route to the products. When a sweet-smelling compound called an ester is formed from an alcohol and an organic acid (as described in Chapter 10), some mineral acid is added to the reaction mixture to act as a homogeneous catalyst.

Enzymes are often referred to as biological catalysts. Enzymes provide active sites in which chemical reactions occur. All enzymes are proteins, so every enzyme has a related gene on a chromosome. Without enzymes chemical reactions in living organisms would be too slow to sustain life. Problems in genes coding for specific enzymes are often responsible for serious illnesses and life-long conditions.

Catalysts:

- can be regenerated and reused after the reaction
- speed up the rate of a reaction by providing an alternative route to the products
- can be poisoned by addition of other substances
- can be enhanced by addition of a substance to act as a promoter
- are often unique to specific reactions
- need to be found by experiment.

Three experiments are described, all based on the decomposition of hydrogen peroxide. The first is a class experiment investigating how different catalysts affect the rate of decomposition. The next two are demonstration reactions.

■ Investigating the effect of different catalysts on the decomposition of hydrogen peroxide

This is a classic reaction, but provides an excellent experiment for illustrating the principles of biological and chemical catalysts using a straightforward reaction suitable for students of all abilities. Hydrogen peroxide decomposes thus:

$$2H_2O_{2(l)} \rightarrow 2H_2O_{(l)} + O_{2(g)}$$

Reaction progress can be tracked by collecting oxygen gas. The best way to do this is over water. This means bubbling gas into a measuring cylinder or burette held upside down in a trough of water and measuring the volume collected over time (Figure 5.10).

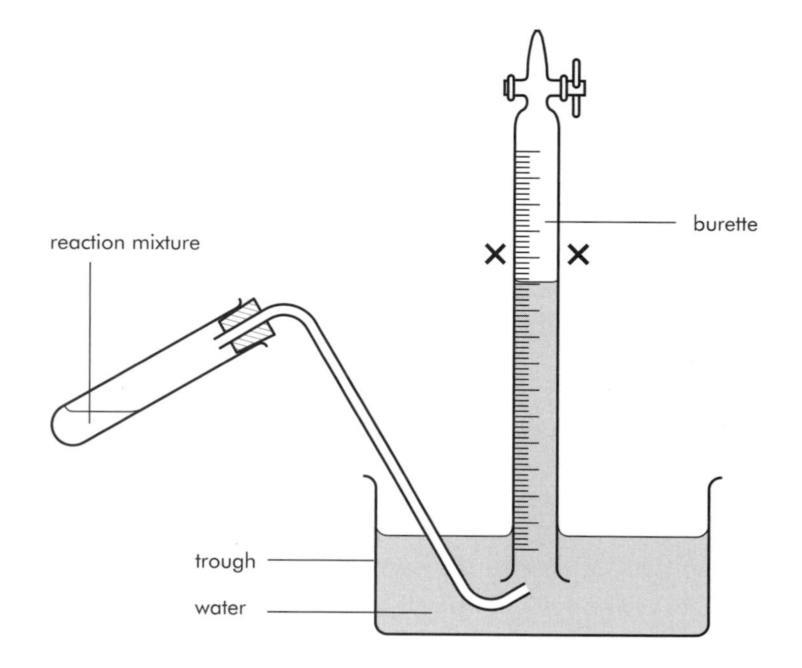

Figure 5.10 Investigating the effects of catalysts on decomposition of hydrogen peroxide

Hydrogen peroxide decomposes naturally over time. The decomposition is speeded up by the enzyme catalase, found in most plant and animal cells, including potato, yeast and fresh liver. Catalase is a 'scavenger' enzyme that provides an active site to which hydrogen peroxide molecules bind. The enzyme–peroxide structure facilitates bond breaking and bond making, so water and oxygen molecules can be made more easily than when hydrogen peroxide molecules break up unaided.

Decomposition is also enhanced by copper(II) oxide, manganese(IV) oxide and lead(IV) oxide. These are heterogeneous catalysts that provide a binding surface for hydrogen peroxide molecules.

PROCEDURE

Student investigation

Materials

- a burette
- a trough or bowl, such as an empty margarine tub
- boiling tubes
- a bung and delivery tube to fit a boiling tube
- a boiling-tube rack
- two clamp stands
- two bosses, two clamps
- a graduated pipette, 5 cm^3
- a measuring cylinder, 10 cm^3
- a stopwatch
- eye protection
- hydrogen peroxide, 5 volume, about 25 cm^3 per group
- yeast suspension, 20 cm^3 (made from 2 g dried yeast in 160 cm^3 water, aerated)

Safety

All students should wear eye protection.

Method

Set up the equipment as shown in Figure 5.10. To do this:

- close the tap on the burette
- fill the trough/tub and the burette with water
- make sure there is sufficient space in the trough to take all water displaced from the burette
- invert the burette in the trough
- clamp the burette in position.

Make the reaction mixtures one by one in the boiling tube according to Table 5.2. To make each reaction mixture:

- measure out the yeast suspension needed into the boiling tube(s)
- measure the hydrogen peroxide solution into the measuring cylinder
- set the stopwatch to zero
- set the delivery tube under the burette
- have the bung ready to fit to the boiling tube
- add the hydrogen peroxide to the yeast
- quickly fit the bung and start the stopwatch
- record the gas volume every 10 seconds for 4 minutes.

Note that each reaction takes about 6–8 minutes to set up and measure. If this is a class experiment, allow all student groups to carry out reaction mixture A, then a selection from B–E depending on the time available.

Table 5.2 Reagent volumes for the hydrogen peroxide experiment

	Mixture				
	A	B	C	D	E
Hydrogen peroxide / cm^3	5.0	5.0	5.0	5.0	5.0
Water / cm^3	0.0	0.5	1.0	1.5	2.0
Yeast suspension / cm^3	2.5	2.0	1.5	1.0	0.5

Handling the data
Plot a graph showing the volume of oxygen released (*y* axis) against the time (*x* axis) for each experiment. Students could then find out the initial rates of reaction using the method shown earlier in this chapter (page 149).

Questions to ask
- How does the rate change as the reaction proceeds?
- How does the initial rate vary with the concentration of the yeast suspension?

How science works: investigation – varying the experiment

Changing the catalyst

Students could compare how the rate varies when:

- catalase from different sources is used, such as a very small amount of liver suspension (the amount will depend on the freshness of the liver) or a piece of freshly peeled potato equivalent to about 1 cm^3
- a chemical catalyst is used instead, such as copper(II) oxide, lead(IV) oxide or manganese(IV) oxide (about 1 g of each).

Different catalysts could be investigated by separate student groups once the basic method has been established.

Changing the reagents: investigating concentration

Note that this experiment could also be used as an illustration of the effect of changing concentration. To do this, decrease the volume of hydrogen peroxide added in each of the five reaction mixtures successively, for example from 5 to 1 cm^3, making up the volume to 7.5 cm^3 with water. Keep the amount of yeast suspension in each reaction mixture constant at 2.5 cm^3.

■ Elephant toothpaste: demonstration

A saturated solution of potassium or sodium iodide catalyses hydrogen peroxide decomposition effectively. This reaction can be used to create an attractive demonstration producing a jet of coloured foam that looks like gigantic toothpaste.

PROCEDURE

Materials
- a large tray with raised sides (such as Gratnells type)
- a 250 cm^3 measuring cylinder (or larger)
- dropping pipettes
- a glass rod
- 30 cm^3 100-volume hydrogen peroxide
- a few drops each of washing-up liquid, food colouring and saturated potassium or sodium iodide solution

Method

1 Set the measuring cylinder in the tray.
2 Pour in the hydrogen peroxide.
3 Add the washing-up liquid and stir with the glass rod.
4 Add a few drops of food colouring and stir.
5 Add a few drops of saturated sodium iodide solution – stand back!

Results

A jet of coloured foam is produced that rises rapidly up the measuring cylinder and spills into the tray.

■ 'Genie of the lamp' demonstration

In this demonstration a tea bag containing manganese(IV) oxide is first suspended and then dropped into 100-volume hydrogen peroxide in a narrow-necked vessel, such as a conical flask. The reaction is so rapid a jet of steam is produced that creates a cloud above the vessel. With imagination, this could be a genie who can be summoned by rubbing a magic lamp!

PROCEDURE

Materials

- a narrow-necked vessel, such as a 250 cm^3 conical flask, with a tight-fitting stopper (or similar 'lamp')
- aluminium foil, enough to cover the vessel completely
- fine, transparent nylon thread
- an empty tea bag
- manganese(IV) oxide to half-fill the tea bag
- 20 cm^3 100-volume hydrogen peroxide

Safety

The teacher and the student to rub the lamp should wear safety goggles or face shields and gloves. Use safety shields between the experiment and the students. The students should be at least 3 m back from the safety shields. A considerable amount of hydrogen peroxide aerosol is produced.

Method

Before the demonstration:

- cover the reaction vessel with foil so it looks like a magic lamp
- place the hydrogen peroxide in the flask
- put the manganese(IV) oxide in the empty tea bag
- tie the tea bag tightly closed with the nylon thread
- suspend the tea bag in the flask above the hydrogen peroxide, leaving a piece of thread hanging outside the flask (Figure 5.11)
- fit the bung tightly so the thread and tea bag are held in place.

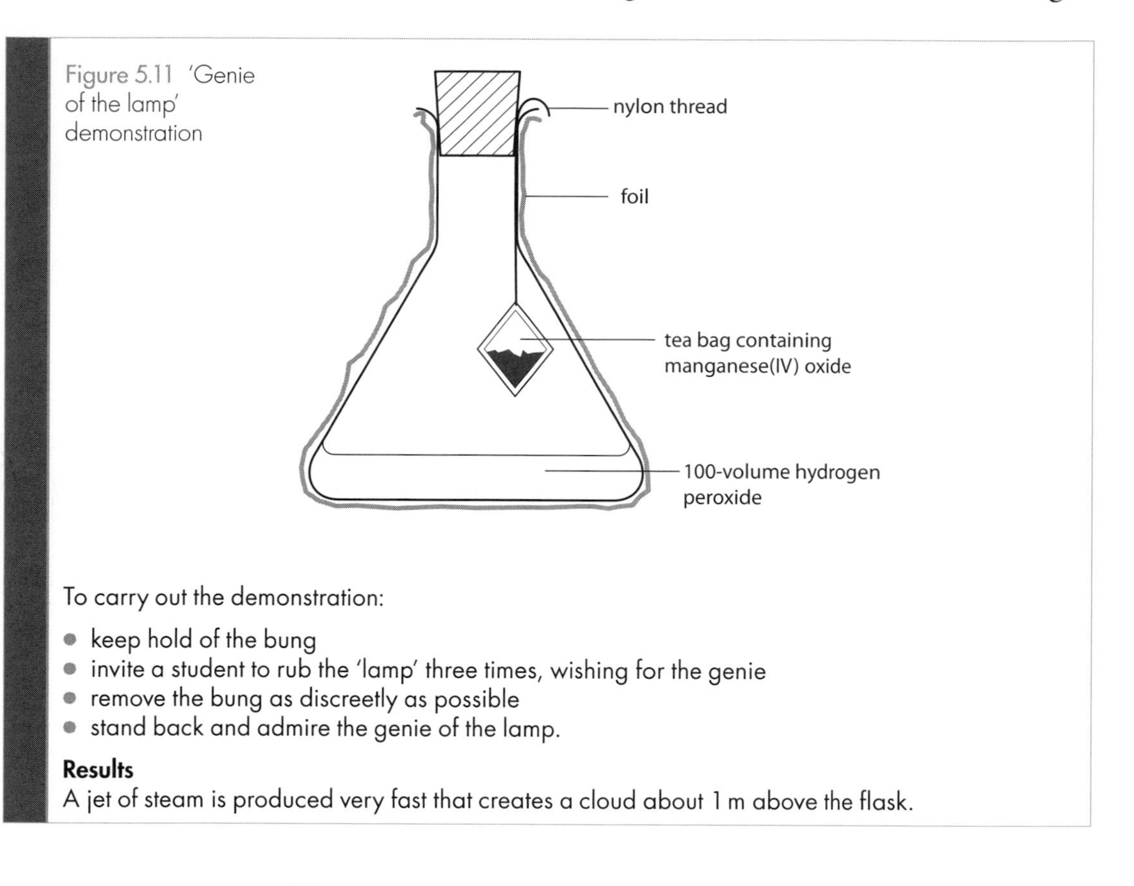

Figure 5.11 'Genie of the lamp' demonstration

To carry out the demonstration:

- keep hold of the bung
- invite a student to rub the 'lamp' three times, wishing for the genie
- remove the bung as discreetly as possible
- stand back and admire the genie of the lamp.

Results

A jet of steam is produced very fast that creates a cloud about 1 m above the flask.

Factors affecting rate of reaction: surface area

Increasing the surface area of a solid-phase reactant in a reaction between a solid and a liquid will enhance the rate of reaction. This is because the reaction takes place at the boundary between the two phases, that is, on the solid surface. Commonly, the effect of increasing surface area in the reaction between calcium carbonate (marble chips) and hydrochloric acid (see equation, page 149) is used as an investigation. Other reactions also illustrate the same point.

■ Investigating how surface area affects rate of reaction

The reaction between calcium carbonate (marble chips) and hydrochloric acid provides a good way of illustrating the effect of increasing surface area.

As a demonstration, the equipment shown in Figure 5.6 (page 150) can be used. For a class experiment, there are various ways of showing the progress of the reaction, such as recording the mass change as the carbon dioxide is produced (this requires ongoing access to balances) or collecting the carbon dioxide over water (as in Figure 5.10, page 161) – although some gas will dissolve, sufficient gas should be produced for measurement.

PROCEDURE

Materials

- a 250 cm^3 conical flask
- a method of gas-production measurement (access to a balance) or gas collection – see Figures 5.6 and 5.10
- a stopwatch
- calcium carbonate in at least three different 'sizes' – such as large marble chips, small chips and powder
- 1 $mol\,dm^{-3}$ hydrochloric acid, about 150 cm^3
- a 50 cm^3 or 100 cm^3 measuring cylinder
- access to a balance
- eye protection

Safety

All students should wear eye protection.

Method

For each sample of calcium carbonate:

1 measure 50 cm^3 hydrochloric acid into the flask
2 measure out a small mass of the calcium carbonate sample, such as 5 g (about 1–3 large marble chips, but this is variable depending on the size of each chip); keep this in a weighing boat for the moment
3 set the stopwatch to zero
4 get the equipment ready to measure the gas production (mass change/gas collection)
5 add the calcium carbonate to the acid
6 start the stopwatch
7 record the gas produced/mass change every 15 seconds until there is no further change.

Rinse the flask then repeat the steps using the alternative sizes of calcium carbonate. Ensure the same mass is used each time. If time is limited, students could complete one or two of the three different reactions.

Handling the data

Plot a graph showing the volume of gas produced (or mass change, *y* axis) over time (*x* axis). All three sets of data should be plotted on the same graph.

Qualitatively, students can observe that different-shaped rate curves are produced for each reaction, noting that the steepest curve is produced using the powdered calcium carbonate.

Quantitatively, students can calculate the initial rates of carbon dioxide produced using the method described on page 149. The relationship between the initial rates can then be questioned. For example, what is the ratio between the initial rate for the powder and that of the large-sized chips? What does this tell us about the difference in surface area between the two samples?

Questions to ask

- What effect does surface area have on reaction rate?
- Why does this effect occur?
- How might chemists wish to use changes in surface area to control reaction rate?

Explaining the effect of surface area

Lego® bricks can be used to illustrate the effect of changing surface area on reaction rate (Figure 5.12). Build about 16 bricks into a cubic-shaped (or cuboid-shaped) block. Measure the exposed surface area by counting the number of brick faces visible externally. Then divide the large block into four smaller ones. Recount the number of brick faces visible. Finally separate out each individual brick – how many faces are visible now?

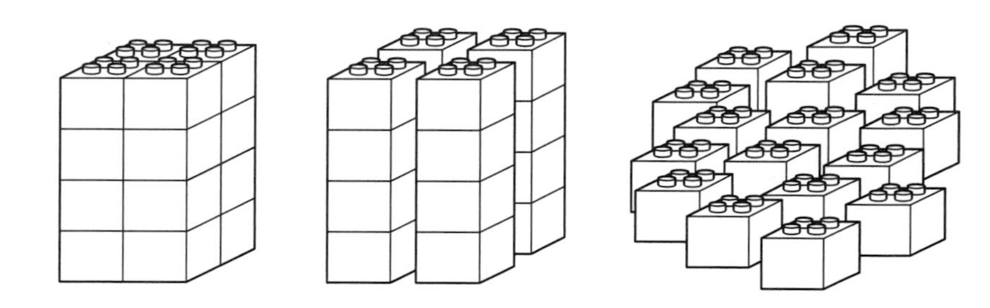

Figure 5.12 To illustrate the effect of changing surface area on reaction rate

How fast can you make mashed potato?

The principle of increasing surface area can be applied to decreasing the time needed to make mashed potato. Although this is hard to represent as a chemical reaction, this is a practical application of an increase in surface area. This can therefore be presented to students as a kind of model or physical analogy. Making mashed potato relies on the cells of the potato losing their structure sufficiently so they can be squashed, releasing starch grains. Exposing more cells to hot water means that the time needed for this is reduced. The effects of chopping potato to different-sized chunks and using different types of potato (Maris Piper, King Edward, red Desiree, standard 'white' potatoes, etc.) can be investigated.

PROCEDURE

Materials

- potatoes
- a potato peeler
- a vessel in which to boil the potato
- a Bunsen burner
- a tripod
- a gauze
- a heatproof mat
- access to a balance
- a stopwatch
- a fork/potato masher
- a knife to chop and prod the potato pieces

Safety
If the mash is to be eaten, the experiment will need to be done in a kitchen rather than a lab.

Method

1 Cut the same mass of potato into different-sized chunks. Test each chunk size separately.
2 Put the chunks in the boiling vessel and cover them with water (starting with water at 60 °C will reduce the time needed to get to mashing point).
3 Start the stopwatch.
4 Bring the water to the boil.
5 Boil until the potato chunks are soft enough to mash.

Discussion points

- When is the potato soft enough to mash?
- Could we boil each chunk size for the same amount of time and then compare softness?
- What happens when different types of potato are used?
- What difference does not peeling the potato make?

5.3 Energetics: why do reactions 'go' and what happens when reactions occur?

Measuring energy changes associated with chemical reactions enables us to understand why they happen and what exactly occurs between reactant particles during a reaction. The main reason chemical reactions occur is that the amount of 'disorder' involved always increases as a result. Disorder can increase in two ways: either energy is more spread out after the reaction than before, or the number of ways the particles in the reaction are arranged is greater than before. Some chemical reactions involve both of these, others just one.

For example, the first reaction in this chapter is:

$$2HCl_{(aq)} + CaCO_{3(s)} \rightarrow CaCl_{2(aq)} + H_2O_{(l)} + CO_{2(g)}$$

The numbers of moles on both sides of the arrow are the same – three moles of reactants and three moles of product are shown. But the state symbols show the reaction produces one mole of gas. There are many more ways of arranging the particles of gas than in the solid state. To understand this, compare the number of ways the same sand grains can be arranged in a sandcastle (one way) and in a random pile of sand (many ways): Figure 5.13.

The particles of carbon dioxide can be arranged in more ways than those of calcium carbonate, which are fixed in a solid lattice structure. So, the amount of disorder is increased in the products.

Figure 5.13 The disorder is greater in the pile of sand than in the sandcastle.

The second way of thinking about disorder relates to energy distribution. Many reactions are exothermic. This means heat is lost to the surroundings as products form. This increases the distribution of the energy. A good example is the combustion of fuel, also shown at the start of the chapter:

$$2C_8H_{18(l)} + 25O_{2(g)} \rightarrow 16CO_{2(g)} + 18H_2O_{(g)}$$

In this reaction, not only is there a large increase in the number of moles (27 reactant compared to 34 product) but energy is given out. This is distributed to the car engine to make movement. This is not 100% efficient, so car engine parts get hot in the process. Energy is released when chemical bonds form between carbon and oxygen atoms, making carbon dioxide, and hydrogen and oxygen atoms, making water. More energy is released when these bonds are made than was required to break bonds between carbon and hydrogen atoms and oxygen atoms in the reactants.

Chemists use the word entropy to describe disorder. Increasing entropy is the most likely reason that chemical reactions occur. Illustrations of this are provided in this section, together with experiments that can be used to show how energy is released when bonds are made and required to break chemical bonds.

Students' misconceptions about chemical reactions

Students' may harbour a number of misconceptions with regard to chemical reactions and energetics. Some of these are highlighted in this section.

■ 1 Students think endothermic reactions cannot be spontaneous

There is a strong association between chemical reactions occurring and energy being released. Students become used to reactions that 'get hotter' when they occur, such as that between hydrochloric acid and magnesium, or acid and carbonate reactions. Finding appropriate spontaneous reactions that get colder is much harder. Hence, students develop the notion that only exothermic reactions occur without extra 'help'.

■ 2 Students think energy is stored in a system ready to be released

The first law of thermodynamics states that energy cannot be created or destroyed. Students may interpret this as meaning that energy is kept in chemicals and is released when they react. Language such as 'fuels are energy stores' or 'fuels contain energy' does not help. The notion is problematic because it sidesteps what actually happens, creating the perception that burning (and by extrapolation, other chemical reactions) is a destructive event in which fuel is 'used up'. This contravenes the law of conservation of mass (Chapter 3). Instead, we want students to realise that chemical reactions involve reactants and products, and that some release energy to the environment.

■ 3 Students think energy is used up when we use fuels

Fuels are used to 'do' something, such as move a motor vehicle, cook a meal or heat a house. There is a strong link between the use of the fuel and the fuel as an energy 'source' for the task. Students may think energy is consumed as the task is done. Calculations of fuel consumption in a car engine or of gas consumed by a boiler lead to this. We need to show students that we transfer energy from usable to non-usable forms. The overall amount of energy in the Universe remains constant.

■ 4 Students think burning is a state change, not a chemical reaction

Students often forget that oxygen from the atmosphere is involved in burning. They perceive changes only to the fuel itself. For example, they often think using petrol in a car engine means turning the petrol to exhaust gas, but without oxygen being involved; similarly, that burning a candle involves wax melting, not reacting with oxygen.

■ 5 Students think that bond breaking releases energy

A natural consequence of thinking of fuels as energy stores is to think that fuel molecules release energy when they burn. The image is rather like cracking an egg – when the shell is broken, the contents are released. Even students who know 'bond breaking requires energy' may think this, on the grounds that a small amount of energy may be needed to start off the break, but the actual breaking releases far more. This misconception makes it very difficult to understand energetics properly.

Introducing disorder – predicting when chemical reactions occur

■ Chance is a fine thing!

Discuss the chances of everyday events occurring. For example, what are the chances or odds of:

- being struck by lightning
- winning a big National Lottery prize
- getting a top mark in a chemistry exam
- a British tennis player winning Wimbledon
- one of the smaller Premier League football teams winning the title
- a smoker getting a smoking-related disease?

The odds of being struck by lightning are about 1/600 000. The chances of winning the UK's National Lottery are about 1/14 000 000. This means most of us (who buy a ticket) have a much bigger chance of being struck by lightning than winning the lottery. However, the odds on either event happening are very small. The most likely events are that we will neither be struck by lightning nor win the lottery. This is because there is a very large number of ways of not winning/not being struck and only one way of winning/being struck.

Three of the other four items listed rely on skills rather than chance. External factors could influence these – for example, getting a better coach or manager may enhance a player's/team's chances of doing well. Working hard at understanding chemistry and revising for an exam may help someone get a top mark. These factors can alter the odds.

The next demonstration illustrates these points.

■ Arranging molecules: demonstration and discussion

PROCEDURE

Materials

- three jars with lids, such as jam jars or instant coffee jars
- two types of dried pulses, such as peas and beans (alternatively two types of any small solid items that would form two distinct layers, such as Lego® bricks of different colours and size, different types of sweets, etc.); enough are needed to make two layers each about one-third the height of the jar
- 200 cm^3 water
- 100 cm^3 ethanol
- 100 cm^3 vegetable oil or similar
- a stirring rod
- molecular models of water, ethanol, oil

Method

1 Pour the peas into one jar. Carefully make a layer of beans on top. Ask students to predict what will happen when the jar is shaken. They are expected to say that the two layers will mix. Ask students to predict the odds, or chance, that when the shaking stops, the two layers will reform unchanged. This should be a very small figure. Ask why the figure should be small. Shake the jar – of course, the beans and peas will mix.

2 Ask students to explain why the beans and peas mixed – based on the earlier discussion, the answer is that this was the most likely event. Introduce the term disorder (or entropy) to describe the number of ways that particles can be arranged. Use this to explain the peas and beans mixing – there are many more ways that the mixing can occur, compared to the number of ways the beans and peas form layers.

3 Next, apply this to mixing particles of liquid. Pour 100 cm^3 water into a separate jar. Ask students to predict what will happen when ethanol is added to the water. They should say that the two liquids will mix (spontaneously). Applying the same reasoning as for the peas and beans, the most likely event is that the two liquids will mix. This is because there are more ways for the particles to exist as a mixture than as separate layers. Reasoning using disorder, the mixing occurs because disorder is greater in the mixture than in two separate layers. Events happen that increase or maximise disorder.

4 Finally, pour 100 cm^3 water into the third jar. Ask students to predict what will happen when 100 cm^3 oil is added to the water. They should say that no mixing will take place. This is strange. If we normally expect mixing to occur, why do oil and water not mix? The answer is that something, an additional factor, must be preventing mixing. Use the molecular models to help arrive at the explanation. Show that the ethanol and water particle models have similar components that permit mixing (technically, these are hydroxyl, or –OH, groups) whereas the water and oil particle models are different (oil particles contain mainly carbon and hydrogen atoms). This means that bonds can form between ethanol and water molecules, but not between water and oil molecules. Differences between the particles act as a factor preventing the most likely event from occurring. Chemists can overcome this by adding something extra or doing something different. In this case, adding soap or detergent to promote mixing would overcome the differences. On other occasions, adding a catalyst or increasing temperature or pressure is used.

■ Stick with it!

This is a good demonstration showing a spontaneous endothermic reaction.

PROCEDURE

Materials

- 32 g hydrated barium hydroxide ($Ba(OH)_2.8H_2O$)
- 10 g ammonium chloride
- a 250 cm^3 beaker
- a –10 to 100 °C thermometer or, if possible, a temperature probe reading down to –30 °C
- access to a fume cupboard
- a heatproof mat/piece of wood large enough to sit underneath the beaker
- water
- eye protection
- access to a fume cupboard

Safety

All students should wear eye protection. Ensure you have access to a fume cupboard.

Method

1. Sprinkle a few drops of water on the mat or piece of wood.
2. Place the beaker directly on the water (the reaction should freeze the water making the beaker stick).
3. Put the barium hydroxide in the beaker.
4. Show the ammonium chloride to the students and ask them to predict what they think will happen when the two chemicals (both white powders) mix. Many may think the temperature will rise.
5. Add the ammonium chloride to the beaker. Stir the chemicals with the temperature probe/thermometer. Do not move the beaker from the mat. As the temperature falls, the water under the beaker will freeze, sticking the beaker to the mat. The temperature may go well below –10 °C.
6. Pick up the beaker without holding on to the mat. Note that ammonia gas is produced in the reaction – after a few minutes place the beaker in the fume cupboard and switch the extractor fan on.

Discussion points

Ask students for evidence that a chemical reaction occurred. Draw attention to the fact that the temperature dropped. Is this to be expected? Did this match their prediction? What was surprising about this reaction?

Students may say that they did not expect the temperature to drop. That the reaction became cold enough to freeze the water was surprising. They may also say that they did not think reactions that got cold would go at all.

The equation for the reaction is:

$$Ba(OH)_2.8H_2O_{(s)} + 2NH_4Cl_{(s)} \rightarrow 2NH_{3(g)} + 8H_2O_{(l)} + BaCl_2.2H_2O_{(s)}$$

The equation shows that three moles of solid (one of hydrated barium hydroxide and two of ammonium chloride) produced eleven moles of product (two of ammonia, eight of water and one of hydrated barium chloride). Also, ten of the moles produced are gas or liquid.

Ask students to compare the level of disorder in the reactants and products. Which is most disordered? The answer is that the products are most disordered. Why did the reaction go? The answer is that the energetics of the reaction made it feasible – although some energy was needed from the environment (measured as a big temperature drop), the formation of products generated increased disorder. There are more ways that the particles of product can be arranged than the particles of reagents. Particles do not 'know' they are reacting (they do not have consciousness). Once reagent particles with enough energy to cross the energy barrier reacted, conditions inside the beaker enabled more to react.

Sharing out energy

Here are two linked demonstrations that show the principle of increasing disorder by redistributing energy.

■ Give us a crisp! Role play

PROCEDURE

Materials

A multipack of crisps – 12-pack of varied flavours (16- or 24-packs are also fine)

Method

Invite two students to 'be' atoms reacting. Give each student six (depending on the size of the multipack) packets of crisps. Explain that the crisp packets represent energy. If desired, use the term 'quantum/quanta'. Explain that energy exists in packets called quanta, that each crisp packet represents an energy quantum and that those with different flavours have differing amounts of energy.

Invite about eight other students to 'be' air particles. They can move around the two reactants.

When the reactants combine (perhaps signalled by holding hands or coming into a smaller space), energy is released. The reactants give away the crisp packets to the air particles. This can be done by throwing them up to be caught. The air particles should try to catch as many as possible. This should ensure a random distribution – some may catch two or more, others none.

Now look at the overall scene. The reactants have less energy than before. The energy they had (six or more packets each) has been redistributed across a larger number of particles. The amount of energy in the environment is now higher than it was before the reaction. Returning to the principle of disorder (entropy), the level of disorder has increased compared to the starting position.

The reaction between aluminium and iodine

This is a very vigorous reaction that creates a brilliant, dramatic effect for little effort.

PROCEDURE

Materials

- a fume cupboard
- a pestle and mortar
- a metal dish or tin lid about 10 cm across
- a tripod
- a gauze
- a heatproof mat
- splints
- a dropping pipette
- water
- 2 g iodine
- 0.3 g aluminium powder
- eye protection

Safety

All students should wear eye protection. The demonstration should be carried out in the fume cupboard.

If the aluminium/iodine reaction does not work, do not remove it from the fume cupboard. Add it to a large beaker of water and add potassium iodide to dissolve the iodine. Then add sodium thiosulfate solution and pour the whole volume down a foul drain with plenty of water.

Method

1. Set the mat, tripod and gauze in the fume cupboard.
2. Grind the iodine to a fine powder using the pestle and mortar.
3. Add the aluminium powder. Do not grind them together.
4. Mix the iodine and aluminium gently using the splint.
5. Ask students to predict what they think might happen when the two elements react.
6. Put the mixture in the metal dish, making a conical pile in the centre.
7. Set the dish on the gauze in the fume cupboard.
8. Squirt a few drops of water on the pile – do not add too much.
9. Close the fume cupboard door and switch on the extractor fan. Wait a few seconds.

Results

The water will vaporise. Purple iodine vapour will be produced. Sparks will be seen and a flame. An exothermic reaction is occurring between the aluminium and iodine. Energy is being released.

When the reaction has finished, a white powder, aluminium iodide, will be present in the dish. The fume cupboard may be spattered with iodine and the powder.

Discussion points

- Ask students if their prediction was correct.
- Ask them where the energy has gone to.
- Ask them what has happened to the amount of disorder – note that there is now one solid rather than two, but point out that the energy has been distributed to the environment.

The equation for the reaction is:

$$2Al_{(s)} + 3I_{2(s)} \rightarrow 2AlI_{3(s)}$$

The number of moles of product (two) is fewer than the number of moles of reactant (five). In this case, the disorder increases because energy is released. The dish, air surrounding the dish and fume cupboard will be hotter after the reaction than before.

What happens when chemical reactions occur?

In this section, fuel–oxygen systems are used to illustrate the principles of bond breaking and bond making in chemical reactions.

■ Burning a candle

PROCEDURE

Materials

- tea lights (small candles in metal casing)
- access to a balance accurate to 0.1 g
- splints
- a Bunsen burner/match
- a heatproof mat
- eye protection

Safety

All students should wear eye protection.

Method

1. Ask students to find the mass of a tea light. Record this mass.
2. Light the tea lights. Dim the laboratory lights to help heighten the effects from the candle flames.
3. Ask students to observe the candle flames closely and to note as many points as possible, such as colour, locations of the colour, the height of the flame, any variation over time, what happens to the wick, the wax, etc.
4. After a few minutes, blow out the candles (What do they notice on blowing out the candles?).
5. Find the mass again and record. (Take care: the metal casing will be hot.)

Questions to ask

- Why do we need to light the candles with a match/splint?
- What mass change was found?
- Why was there a change in mass?
- Why is part of the flame yellow, part transparent and part blue?
- What happens to the wick?
- What happens to the wax?
- What materials are suitable for candles?

Discussion points

Observations may include: a blue colour at the base of the flame, transparent regions, a yellow flame towards the top, a red-tipped wick, a pool of melted wax, heat and light, vaporised wax when blowing out the candle.

The wick acts as a conduit for melted wax. The wick does not alter, other than to char slightly. The original match flame is needed to start the reaction – heat from this flame melts enough wax to create flow up the wick and some molecules vaporise. These react with oxygen molecules in the air. Energy is released. Enough energy is released to keep the reaction going, so the candle lights. Some of the energy is used to melt and then vaporise the wax. To see the wax vapour, blow out the candle then immediately bring a lighted splint close to the wick. The candle will relight without the wick being touched.

Combustion, the reaction between the wax and oxygen, is occurring in the blue and transparent parts of the flame. The yellow part contains unreacted hydrocarbon particles from the wax. This can be shown by holding a cold piece of metal in the flame – carbon particles are deposited (condense) on the metal.

The reaction occurs as bonds are broken in the wax and oxygen molecules. Energy is used to break these bonds. Energy is released when water (as vapour) and carbon dioxide molecules are made. Overall the reaction is exothermic.

The flame is luminous because the unreacted hydrocarbon particles absorb heat and emit this as radiation. The temperature at the centre of the flame is around 1000 °C.

Over time, substances such as beeswax, tallow (animal fat) and vegetable fat have been used in candles.

■ Burning a fuel

This class experiment shows how much energy is released when a fuel burns.

PROCEDURE

Materials

- fuels – such as portable LPG burners, candles, small glass fuel burners
- alternative fuels, such as cooking oil, sugar cubes
- watch glasses (on which to burn the alternative fuels)
- a calorimeter – for example, a small, cleaned food tin
- a heat shield – such as a large tin can or heatproof mat
- 50 cm^3 water
- a 50 cm^3 measuring cylinder
- a −10 to 100 °C thermometer
- two bosses and two clamps
- a clamp stand
- a heatproof mat
- splints/matches and a Bunsen flame to start the fuels burning
- access to a balance accurate to 0.1 g
- a ruler
- eye protection
- molecular models of fuel particles and oxygen particles

Safety
All students should wear eye protection.

Method
Allocate fuels around the class. Gather data from the whole class. The equipment should be set up as shown in Figure 5.14.

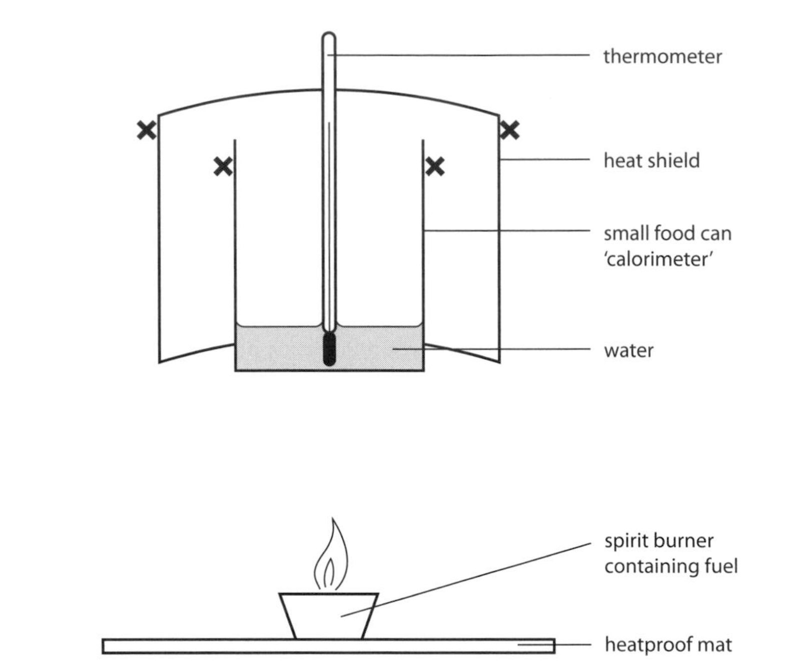

Figure 5.14
Burning a fuel

To find out how much energy is released when a fuel burns:

1. Place 50 cm^3 water into the calorimeter.
2. Record the temperature of the water.
3. Find the mass of the fuel and its container. Record this value.
4. Use the ruler to set the calorimeter 10 cm above the fuel burner.
5. Ignite the fuel. Place the heat shield so as much heat from the fuel goes into the calorimeter as possible.
6. Stir the water using the thermometer.
7. Put out the flame from the fuel when the temperature of the water has risen by 10 °C.
8. Find the mass of the fuel and its container again. Record this value.
9. Calculate the mass of fuel burned in grams.
10. Repeat the experiment with a different fuel.

Handling the data

There are two ways of handling the data from this practical. One is easy; the other is more advanced and may not be suitable for all students.

1 Easy: students can calculate the amount of energy per gram of fuel burned using the simple equation:

$$\text{temperature change per gram of fuel} = \frac{\text{temperature change (°C)}}{\text{mass of fuel burned (g)}}$$

2 More advanced: students can calculate the amount of energy per mole of fuel burned by converting the mass in grams of fuel burned to moles. To do this they will need the relative molecular mass of each fuel. To find the number of moles:

$$\text{amount of fuel (moles)} = \frac{\text{mass (g)}}{\text{relative molecular mass}}$$

Questions to ask

- Which fuels gave the highest and lowest temperature changes?
- Do all fuels give out energy when they react?
- Write a chemical equation to represent what happens when a fuel burns. What are the products of the reaction?
- Where does the energy come from when a fuel burns?
- Why is a spark or match needed to start a fuel burning?
- In what way is sugar a fuel?

Discussion points

Use the molecular models to represent what happens when a fuel burns. Show that in order to start a reaction, energy needs to be put in to break bonds between atoms in the fuel and oxygen molecules. Show this by physically breaking apart some models. Explain that in the reaction, new molecules form – these are carbon dioxide and water. Energy is released when bonds are formed (as shown in the next demonstration). The name for the chemical reaction in which fuels burn in oxygen is combustion (Chapter 7).

Fuels vary in the amount of energy produced on combustion. This is because the number of water and carbon dioxide molecules produced differs between them. Notice that some fuels burn with a lot of 'waste' – carbon that does not combust. Others burn 'efficiently' with little waste. This can be discussed.

■ Energy is released when chemical bonds form

PROCEDURE

Materials

- 125 g hydrated sodium ethanoate, plus a few extra crystals
- 250 cm^3 beaker
- a watch glass or cling film to fit over the beaker
- a crystallising dish
- a chemical hand warmer (can be bought from camping/outdoor shops/websites)
- a heatproof mat

- a tripod
- a gauze
- a Bunsen burner
- a measuring cylinder
- eye protection

Safety

All students should wear eye protection.

Method

About 90 minutes beforehand, make a saturated solution of hydrated sodium ethanoate by placing the 125 g solid in the beaker and adding 12.5 cm^3 water. (Note: this is a very small amount of water relative to the amount of solid.) It is best to do this in the location where the demonstration will take place. Heat the mixture gently until all the solid dissolves and a clear, colourless solution has formed. Turn off the heat. Cover the beaker with cling film or a watch glass and leave it undisturbed.

The solution is ready when the beaker is warm to the fingers. As a precaution, make two solutions – inevitably, something will go wrong with the first. If it does not, the demonstration is worth repeating anyway.

To do the demonstration:

1. Have the crystallising dish ready.
2. Put one or two small crystals of hydrated sodium ethanoate in the dish.
3. Remove the cling film/watch glass from the beaker containing the solution.
4. Hold the crystallising dish in one hand and the beaker in the other.
5. Pour the solution into the dish, slowly. Try to pour on exactly the same spot.
6. The solution should solidify as you pour, creating a stalagmite.
7. Keep pouring until all the solution is in the dish.

Now show the result to the students. Heat should be felt radiating from the stalagmite.

Questions to ask

- Where does the heat come from?
- What has happened to the disorder?
- Why has the change occurred?

Discussion points

The heat is produced when chemical bonds form between the particles in the solution. The product, the solid, is more ordered (less disordered) than the solution. But energy is given out to the environment, so the principle of increasing disorder is retained.

This shows what happens when a fuel combusts – bonds are formed and energy is released. Although the products are less disordered than the reactants, energy is redistributed to the environment.

Other resources

Books and journals

Barker, V. (2002). *Building Success in GCSE Science: Chemistry.* Dunstable: Folens

Kind, V. (2004). *Contemporary Chemistry for Schools and Colleges.* London: Royal Society of Chemistry. This text includes activities relating to hydrogen fuel cells.

Lister, T. (1995). *Classic Chemistry Demonstrations.* London: Royal Society of Chemistry Experiments 22 and 23 (pages 48–52). These provide details for two more 'clock' reactions, called the 'blue-bottle experiment' and the 'Old Nassau clock reaction'.

Van Driel, J.H., de Vos, W. & Verloop, N. (1998). Developing secondary students' conceptions of chemical reactions: the introduction of chemical equilibrium. *International Journal of Science Education*. **20**(4): 379–392.

Websites

Beyond Appearances: Students' Misconceptions about Basic Chemical Ideas (Kind, V., 2004) can be freely downloaded from the Royal Society of Chemistry website: www.rsc.org/Education/Teachers/Resources/Books/Misconceptions.asp

YouTube features clips that show the cobalt(II) chloride equilibrium reaction, search for 'equilibrium in copper chloride'. One clip, www.youtube.com/watch?v=BGfYf8OQzuk, shows what happens when a solution of cobalt(II) chloride is heated in a U-tube.

To enhance the 'genie in the lamp' experiment, you could try using a 'genuine' Aladdin's lamp, such as the one available at www.sillyjokes.co.uk. However, because it is plastic, test it for durability with the 100-volume hydrogen peroxide carefully.

6 Acids and alkalis

John Oversby

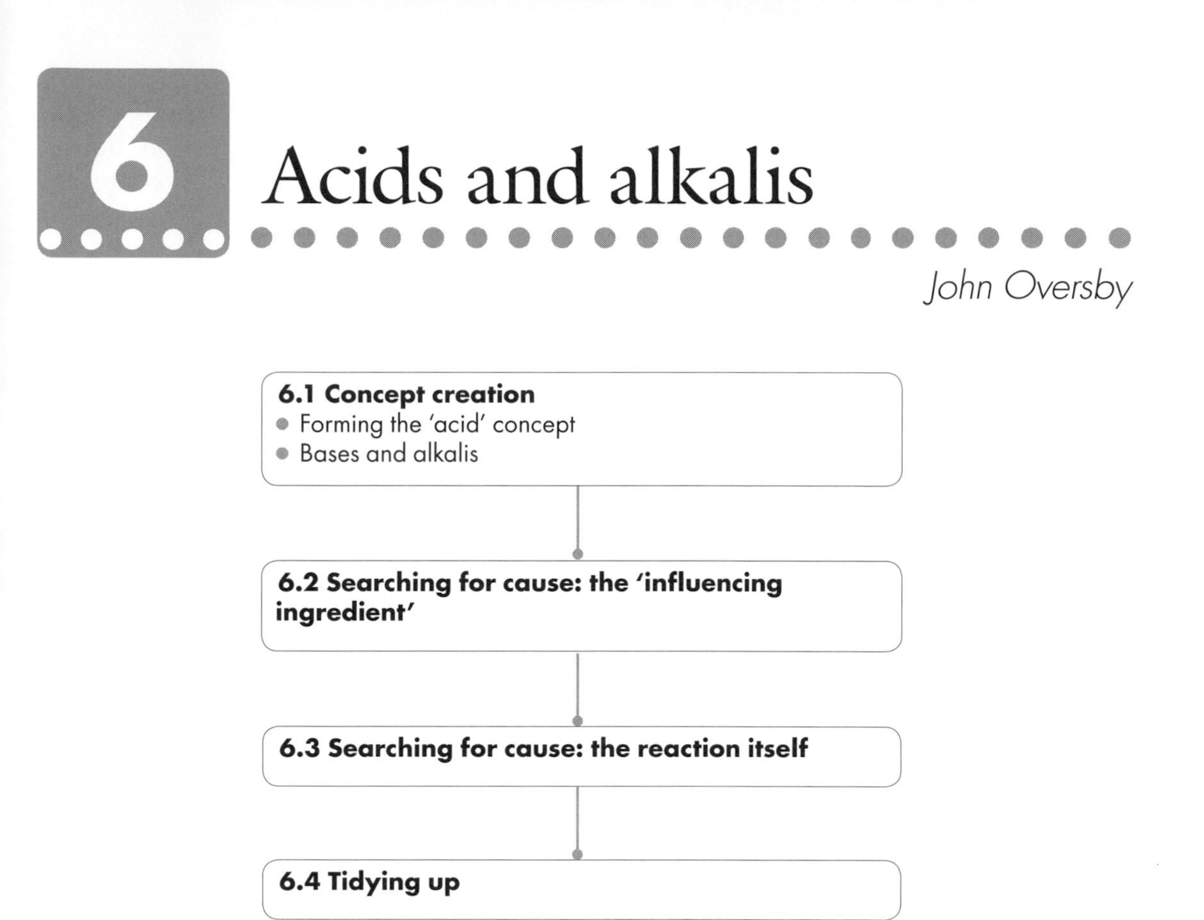

Chapter overview

The Introduction (overleaf) sets the scene for the approach I am taking in this chapter. I use a strong philosophical framework to sequence the topic in an order to develop abstract notions after establishing the chemical phenomena. The chapter is based on the outcomes of a European project, History and Philosophy in Science Teaching (HIPST), in which I created and elaborated the ideas, working with a group of teachers and their classes.

The first section is based on a historical progression, although it will be familiar to experienced teachers as the normal start of this topic. The distinctive contribution of this section is an explicit regard to the work of past chemists which is largely lacking in textbooks today. Such an approach is strongly humanistic and aims to portray the chemists in the contexts of their time.

The second section, 6.2, also follows a typical progression, but the explanatory discussions are given in a cause-and-effect context, linking closely to recent work on the idea of 'how science works', but rather focusing on historical material.

Section 6.3 covers the historical development of the acid–alkali reaction itself. In historical terms, this development took place some centuries after the concept of an acid was first established. The effort expended by the chemists involved is too often omitted or glossed over in textbook accounts. By including this history, some recognition of their hard work and intellectual prowess is acknowledged and integrated into this abstract study.

I have rather swept through the sequence described here and left out some significant issues that could obstruct the flow if attended to earlier in the chapter. They are collected in the final section for completeness.

Introduction

Initial teaching of acids and alkalis appears hardly to have changed over the decades, with emphasis on the properties of these classes of compounds and their reactions with other materials and between themselves. It is at higher levels of explanation that there have been some significant changes. In this chapter, I intend to provide a rationale for the progression I am recommending. I will also provide links to the history of chemistry, in an attempt to demonstrate humanity in chemistry, and to some philosophical aspects, as part of the present focus on the nature of science, or 'how science works'.

While this topic may seem to be very familiar, there are some long-standing challenges to effective learning, including the provision of engaging and manageable practical work. While chemical educators will never be able to suggest ways to overcome all of these challenges, some of the ideas I propose here have worked well in many circumstances. As with many topics that have been closely intertwined with day-to-day life, there are everyday understandings that can get in the way of useful learning. These must be identified and I will suggest some successful strategies, not least direct confrontation on occasions.

Choosing a route

I begin the topic with an approach based on readily observable phenomena, as an attempt to carve out the concept of acidity from the landscape of chemical reactions. (There is a well established history of the topic that lends itself to such an approach, based on the mining industry of ancient times and kitchen medicine.) This is then followed with an attempt to explain acidity on a causal basis, mirroring closely the historical timeline, incorporating some features of the characters involved.

At higher levels, mathematical considerations are at play, involving the notions of equivalents and chemical equations, as well as the rather enigmatic pH scale.

Previous knowledge and experience

Students are likely to be familiar with the term 'acid' and may assume that all acids are very dangerous and 'eat through' materials readily. They will know the word 'base', but may not be familiar with its chemical meaning. Students will be familiar with many everyday materials that are acids or bases (fruit juice, cleaning materials, vinegar, baking soda, etc.), but probably without appreciating how they would be classed chemically.

6.1 Concept creation

Studying aspects of chemistry often includes grouping together some chemical materials with similarities of behaviour. This is what we mean by concept creation. The concept of acidity is one of these. In chemistry teaching at this stage, teachers do not normally use the acids directly, but make use of their aqueous solutions. Studying similarities of acidic behaviour means that we must include examples of 'non-acids' to be sure that the learners can recognise the difference between the observed behaviour and materials that do not behave in this way. The two reactions chosen are:

- reaction of the acidic solution with a reactive metal such as magnesium
- reaction of the acidic solution with a carbonate such as powdered sodium carbonate.

The choice of materials as 'non-examples' is not so obvious since many non-acids, such as copper(II) sulfate, dissolve in water and then react with the water to make acidic solutions. These must be avoided. I recommend aqueous solutions of sodium chloride, magnesium sulfate and sodium nitrate, which will suit the purpose of acting as non-acids. The choice of acids is also not so obvious but, if magnesium ribbon is used for the reactive metal, it will react appropriately with dilute nitric acid, forming hydrogen. The mineral acids – hydrochloric, sulfuric, nitric and phosphoric acid – in dilute solution are all relatively safe for this activity with the usual safety precautions. These acids can be used alongside a variety of fruit acids, as available, such as citric, malonic, as well as ethanoic acid in the form of vinegar.

The reactions of magnesium metal with the aqueous solutions of acids provide a distinction between the mineral (strong) acids, where heat is evolved which can be detected through the wall of the test tube or with a thermometer, and the natural (weak) acids, where little heat is detectable. Hydrogen also fizzes from the solution and can be tested as usual with a lighted splint to give the characteristic pop sound. Attention to normal safety precautions is required, such as wearing eye protection and washing off accidental spills immediately, but it is too easy to overstress the dangers, resulting in undue anxiety. The students should be sufficiently at ease to focus on the chemical reactions and not the fear. Fruit acids and vinegar are hardly dangerous in these situations. (You may wish to discuss storage of these solutions in glass (traditional) or plastic bottles, but not metal containers such as aluminium or steel, because of the reactions observed.) The rate at which hydrogen is evolved can provide some information about whether the acidic solution is strong or weak, and this can be discussed by more able students.

A typical reaction equation (discussed more fully in section 3.2 of Chapter 3), using a full formula equation, for the reaction of magnesium with dilute sulfuric acid is:

$$Mg_{(s)} + H_2SO_{4(aq)} \rightarrow MgSO_{4(aq)} + H_{2(g)}$$

The same reaction, using an ionic equation, is:

$$Mg_{(s)} + 2H_3O^+{}_{(aq)} \rightarrow Mg^{2+}{}_{(aq)} + H_{2(g)} + 2H_2O_{(l)}$$

Here is an example for a carbonate reaction:

$$CuCO_{3(s)} + 2HCl_{(aq)} \rightarrow CuCl_{2(aq)} + CO_{2(g)} + H_2O_{(l)}$$

Reaction with a carbonate is another identifier of acidity. In medieval times, when searching for suitable ores for smelting, it was already known that carbonates (or ‘earths’ as they were then called) were easy to smelt. These earths could be identified by the fizzing they gave when a drop of hydrochloric acid fell onto the rock. Looking at the reverse logic provides a way of classifying materials as acids or non-acids. It is possible to test the gas produced in this reaction with lime water, which goes cloudy with carbon dioxide. Identifying the gas, and thus the presence of carbonate, is satisfying for able students, but may be too much information for some. The rate at which the bubbles form provides evidence for whether the solution is weakly or strongly acidic, confirming the metal reactions.

Forming the 'acid' concept

This section will discuss the notion of forming a concept, using the example of 'acid'. A central activity of chemists is to search for regularities in behaviour, since explanations then prove to be more useful than for a single behaviour. Recognition of regular behaviour depends on how chemists recognise that behaviour, of course. Modern chemists have the advantage of a wide range of advanced instruments, some collecting spectroscopic data, such as NMR (nuclear magnetic resonance). Students aged 11–16 years do not, and in that sense are driven back to phenomenological processes with direct observation as a major data-gathering tool. They are similar to the early chemists in their methodology. Students should be helped to understand that ideas such as 'acid' are human constructions: a way of imposing understanding on the whole variety of the Universe in which we live. This is normal in chemistry.

The two identification processes mentioned earlier, known to medieval alchemists, enable students to describe the solutes as acids or not. Discussion of the notion of 'an ideal acid' from the examples provided takes the form of induction – that is making a generalisation from specific instances. It suffers, as do all inductive processes, from the possibility of finding an example of an acid that does not fit this characterisation. Nevertheless, induction is a valuable tool for practising chemists, as long as they are cautious about over-generalising. Idealisation is also involved in concept creation, since the ideal acid cannot really exist. We can, however, behave as though it does and predict reactions of acids yet to be investigated; the process of deduction. As chemists confirm that all the acids they investigate behave in similar ways, they gain confidence that the whole class of substances does, indeed, behave this way. They may even suspect that a material called in everyday usage an acid, but that does not behave in these ways, is not really an acid.

■ Extending the identification of an acid

Litmus is believed to have been used as an indicator by about AD1300 by Spanish alchemist Arnaldus de Villa Nova, although the evidence that he was the first user is rather uncertain. This indicator is extracted from lichens and turns red with acidic solutions and blue with alkaline solutions. Historically, extracts of plants were widely used as household medicines, a favourite one being syrup of violets. It was made by steeping violet petals in syrup and was used to cure a variety of ailments, including insomnia, anger and stomach upsets. In the Middle Ages when visiting a hospital was rather risky, household self-medication seems to have been common and containers of these syrups, preserved by their sugar content, would readily have been available in many kitchens.

Having already established the category of acids, the use of indicators that have only two colours provides further confirmatory evidence for the validity of identifying acids as a special case of materials. This property of a theory, that in this situation there is a group of materials that have a chemical claim to be distinct from other groups, and that further investigations can be carried out to consolidate this claim, is known as fertility. The theory that there are acids is a fertile one. One way of testing the value of this theory is to use an indicator that distinguishes between acids, alkalis and non-acids. Litmus is very useful here since it does not change with neutral solutions. (Be careful to check this with both blue and red litmus, since red litmus would not change with acidic solutions!) We should avoid universal indicator here since anecdotal evidence suggests that students find it hard to appreciate that red, orange and yellow really are the same in this context and this may be confusing.

The next stage is to investigate the idea of neutralisation. Early learning in chemistry is full of dichotomies, such as metals vs non-metals, soluble vs insoluble, conductors vs insulators and solutes vs solvents. The acid–base dichotomy is one of these and is, of course, a simplification as most dichotomies are. Nevertheless, they serve a valuable purpose in not overloading the brain's working space. Medieval alchemists interpreted these dichotomies in their dualistic theories when trying to make sense of material behaviour. It appears that this dualistic approach was widespread, such as good and evil, yin and yang. It is obviously a powerful approach and one which is adopted generally throughout lower school science.

Bases and alkalis

There is a common cause of confusion while talking about the opposite of acids in solution, alkalis, which term is often confused with the term base. The French chemist, Rouelle (1703–1770), created the modern idea of a base as the solid material that was the basis for the action of an acid to form a salt. At that time, many acids were volatile liquids and salts were crystalline. Hence it was thought that there should be a solid base for the salt to grow out of, giving rise to the term. Many bases are insoluble but the salts they form with acids are soluble, making it appear that the base dissolves in the acid. Some bases are soluble in water. The solutions are then described as alkaline (see below for derivation of the term 'alkali'). I recommend the term 'base' for any substance that reacts with an acidic solution to make a salt and water only. I recommend the term 'alkali' for a soluble base. As bases are dependent, at this stage, on the concept of acid, being the opposite, any discussion of the entities in solution should be avoided, since a descriptive empirical approach is recommended.

There is an issue with some materials such as ammonia gas. Ammonia is soluble in water making a solution with pH > 7, i.e. alkaline. Traditionally this has been rationalised using the label 'ammonium hydroxide solution'. However, the solution is mainly neutral ammonia molecules in water, with only a small proportion reacting to give ammonium and hydroxide ions, depending on the concentration. Aqueous ammonia solution is truly alkaline (pH > 7) but ammonia is not a base like sodium or magnesium oxides or hydroxides. I recommend avoiding this potential confusion by avoiding the use of aqueous ammonia solution at this level.

Metal carbonates also react with acidic solutions forming a salt and water, but also carbon dioxide. It is tempting to refer to these carbonates as bases, but this can only serve to confuse. I recommend just calling them metal carbonates.

■ Salt formation

A characteristic of acids is that they form salts with bases, carbonates and metals. The term 'salt', which appears in a similar form in many languages, is derived from an Old English word for the sea, 'sealt'. Many salts can be extracted from sea water, including sodium chloride, or common salt, widely used in the past as a food preservative. 'Salary' is also related to this use, with Roman soldiers thought to be paid part of their wages in salt, a valuable commodity that could be exchanged readily in markets.

Many metals can be extracted from ores that are mainly carbonates. As mentioned earlier in the chapter, even in the Middle Ages mining engineers were able to test rocks for the presence of carbonates using hydrochloric acid. In school chemistry, many salts can be made simply from carbonates that are insoluble in water. To a dilute solution of the acid, the carbonate is added in small portions until there is no more gas given off and an obvious excess is seen. The liquid can be filtered or simply poured off (decanted) with care. If necessary, the solution can be concentrated by gentle heating in a container. Slow evaporation provides crystals of the salt. Many carbonates produce coloured salts (beautiful green or blue crystals from copper(II) carbonate, see Figure 6.1), depending on the acid.

Making salts from the respective carbonates is a fulfilling activity, developing manual dexterity and basic skills such as filtering, and encouraging patience while the solution evaporates over a few days. The crystals can be sandwiched between two pieces of sticky tape for preservation and then attached to a notebook as evidence of success.

Figure 6.1 Copper sulfate and copper chloride crystals

Investigating bases

The chemistry of acids and bases devotes more attention to the former, not least because acids taste sour or sharp and this is an obvious feature, although students should be strongly dissuaded from experimenting with this test for safety reasons! The term base, for the opposite of an acid, is very much less well known. The term alkali is said to come from the Arabic term for the ash that is left in the roasting pan and is related to the term 'potash', literally the ash in the pot!

> 'Sodium hydroxide history can be traced as far back as Ancient Egypt when it was used in early soap making. In Babylonian times it is believed that soap-like substances were in use for general bathing. Indeed, evidence has been found on a clay tablet dating back to around 2800BC that a soap-like material made from water, oil and lye was used for bathing, 'lye' being another name for sodium hydroxide. An interesting discovery about sodium hydroxide history concerns the Ebers Papyrus dating from around 1550BC which indicates that animal or vegetable fats were mixed with lye to produce a type of early soap. References in the papyrus indicate that this substance was used for bathing but that it was also used for washing wool, probably before it was woven into cloth.'
> Source: adapted from www.sodium-hydroxide.co.uk

Investigations on cooled wood ash left over from barbecues or the ash derived from dried vegetable or fruit skins can be carried out, with suitable precautions for dealing with alkalis. Dried fruit skins, such as bananas, often burn with a lilac flame indicating the presence of potassium. The ash is partly soluble in water, and both the dry ash and its solution fizz readily with mineral acids, indicating that they contain a carbonate. I understand that the dissolved part of the wood ash contains potassium carbonate and sodium carbonate in the ratio of 10:1. Heating the wood ash to much higher temperatures (I used a pottery kiln) produces a much more alkaline solution on leaching. It may be that some of the calcium carbonate in the ash is decomposed at the higher temperature to calcium oxide, which partly dissolves, when cold, to give a more alkaline solution. The ash solution is highly alkaline and can be used to neutralise acids. If the resulting neutral solutions are slowly evaporated, crystals of potassium salts form.

In Georgian times, the wood ash was used as a detergent to wash clothes ('lye' is said by various dictionaries to come from Germanic words for lather or bath, i.e. washing). Presumably, it converted any fatty or oily material on the clothes into a kind of soap which would then emulsify the dirt. The emulsified dirt could then be washed away. Use of the lye to make soap was also carried out in Wales using bracken ash.

A historical experiment to make soap using wood ash or any of the other ashes is an engaging exercise for students. Any oil or fat can be used and the normal vegetable oils for cooking are suitable. There are many recipes for making soap which are widely available on the internet. One example is at www.ehow.com/facts_7279261_make-soap-base-scratch.html.

Please note that lye (usually sodium hydroxide these days but potassium hydroxide can be used) is a caustic material both as the solid or as a concentrated solution. It is corrosive and appropriate safety precautions must be employed. Gloves and goggles must be used to protect the skin. Any spills must be thoroughly washed with cold water. Medical assistance should be sought if necessary.

Figure 6.2 The apparatus needed to make soap in the lab

6.2 Searching for cause, the 'influencing ingredient'

So far, the emphasis has been on concept creation, roughly following the route of history. This brings us to around the eighteenth century. French chemist Nicolas Lémery (1645–1715) was born in Rouen and was one of the first scientists to develop theories on acid–base chemistry. By a theory, I mean a causal explanation that makes sense of all, or as much as possible, of the data that has been collected to date. The theory should be able to predict new information that can be collected and tested for compatibility. Earlier in this chapter I called this 'fertility'. A theory of acidity should explain why all the members of the class of acids behave as they do and why the non-members do not. When chemists construct a theory, they may find that minor alterations are required. This is called accommodation and is felt to be natural in the development of a theory. If, however, there comes a point where accommodation cannot explain a large portion of the new evidence, a new explanation is created. This is called a paradigm shift and the topic of acidity offers examples of it that will be explored later.

In the late 1600s and early 1700s, at the time when Lémery was working, chemists (scientists) were developing techniques of investigation that many might now view as being obvious. One example was the handling of gases, which depended on the manufacture of appropriate equipment. Lémery, whose theory I shall shortly discuss, pioneered the use of leather containers to collect gases. Before this, gases were so elusive that chemists were unable to study them properly and referred to them as different kinds of 'airs'. Antoine Lavoisier (1743–1794) was another chemist whose equipment enabled him to handle gases with ease. He invented a large wooden trough filled with mercury and used upturned glass containers to hold and observe gases and measure their volumes. Unfortunately, the changing humidity and temperature in his laboratory made the wooden joints leak, losing the mercury. He solved this by constructing troughs from solid blocks of marble, 2 metres long, which could not leak. We can only imagine how heavy they must have been. Similar troughs were used by most chemists of the time to handle gases.

In 1680, Lémery constructed a theory to explain the sharp taste of acidic materials. He imagined them to be made of particles with little points that pricked the tongue. He imagined bases to have holes or pores into which the points fitted to neutralise this effect, so explaining the formation of salts. Of course, he had no way of seeing these particles or 'corpuscles' as he called them.

The scene was set for the eighteenth century to find a causal explanation for acidity. Lavoisier was credited with this in 1776. Even at that time chemists knew that various non-metals burned to give new materials with the characteristics of acids. Lavoisier created the idea that there was a specific part of air involved, which he named acid-maker or 'oxygen' from the Greek. His equipment for handling gases, mentioned earlier, was instrumental in creating this theory. Lavoisier's discovery was all the more remarkable because most gases are colourless, making it very difficult to distinguish them. Another secret of his success was the construction of a very accurate and sensitive balance by Fortin, the French instrument maker. This enabled Lavoisier to demonstrate accurately the small changes in weight that took place on burning. Lavoisier named the oxides of the non-metals 'acids'. So, SO_3 was acide sulfurique, SO_2, acide sulfureux, and CO_2, acide charbonique. He knew that the strength of the acids increased with more oxygen, so he invented a naming system precisely to reflect this. Like many chemists of the time, Lavoisier did not appreciate the role of water in acidity, except as a medium for dissolving. However, given the limitations of his equipment, his theory was most creative and is the basis for naming acids, even today.

Modern safety regulations restrict the burning of non-metals in oxygen to teacher demonstration, with the extra precaution of a fume cupboard restricting observations. There are videos of various elements burning at www.periodicvideos.com. Burning metals in oxygen provides the contrast of basic oxides. Some basic oxides are soluble in water and are described as alkaline, as mentioned above. They all neutralise acidic solutions.

One significant issue of confusion for students is the leap from the 'metal/non-metal' dichotomy to the 'basic/acidic solutions of their oxides' dichotomy. Many find it difficult to connect the two dichotomies (Table 6.1).

Table 6.1 Examples of the oxides of some elements

Element	Nature of element	Nature of oxide
Sulfur	Non-metal	Acidic
Sodium	Metal	Basic and alkaline (it is soluble in water)
Copper	Metal	Basic only (it is insoluble in water)
Magnesium	Metal	Basic and alkaline (it is soluble in water)

This principle can be demonstrated by asking students to use an indicator to test the nature of 'solutions of different oxides', as shown in Table 6.2.

Table 6.2 Testing 'solutions of different oxides'

Solution provided	Chemical label	Test with litmus indicator	Conclusion
Dilute sulfuric acid	Solution of sulfur oxide	Indicator turns blue to red	Solution is acidic, so substance is a non-metal oxide
Soda water	Solution of carbon oxide	Indicator turns blue to red	Solution is acidic, so substance is a non-metal oxide
Dilute sodium hydroxide solution	Solution of sodium oxide	Indicator turns red to blue	Solution is alkaline, so substance is a metal oxide
Lime water (calcium hydroxide saturated solution)	Solution of calcium oxide	Indicator turns red to blue	Solution is alkaline, so substance is a metal oxide

Lavoisier had adopted the dualistic theory of neutralisation: that the base and the acid (the non-metal oxide) simply combined with each other to make a salt. So, the reaction between copper(II) oxide and sulfuric acid would be:

$$CuO + SO_3 \rightarrow CuO,SO_3$$

CuO,SO_3 is what chemists would now write as $CuSO_4$ or copper(II) sulfate. However, the dualistic formula does show the composition of the salt from a base and an acid. It may be strange to think of chemical formulae in this way, but consider these two alternative versions of the thermal decomposition of copper(II) carbonate:

$$CuCO_3 \rightarrow CuO + CO_2$$

$$CuO,CO_2 \rightarrow CuO + CO_2$$

Which one is a clearer explanation for what is going on?

The idea developed by Lavoisier constitutes an explanation based on an 'influencing ingredient' of acids and was very compelling because of its simplicity. Sir Humphrey Davy (1778–1829), later, was more concerned with unravelling the hydrochloric acid problem; he was able to show that there was no oxygen present at all in hydrogen chloride. In 1839, using new formulae for the acids, such as H_2SO_4, he proposed that all acids contained hydrogen that could be displaced by a metal. His theory, though, was in the same camp as that of Lavoisier, since it involved an 'influencing ingredient' of acids – hydrogen. Davy's explanation gave a straightforward view of the reaction of an acid with a reactive metal. For example:

$$Mg + H_2SO_4 \rightarrow H_2 + MgSO_4$$

The reaction involves the simple displacement of the hydrogen by magnesium. We now have an explanation for the reaction of an acid with a reactive metal, a defining feature of an acid, and for the consequent formation of a salt, magnesium sulfate.

What about the reaction of an acid with a carbonate? This was construed as a two-step reaction, for example:

$$H_2SO_4 + MgCO_3 \rightarrow H_2CO_3 + MgSO_4$$

$$H_2CO_3 \rightarrow H_2O + CO_2$$

In the first part, one acid, sulfuric acid, is displacing another, carbonic acid. In the second part, the carbonic acid simply breaks up, giving off the carbon dioxide.

6.3 Searching for cause: the reaction itself

This section focuses firstly on Svante Arrhenius (1859–1927), a Swedish chemist, and covers material which is more appropriate for students aged 14–16. Working in the late 1800s, Arrhenius stumbled on a new principle of acidity, while investigating a quite different problem – this is known as serendipity. Arrhenius was interested in physics, not chemistry, and chose to work on the conductivity of electrolytes. In 1840, Kohlrausch, a German physicist, had invented a method of measuring very low conductivities using a kind of alternating current to detect the position of breaks in underground telephone wires. In the 1880s, Arrhenius adopted this instrument for his investigation. From his data, he came up with the idea that some compounds existed as charged bodies when dissolved in water. These bodies eventually became known as ions. Arrhenius' professors were surprised, but unimpressed, by this and he was graded at a very low level for his doctorate. Later, he was to win the Nobel Prize for this work. In the part of his study relating to acids, Arrhenius proposed that acids form H^+ ions in solution and that it is this which then reacts. It is no longer part of the acid, but is produced in solution, and then itself reacts with bases and alkalis. The converse of this idea is that the alkaline component is the hydroxide ion, OH^-. The neutralisation reaction is, essentially:

$$H^+ + OH^- \rightarrow H_2O$$

This reaction is the production of water from the two ions. Arrhenius' ideas were focused specifically on aqueous solutions.

This principle can be shown in class by measuring the electrical conductivity of acid solutions with a modern conductivity meter. The traditional mineral acids will have a relatively high conductivity, while fruit acids have a much lower conductivity. Chemists now have a modification of the theory that an acid produces a specific species, the H^+ ion, and they can now identify how much is produced – a quantitative measure of acidity. Changes in conductivity when an acid is neutralised by an alkali, or vice versa, can be followed during titrations.

The concentration of hydrogen ions in aqueous solution varies over a wide range, from about 1 mol dm^{-3} for strongly acidic solutions to around 10^{-14} mol dm^{-3} for strongly alkaline solutions. Soren Sorenson, a Danish chemist, created the pH scale to deal with this wide range in 1909. The origin of the term pH appears to be 'potens hydrogen' or 'the potential of the hydrogen'. Sorenson defined pH as:

$$\text{pH} = -\log\text{ (concentration of hydrogen ions)}$$

The function log is not so widely used or understood these days so an alternative form of this is:

$$\text{concentration of hydrogen ions} = 10^{-\text{pH}}$$

Some textbooks state that the pH scale runs from 0 to 14, while others that it runs from 1 to 14. They both state that 7 is in the middle! It seems that the pH scale can run from a practical value of around –1.6 to around 15.6. Typical wide-range indicator paper available in schools runs from 3 to 11, and the same is true for universal indicator solutions. The colour of pH 7 is green for some universal indicator solutions, but yellow for others. The recipe used for the mixture of indicators determines the actual colour seen. Yamada created and patented the first recipe (water, propan-1-ol, phenolphthalein sodium salt, sodium hydroxide, methyl red, bromothymol blue monosodium salt and thymol blue monosodium salt) in 1923.

Also in 1923 Brønsted and Lowry were working to extend the notion that H^+ is the agent of acidity. They noted the similarity of the reaction of ammonia and hydrogen chloride gases (shown as Figure 2.9 on page 69 of Chapter 2) to make ammonium chloride with the parallel reaction in aqueous solution. They took the idea of an acid one stage further to include reactions where a hydrogen ion is transferred from one chemical species to another:

$$HCl + NH_3 \rightarrow NH_4^+Cl^-$$

In this equation, the focus is on the reaction that defines the acid as a proton donor. Conversely, a base is a proton acceptor.

6.4 Tidying up

1 The nature of the hydrogen ion in solution is not as easy as it seems. The simple explanation is that it is H^+. However, there is evidence that the hydrogen ion cannot exist on its own in aqueous solution and many books describe it as a hydrogen ion combined with a water molecule, or H_3O^+. Some research chemists have found evidence for the species $H_9O_4^+$ in some solutions, where four water molecules are bonded. I prefer to use the simplest form wherever possible. The neutralisation reaction using H_3O^+ is:

$$H_3O^+ + OH^- \rightarrow 2H_2O$$

2 Should we use the term 'strong' for an acid or an acidic solution? For students aged 11–14, I recommend a strict use of the idea of acidity only in solution and to use 'strongly acidic solutions' for those with pH less than 3. For older students, who understand the idea of partially dissociated acids in solution (such as the weak acid, ethanoic, in vinegar), then strong acid can refer to those more fully dissociated, such as the mineral acids.

3 For students aged 11–14, I recommend that pH be seen as a refinement of the colours attached to indicator solutions. So, pH 5.5, seen in advertisements for shampoos, refers to somewhere in the yellow region of the indicator. It is simply a more precise way of identifying a position on the acidity scale. Later on, for those students capable of working with logs or powers of 10, the mathematical idea of pH can be used.

4 The problem of the pH scale requiring increasingly acidic solutions for lower pH can be introduced as simply a quirky outcome of having to have a scale for something that is not easy to measure.

5 I mentioned earlier the place of metal carbonates in this concept of acidity. If metal carbonates are seen as a dual compound, e.g. magnesium carbonate is 'magnesium oxide, carbon dioxide' (MgO,CO_2) where the comma means combined as in nineteenth century chemistry, then the chemical reaction is between the base part of the carbonate (MgO) leaving the carbon dioxide to bubble off. A typical reaction is then:

$$MgO,CO_2 + H_2SO_4 \rightarrow MgSO_4 + H_2O + CO_2$$

I recommend, though, simply treating reactions of metal carbonates to make a salt and water (and carbon dioxide) as a special case and not mixing them up with a general discussion of reactions of acids and bases. Readers may find this in contrast with more traditional textbooks.

Other resources

Beyond Appearances: Students' Misconceptions about Basic Chemical Ideas (Kind, V., 2004) can be freely downloaded from the Royal Society of Chemistry website. It details common student conceptions across a range of chemical topics. www.rsc.org/Education/Teachers/Resources/Books/Misconceptions.asp

The Royal Society of Chemistry resource *Chemical Misconceptions – Prevention, Diagnosis and Cure* (Taber, K.S. 2002. London: Royal Society of Chemistry. Volumes 1: Theoretical background; Volume 2: Classroom resources) includes two activities relating to acids. A revision activity for lower secondary is built around concept mapping ('Acid revision map'). Another activity 'Explaining acid strength' is for more advanced students, considering the difference between acid strength and concentration. Both of these activities can be downloaded from the Royal Society of Chemistry's website: www.rsc.org

7 Combustion and redox reactions

Vicky Wong, Judy Brophy and Justin Dillon

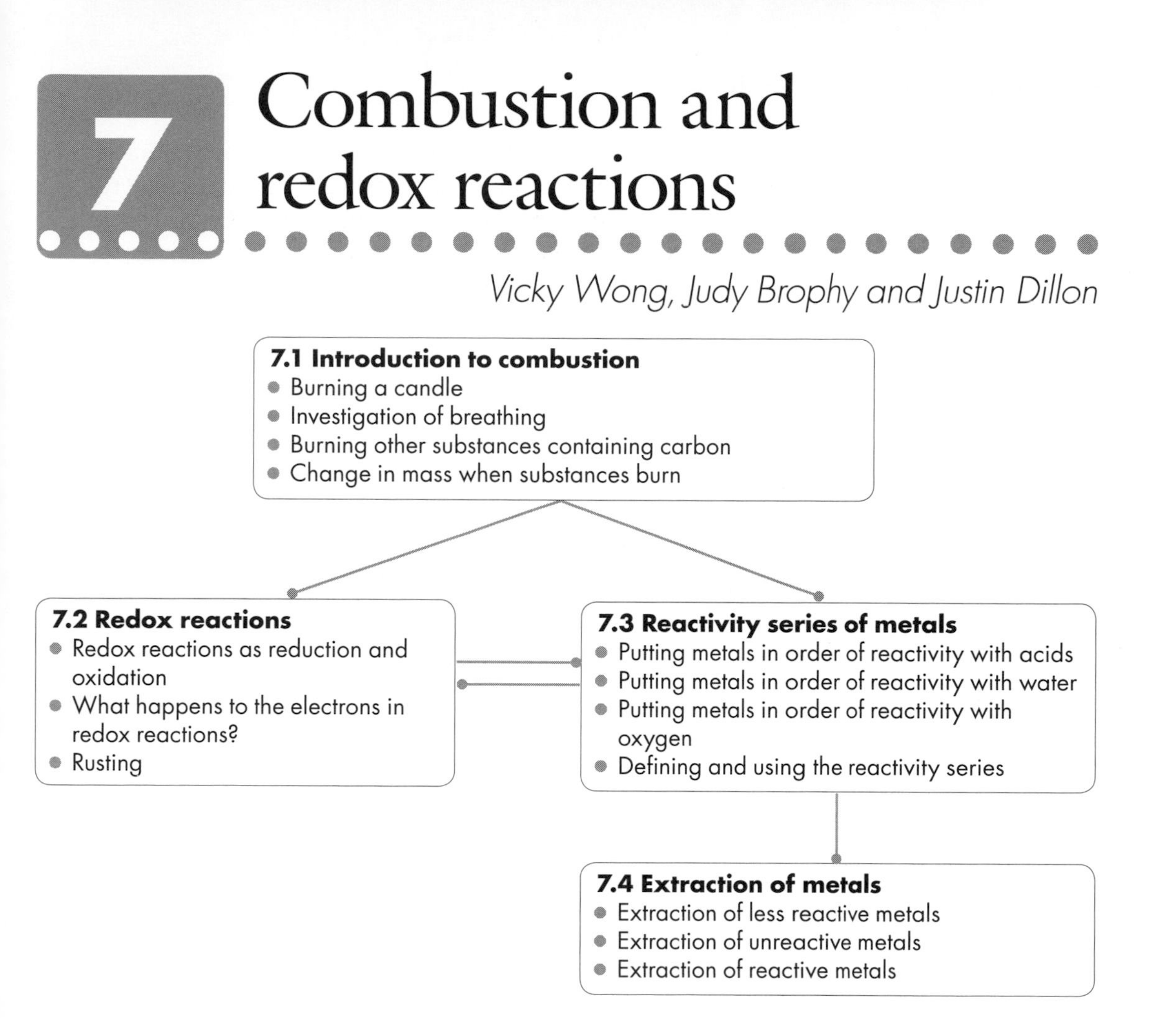

Introduction

Students are always keen to investigate the burning of substances. When they arrive in secondary school there is great enthusiasm for the Bunsen burner and students are happy to use it as often as possible. The composition of air, together with various experiments to investigate the nature of combustion, is a good starting point to develop observational and thinking skills. Investigations of burning, breathing and rusting to find the similarities will help to further understanding. Comparing the ability of different metals to burn, and more generally to react, leads to the idea of a reactivity series. Testing the products of combustion reinforces ideas about differences between metal and non-metal elements and can introduce or revisit ideas of acidity and alkalinity.

Students can end up thinking about chemical reactions with any number of faulty models – leading to confusion, misconceptions and frustration. In designing a teaching sequence it is important to ensure that students understand what has gone before and have clear thinking about chemical reactions (covered in greater detail in Chapter 3) in order to ensure that progress is made. In particular, throughout the work in this chapter, students could consider particle pictures of the states of matter (covered in Chapter 2) before them and be asked to draw diagrams to show their ideas explicitly. This helps them to express their ideas about what is going on and the teacher to see where students' misconceptions lie.

7.1 Introduction to combustion

Previous knowledge and experience

The topic of combustion is likely to come early in a secondary school course and students are unlikely to have much experience of chemical reactions from their primary school science. If the topic is covered later in the course, then students may have done:

- some chemical reactions but may not have been asked to make careful observations and explain them; they may have written word equations and/or formulae equations
- tests for oxygen, nitrogen and carbon dioxide
- tests for water using cobalt chloride paper and anhydrous copper sulfate (if they have not come across these tests before, they could be introduced here)
- particle pictures of solids, liquids and gases.

In addition, all students will have some everyday experience of combustion from watching bonfires, home fires, candles and fireworks, among others.

A teaching sequence

The following list represents a suitable order for covering the required parts of a study of combustion:

- burning – the burning of a candle
- products of combustion – particularly showing that products can be gases
- respiration as a form of combustion
- products of combustion of a range of carbon-containing materials
- combustion of other materials – change in mass
- combustion of other materials – naming the products
- rusting.

This topic will probably not be taught as a single sequence but may be spread through the school curriculum. It is important to revisit and recap the ideas covered earlier in the sequence each time it is picked up.

All of these ideas could be taught to students between the ages of 11 and 13. The products of combustion and ideas of change in mass and naming products are likely to be revisited throughout the secondary school science curriculum.

Burning a candle

The aim here is for students to understand that air is a mixture of several different gases including oxygen. When substances burn in air, they combine with oxygen and give out hot gases. These are usually seen as flames. The process is known as combustion. The investigation into the burning of a candle given here can be done at many different levels with students of a wide range of ages.

PROCEDURE

Investigation of the burning of a candle

Materials

Each group of students will need:

- a candle
- a large beaker
- cobalt chloride paper
- anhydrous copper sulfate, one spatula
- universal indicator paper
- a watch glass
- splints
- a temperature probe (a thermometer is not suitable for this experiment)
- beakers of different sizes
- gas jars with covers
- lime water
- eye protection

Safety

Eye protection should be worn. Students known to have skin allergies should wear nitrile gloves when using cobalt chloride paper. Hands should be washed after their use.

Method

1. In groups of three or four, students should light a candle.
2. Students should make as many observations about the candle as possible in about 8–10 minutes.
3. Ask groups to compare their observations with those of another group.
4. From the discussion with another group, students should choose one finding and try to explain it.
5. Ask the groups to make a poster of the information/explanation.

Observations/questions about the flame may include:

- Are there different colours in the flame at the top, middle and bottom?
- What happens to the wax as the candle burns?
- What happens to the wick as the candle burns?
- Is there any smoke while the candle burns?
- Which is the hottest part of the flame?
- Can the temperature of the hottest part of the flame be measured with a temperature probe?
- Does the type of wax alter the temperature of the flame? (To answer this, students would need access to more than one type of candle.)
- What should be done in each investigation to ensure the results are reliable?
- How are the comparisons a fair test?

Make a display of all the posters. The posters can be used to probe students' understanding of what is happening as the candle burns and to challenge inconsistencies in their explanations.

Students' misconceptions about burning

Students have many misconceptions relating to burning. These may include some of the following: air is not involved with burning; burning does not produce new substances; the evaporation of a substance is the process of burning; no molecules exist in a burning flame.

Many students think that a candle burning is a state change, often describing the wax as melting or evaporating. This may result from the fact that the oxygen is invisible, so it may be easy for some students to ignore the role it plays. Some students think that the candle flame is caused by the wick burning rather than the wax. They may reason that the heat from the flame (which is the wick burning) causes the candle to melt.

Students' poor understanding of the particle model does not help as many do not recognise that the wax and the flame are particulate. Very few students understand that the flame includes particles of hydrocarbon. This is obviously going to be true of younger students who have not yet met the ideas of hydrocarbons, but is also prevalent among older students who have.

While most students know that oxygen is needed for burning, they cannot explain how it is involved.

There are a number of aspects of combustion which are often absent from students' thinking. These include the idea that gases can be produced in the reaction, that gases might have mass, and the existence of atoms or molecules.

An important role for chemistry lessons is, therefore, to extend the range of observations that students make about burning to include an emphasis on the involvement of gases in the reactions. An understanding of conservation of mass does not necessarily follow from an understanding of the role of gases in burning. Many students think that a change in form of a substance can cause a change in mass and in particular that gases have zero or negative weight. Without an understanding of conservation of mass it is impossible for students to make sense of experiments involving weighing, particularly reactions involving gases such as combustion.

■ Further investigations of burning candles

PROCEDURE

Materials

- candles
- beakers of various sizes
- cobalt chloride paper
- universal indicator paper
- lime water
- splints
- watch glass
- gas jar

Method

Burn a candle on a watch glass in a confined volume of air under upturned beakers of different sizes. Ask students to measure the time taken for the candle to go out. They could also try raising the beaker slightly to let more air in as the candle starts to go out.

Can the class explain why the flame lasts longer under a bigger beaker? Can they explain the role that oxygen plays in burning?

Later investigations (of the products of combustion of a candle and of breathing) will involve tests for water and carbon dioxide. If students have not yet been introduced to these tests, now would be a good time to demonstrate them.

Draw your students' attention to the formation of mistiness on the side of the beaker. This can be tested with cobalt chloride paper. What is the result? What does it mean?

Next, burn the candle with a gas jar over the top. When the candle goes out, put a cover on the gas jar. Ask students to add a small amount of lime water to the jar and shake it carefully. If it goes milky then students should understand that this is a positive test for carbon dioxide. As a control, repeat the test with a gas jar which has not had a candle burned in it.

For younger students, this last investigation can be used to show that the air in the gas jar is different after burning – the gases in the jar have changed. This is a good way of probing the ideas that there are gases involved in this reaction – both oxygen being needed for the burning to take place and also gaseous products. This also shows that air is a mixture of gases and that the precise composition of air varies.

Change in mass when a candle burns – teacher demonstration

This demonstration can be used to further probe students' understanding of the products formed when a candle burns. Key to this is the idea that gases have mass. The soda lime traps the gaseous carbon dioxide and water vapour and allows them to be weighed. The idea that the gaseous products of combustion might weigh more than the part of the solid candle that has burned may well challenge students' thinking and it is worth taking some time to discuss exactly what they think is going on before presenting the scientific viewpoint.

PROCEDURE

Materials

- a thistle funnel with a bent stem
- a U-tube, 10 cm, limbs fitted with bungs and delivery tube
- a filter pump
- a stand and clamp
- a watch glass
- a candle
- mineral wool and tweezers
- soda lime, granular form
- a balance (to two decimal places if possible)
- eye protection

Safety

Students should wear eye protection. The demonstrator should wear gloves when handling soda lime. It may be advisable to have the soda lime packed into the U-tube in advance of the demonstration.

Method

1. Connect the apparatus as shown in Figure 7.1.
2. Pack the U-tube loosely with soda lime granules to a depth of 5 cm in each limb.
3. Use some mineral wool to hold the granules in place.
4. Weigh the candle and the watch glass together.
5. Weigh the rest of the apparatus.
6. Clamp the apparatus and connect it to a filter pump.
7. Light the candle and turn on the pump.
8. Allow the candle to burn for 2–3 minutes, noting the length of time of burning.
9. Extinguish the candle and at the same time disconnect the pump but do not turn the pump off.
10. Reweigh the apparatus when it has cooled.
11. Reweigh the candle and watch glass together.
12. Repeat the experiment without lighting the candle but by reconnecting the filter pump for the same length of time and with the same rate of flow.

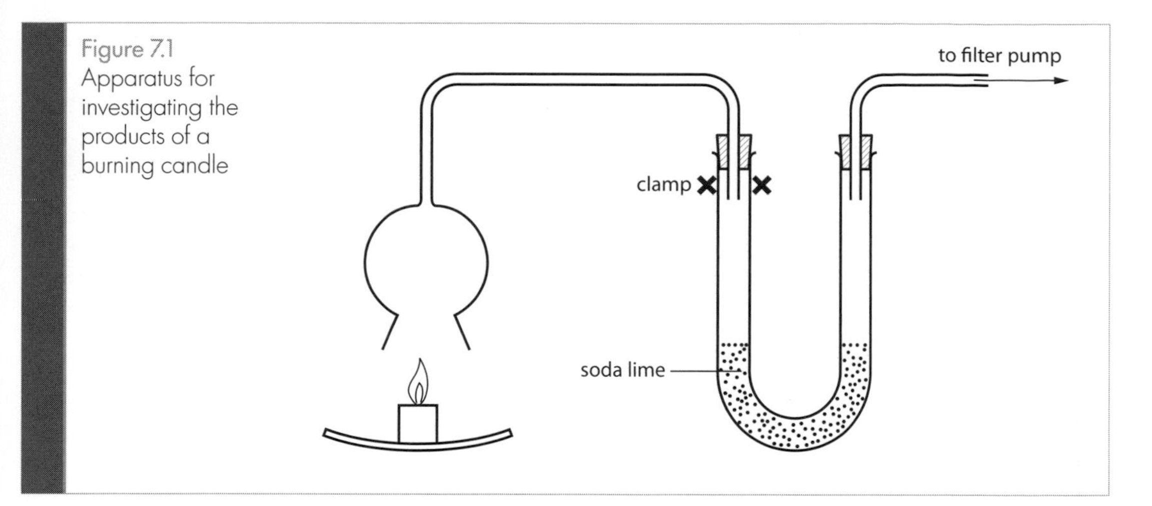

Figure 7.1 Apparatus for investigating the products of a burning candle

The results will show that the increase in mass of the U-tube apparatus is more than the loss in mass of the candle. The products of combustion weigh more than the part of the candle that has burned. This is because oxygen has been taken from the air. You might wish to initiate a discussion at this point on whether there are other feasible interpretations of these results. Can we be sure that the increase in mass is due to the burning process? Is there a way to set up a control to make sure we are doing a fair test? Part 12 of the procedure will also give an increase in weight, which will show the amount of moisture and carbon dioxide absorbed from the atmosphere during the time of combustion of the candle.

■ Combustion products of a candle

Some of the ideas given in this section could be covered by the practical work already described, but this is a very commonly used school demonstration so is included here for completeness. To understand why the candle releases carbon dioxide and water when it is burned requires students to understand that the candle is made of a hydrocarbon and that this is composed of carbon, which burns in oxygen to make carbon dioxide, and hydrogen, which burns to make hydrogen oxide – more commonly called water. The formation of carbon dioxide and water can then be used as evidence that the candle did indeed contain a hydrocarbon.

PROCEDURE

Materials

Version 1:

- lime water
- anhydrous copper(II) sulfate
- universal indicator solution
- test tubes fitted with bungs and delivery tubes
- a candle
- a suction pump

Version 2:

- a glass funnel
- glass and rubber tubing
- a U-tube in an ice bath
- a side-arm test tube fitted with a delivery tube
- lime water
- a candle
- a suction pump

Safety

Students should wear eye protection.

Method

1 Connect the apparatus as shown in Figure 7.2 or 7.3 (depending on which version is being demonstrated).
2 Turn on the pump.
3 Light the candle and allow it to burn for about 10 minutes.
4 Observe the lime water during this time.

(Note: In version 1 it is important to connect up the anhydrous copper sulfate first. This is because both the universal indicator and the lime water contain water, so if the copper sulfate comes after them it is not showing that the gases from the candle contain water.)

Figure 7.2 Version 1 of the apparatus

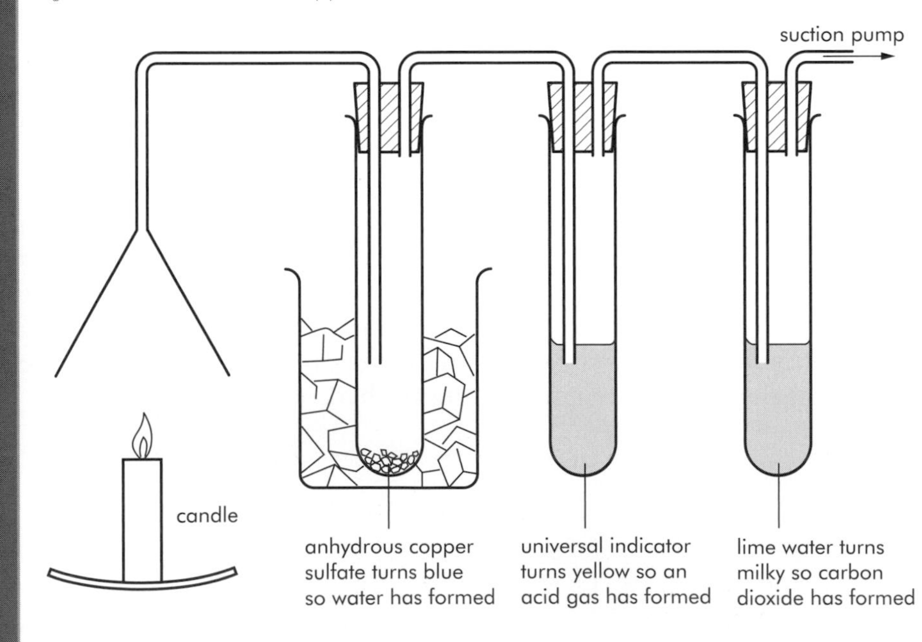

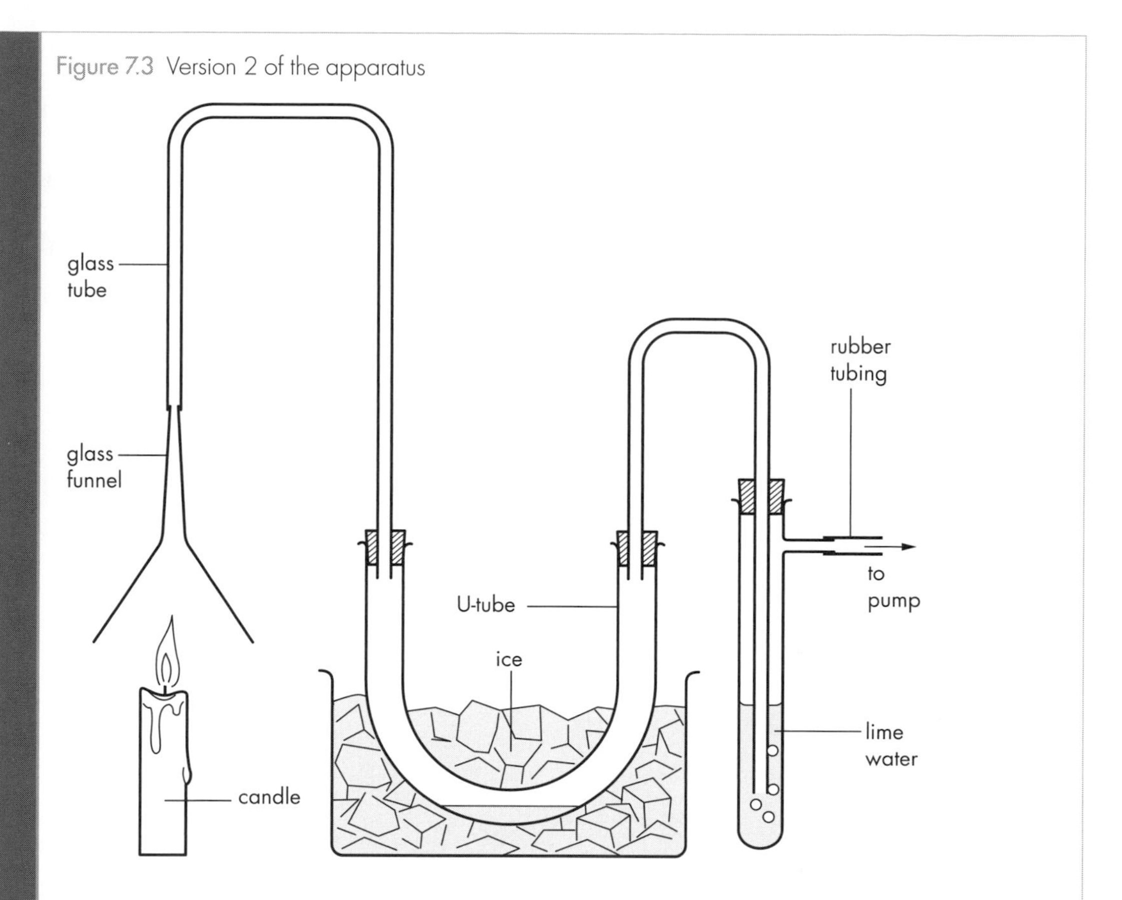

Figure 7.3 Version 2 of the apparatus

Observations

- Lime water changes from colourless to milky (white precipitate) showing the presence of carbon dioxide. (Note: as more carbon dioxide is passed through the solution it becomes clear again. This is to do with the chemistry of the reaction of carbon dioxide with lime water and does not show that no more carbon dioxide is formed or that it has all been removed.)
- The liquid in the U-tube in version 2 may be tested with anhydrous copper(II) sulfate or anhydrous cobalt chloride paper to show the presence of water.
- The glass funnel and tubing will become black due to the soot formed from the incomplete combustion of the candle wax.

Further teaching suggestions

- Include the idea of the 'fire triangle' alongside combustion.
- Demonstrate the fat-pan fire: www.nuffieldfoundation.org/practical-chemistry/fat-pan-fire
- Use carbon dioxide to put out a fire. This can be done very simply using a gas jar or beaker of carbon dioxide generated in the lab with acid and carbonate. The gas jar or beaker of the gas can be poured over a candle flame to show that it will extinguish

it. Ensure that the beaker or gas jar is dry so that there are no drops of water to put out the flame.

- Many students have the impression that 'chemicals' and 'chemical reactions' are artificial and nothing much to do with life. Including respiration in a sequence of lessons about combustion helps to correct this. Respiration is the reaction of material containing hydrogen and carbon with oxygen giving identical products to those of combustion: carbon dioxide and water.

Investigation of breathing

PROCEDURE

Materials

- two boiling tubes
- two bungs with holes to fit the boiling tubes
- plastic tubing
- sterilising liquid (such as Milton® sterilising fluid)
- a plastic T-bar to ensure correct flow through equipment
- lime water

Figure 7.4 Apparatus to test the products of respiration

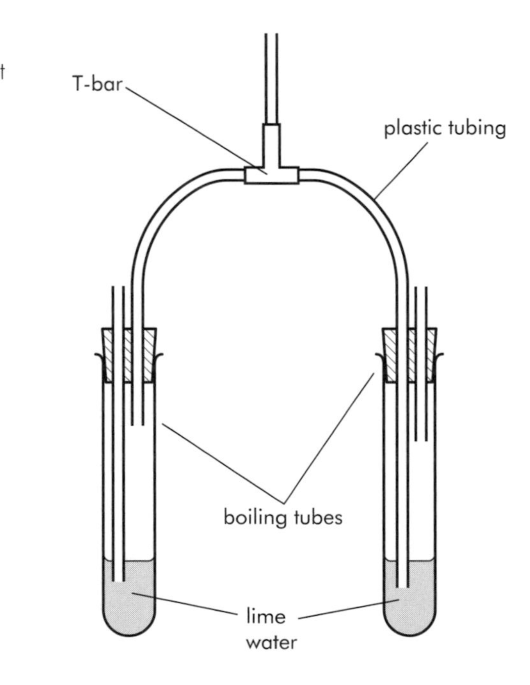

Method

Half-fill the two boiling tubes with lime water, ensuring that the lower glass tube is covered but the upper one is not. Make sure that when the equipment is breathed through, students will not suck in lime water. Ask a student to breathe through their mouth using the apparatus for about a minute. The lime water will go milky with the exhaled air but not with the inhaled air.

A similar set-up can be used with empty and dry tubes to demonstrate that water is another product of respiration. After the student has finished breathing, the gas in the boiling tubes can be tested with cobalt chloride paper or anhydrous copper sulfate. The exhaled air should contain moisture which can be detected as water by either of these methods.

Burning other substances containing carbon

PROCEDURE

Student investigation

Materials

Each group of students will need:

- four test tubes, 100 × 25 mm, fitted with corks
- a combustion spoon
- wood
- paper
- sugar
- starch
- a wax taper
- lime water
- cobalt chloride paper
- eye protection.

(Note: if combustion spoons small enough to fit into the test tubes are not available, this could be a teacher demonstration using gas jars instead.)

Safety

All students should wear eye protection. Take care when using ethanol in the teacher demonstration.

Method

1. Take a small amount of one substance (wood, paper, sugar or starch), ignite it on a combustion spoon and put it in a test tube.
2. As soon as the flame is extinguished, withdraw the spoon and cork the tube.
3. Test the mist on the inside of the tubes with cobalt chloride paper and then use the lime water to test for carbon dioxide. (Students may need to do each one twice to be able to carry out both of these steps.)

When all students have finished the practical, initiate a discussion focusing on the following points:

- which substances burned
- which did not burn
- which changed permanently
- which changed temporarily
- which did not change
- which products can be collected and tested for.

4. As a further teacher demonstration, use a sample of ethanol (industrial methylated spirits) as the substance to be tested.

The 'whoosh' bottle – products of combustion of alcohols

This is a very exciting demonstration and can be used to show the products of combustion of alcohols, the variation in combustion of a series of alcohols and the energy stored in a small sample of an alcohol/oxygen mixture.

The demonstration requires careful preparation and it is advisable to consult a model risk assessment before carrying it out. Under no circumstances use a cracked or otherwise damaged bottle and do not use a bottle of any type other than described overleaf.

PROCEDURE

Materials

- eye protection (for the whole class)
- reaction vessels (see Safety notes below)
- rubber stoppers or plastic caps (to fit the reaction vessels)
- a 250 cm^3 beaker for each alcohol used
- wooden splints, as needed
- a metre rule
- one or more of the following alcohols, 40 cm^3 of each one used (see Notes below):
  - methanol
  - ethanol (IDA or Industrial Denatured Alcohol)
  - propan-1-ol
  - propan-2-ol

Safety

Each reaction vessel should be a large polycarbonate (marked PC) bottle such as those used in workplace water dispensers. These have a volume of 16–20 dm^3. After use, the bottle will need to be cleaned and dried. The easiest way to dry it is to invert it to allow any remaining water to drain away and then leave it upright for several days to dry.

Select a safe, level place for the demonstration, with at least 2.5 m clearance above the top of the vessel to the ceiling above and no flammable materials above it. If the laboratory bench does not allow for this, four stable laboratory stools supporting a large wooden tray may give sufficient clearance and stability. Alternatively, stand the bottle on the floor with a fire blanket or similar underneath it.

Students should be 4 metres away from the demonstration and should be wearing eye protection.

Method

1. Pour 40 cm^3 of one of the alcohols into the polycarbonate bottle. Put the bung in and then swirl the bottle around to coat the inside of the bottle with the alcohol and encourage evaporation.
2. Tip the remaining alcohol into a separate beaker and move it to at least 1 m from the reaction. Replace the bung.
3. Attach a splint to the end of the metre rule (an elastic band is the easiest way) and bend it down 3 cm from the end to allow this piece to be poked easily into the bottle.

Remove the bung, light the splint on the metre rule and light the vapour in the top of the bottle. There will be a whooshing noise and flames will shoot from the top of the bottle.

Notes

If you are carrying this demonstration out with a number of the alcohols, it is probably best to start with the propanols and work your way up to the most exciting – methanol. It is possible to observe changes in the flame colour, amount of soot produced and the length of the reaction as the series of alcohols is burned.

To show that the reaction produces carbon dioxide: allow the reaction vessel to cool a little, tip out the water which is present and pour in some lime water. Swirl the lime water around the vessel and pour out again. It will be seen to be cloudy indicating the presence of carbon dioxide. The liquid poured out will not give a positive result for water with the usual tests.

Combustion of other materials

The discussion of combustion has concentrated up until now on burning substances that contain carbon. Most students begin studying combustion in this way and it can lead them to assume that whatever is burned must produce carbon dioxide and, perhaps, water. Some students assume that any black product produced by combustion (such as copper oxide) must be soot. Some can correctly transfer their existing chemical knowledge of reactions to new ones but many require some help to do so. As with burning candles, there is widespread confusion about ideas of conservation of mass and the involvement of oxygen.

Change in mass when substances burn

The combustion of iron wool

This is a simple yet elegant practical to show that the mass of substances can increase on burning. Students can predict what they expect to happen and try to explain their predictions on seeing the results.

PROCEDURE

Materials

- eye protection
- Bunsen burner
- a heat-resistant mat
- a wooden metre rule
- aluminium cooking foil, about 10 cm × 10 cm
- sticky tack, a few grams
- a triangular block or something similar
- steel wool, about 4 g

Safety

All students should wear eye protection.

Method

1. Cover the end of the metre rule with cooking foil to protect it. Place the rule on the triangular block.
2. Tease apart some iron wool and place it on the end with the cooking foil. Put enough sticky tack on the other end to just tip the balance over on to that side.
3. Heat the steel wool with a Bunsen burner on a blue flame until it catches fire. The wool will glow and burn and the balance will tip towards the end that held the iron wool.

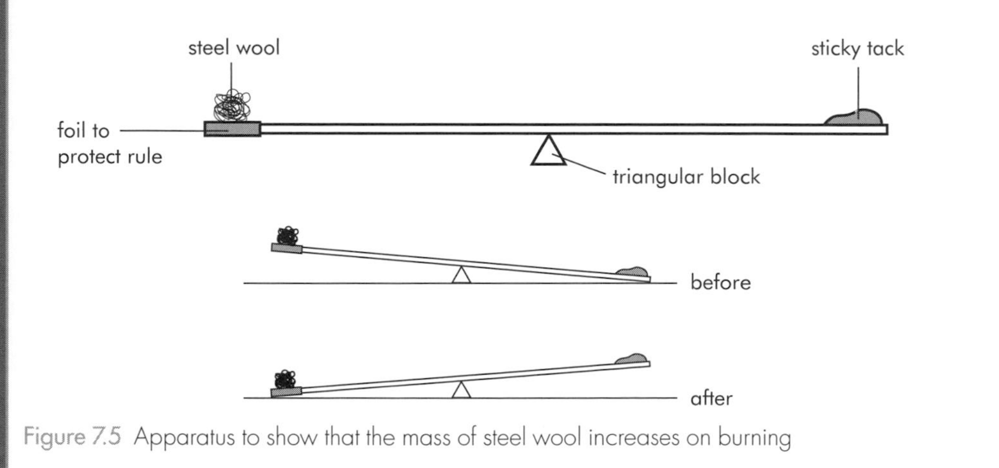

Figure 7.5 Apparatus to show that the mass of steel wool increases on burning

■ Do all substances change mass when heated in air?

Students can be asked to make predictions about whether or not certain substances will gain or lose mass on heating. They can try to predict the products and explain the reasoning behind their predictions.

The change in mass when magnesium burns

PROCEDURE

Materials

- eye protection
- a crucible and lid
- a pipe clay triangle
- magnesium ribbon, about 15 cm
- some emery paper
- tongs
- a Bunsen burner
- a heatproof mat
- access to a two-decimal-place balance

Safety

All students should wear eye protection. Be aware that tongs can be unreliable and difficult to use. If students are unfamiliar with them, it is a good idea to get them to practise lifting the lid on and off the crucible while it is cold. This also tests that the tongs are working.

Magnesium is highly flammable.

Method

1. Weigh a crucible and lid.
2. Take about 15 cm of magnesium ribbon, clean the surface with emery paper and make it into a loose coil.
3. Place the magnesium in the crucible and reweigh it with the lid. Place the pipe clay triangle onto the tripod in a 'star of David' configuration, ensuring that it is secure.
4. Place the crucible with lid on the pipe clay triangle and heat it strongly with the Bunsen burner.
5. Raise the lid periodically using the tongs to allow oxygen to enter the crucible. The lid needs to be replaced to minimise the loss of product. Heat until no more changes occur.
6. Allow the apparatus to cool and reweigh.
7. If time allows, reheat it to a constant weight to make sure the reaction is complete.
8. The magnesium will have gained mass. Students can explain why and write equations for any reactions that have occurred.

Extension

As an extension activity, this experiment can be used to calculate the formula of magnesium oxide.

This can be done by plotting a graph of mass of magnesium in grams against mass of oxygen in grams for all the class results. Then superimpose on the results the graph lines which would be expected for Mg_2O, MgO and MgO_2. The results almost always cluster along the MgO line. This approach is most successful if you ensure that there are several different masses of magnesium used by supplying cut pieces of different lengths.

Heating zinc dust in air – teacher demonstration

PROCEDURE

Materials
- zinc dust
- a steel needle or spatula
- a tin lid
- a tripod
- a Bunsen burner
- eye protection

Safety
All students should wear eye protection.

Method
1 Place a spatula of zinc dust on a tin lid on a tripod.
2 Heat it strongly in one corner. When the zinc catches fire and starts to burn, remove the Bunsen and, using a needle or spatula, gently prise up the yellow crust that forms to expose the metal underneath.

Observations
When exposed to the air, the fresh metal will burn forming a white fluff known as Philosopher's wool. There are lots of opportunities here for students to come up with explanations of what is happening and ideas about whether the zinc will have gained or lost mass.

As an extension activity, students could plan their own experiments to see whether (a) copper and (b) zinc oxide change mass on heating. To make these genuine experiments, students should work in small groups to first make a prediction and then try to support it with their reasoning – in other words, they should develop a hypothesis to test. On no account allow students to mix these two together and heat them.

■ What happens when substances are burned in air and oxygen? – teacher demonstration

PROCEDURE

Materials
- combustion spoons
- tongs
- gas jars of oxygen
- samples of the following materials:
  - wood charcoal
  - iron wool
  - copper
  - magnesium ribbon
  - powdered sulfur
- universal indicator solution

Safety

All students should wear eye protection.

No more than 0.3 g of sulfur can be burned in the open laboratory. Any more than this requires a fume cupboard. Be aware that burning sulfur produces sulfur dioxide which can exacerbate pre-existing breathing problems such as asthma. Effects of the exposure can be delayed for some hours.

Warn students not to look directly at burning magnesium. Looking through blue glass or partially opened fingers in front of the eyes offers protection.

Note that gas jars of oxygen can be produced using a gas cylinder or an oxygen generator. See: www.nuffieldfoundation.org/practical-chemistry/generating-collecting-and-testing-gases for details of how to generate and collect oxygen.

Method

1 Heat a small amount of carbon (charcoal) in air and then plunge it into a gas jar of oxygen.
2 Repeat with iron (wire wool), copper, sulfur and magnesium ribbon. Sulfur and magnesium will catch fire, whereas the others just glow. (For some substances, such as sulfur, it is easier to use a combustion spoon; for others, such as magnesium ribbon, tongs are preferable.)
3 Students can make observations of the substances before, during and after burning and compare the burning in air and in oxygen.
4 Add water to the gas jars, find the pH using universal indicator (discussed further in Chapter 6) and note any patterns.

Results

The products of the combustion reactions are called oxides and students could be asked to name them. These are chemical changes and a glow or a very bright flame will be seen. Students can observe that the reactions are much more vigorous in oxygen compared to air. They can be asked to write word and/or symbol equations.

This set of reactions produces two sorts of oxides – acidic and basic. Metal oxides are basic and non-metal oxides are acidic (as discussed in Chapter 6). This means that it may be possible to classify elements by their chemical properties, not just their physical ones.

Other suggestions

- The copper envelope is a simple and straightforward reaction that students can carry out: www.nuffieldfoundation.org/practical-chemistry/copper-envelope
- Video material about fire and flames is available from the STEM centre website: www.nationalstemcentre.org.uk. Search for 'flames and chemical change' and 'combustion or decomposition'.
- Video and animation material about reactions, including the reaction of magnesium and copper with oxygen, is also available from the STEM centre website, search for 'substances changing reaction on heating'. You will need to register to view the material, but registration is free.

7.2 Redox reactions

Previous knowledge and experience

Students will have experience of combustion reactions as reactions with oxygen. They may have called these oxidation reactions.

Students' misconceptions

Students' difficulties with these reactions and their alternative conceptions are similar to those experienced with other classes of chemical reactions (as discussed in Chapter 4).

These reactions also provide a good opportunity to reinforce the teaching of chemical equations. Several studies have explored the causes of students' errors in balancing equations. Interestingly, mathematical errors and violations of mass conservation were seldom found to be reasons for incorrectly balanced equations. More often, students did not understand the methods of writing formulae – subscript numbers, brackets and coefficients. Introducing these concepts early on in a student's school career allows time for plenty of practice and familiarisation with protocols for writing equations. For ionic equations (such as those often found in redox reactions) the concepts of spectator ions and ionic charge posed the greatest obstacles for students. One way of helping students understand about spectator ions is to use electrolysis reactions (see Chapter 8). Writing half equations showing exactly which ions are reacting makes it clearer which ones are not. These ideas can then be transferred back to redox equations. Using models such as chemical jigsaws can assist students in understanding ionic formulae, which makes it easier for them to use such formulae in equations.

Students' misunderstandings of chemical equations are also due to their failure to associate the symbols and numerical answers with real objects and events. Thus, students' difficulties in understanding chemical reaction equations have been attributed to misunderstanding the symbols themselves, to misunderstanding concepts related to the atomic and electronic structure of matter, and to not understanding the relationship between the equations and the macroscopic level.

This topic can involve students carrying out many reactions and writing equations for them. Careful guidance should help them to make the connection between the symbols and the actual reaction. They will also need help in understanding the conventions of writing formulae because without correct formulae, it is impossible to write a balanced chemical equation.

Additionally, considerable evidence indicates that a key issue is the language of chemistry. Chemists' meanings for words such as 'substance', 'element' and 'pure' differ significantly from everyday meanings. Students need to be given opportunities to learn the chemical meanings, to appreciate what they really mean and to use them correctly.

A teaching sequence

A possible teaching sequence for redox reactions is given here:

- What is oxidation?
- What is reduction?
- Redox reactions – both reduction and oxidation
- Oxidising and reducing agents
- Electrons in redox reactions
- Investigating redox reactions
- Rusting

Although they are separated out in this chapter, the teaching of redox reactions is likely to be intertwined with the reactivity series of metals (section 7.3) and work on the extraction of metals (section 7.4). This is because all these reactions are redox reactions, as are the combustion reactions met earlier.

By the age of 14 students should have some experience of combustion reactions and reactions with oxygen. They should know the term oxidation. The reactivity series of metals may have been covered by this stage too.

By 16 students should have experienced several oxidation and reduction reactions. They should understand that these occur together (oxidation is always accompanied by reduction, and vice versa) and be able to write equations for some of them. More able students may cover these in terms of electrons, which provides a good link to electrolysis (see Chapter 8).

Redox reactions as reduction and oxidation

One very important type of chemical reaction is a redox reaction which involves both reduction and oxidation reactions together. To understand what is happening, it is useful to look first at some reactions illustrating oxidation and some illustrating reduction – although it is important that students understand that oxidation and reduction always occur together.

■ Oxidation reactions

Oxidation at a simple level can be looked at as the addition of oxygen. When magnesium reacts with oxygen it is oxidised to magnesium oxide:

$$\text{magnesium} + \text{oxygen} \rightarrow \text{magnesium oxide}$$

$$2Mg \quad + \quad O_2 \quad \rightarrow \quad 2MgO$$

The magnesium gains oxygen – it has been oxidised. Oxidation is also the removal of hydrogen.

Generating chlorine – teacher demonstration

This demonstration will mean nothing to students unless they understand that the hydrochloric acid contains chlorine bonded to hydrogen. The potassium manganate(VII) 'removes' the hydrogen and chlorine gas is formed. The obvious yellow-green colouration of the chlorine gas allows it to be clearly seen as it is formed and fills up the gas jar.

PROCEDURE

Materials

- concentrated hydrochloric acid
- potassium manganate(VII)
- apparatus as shown in Figure 7.6
- a funnel
- two gas jars and covers
- Vaseline®
- eye protection
- gloves

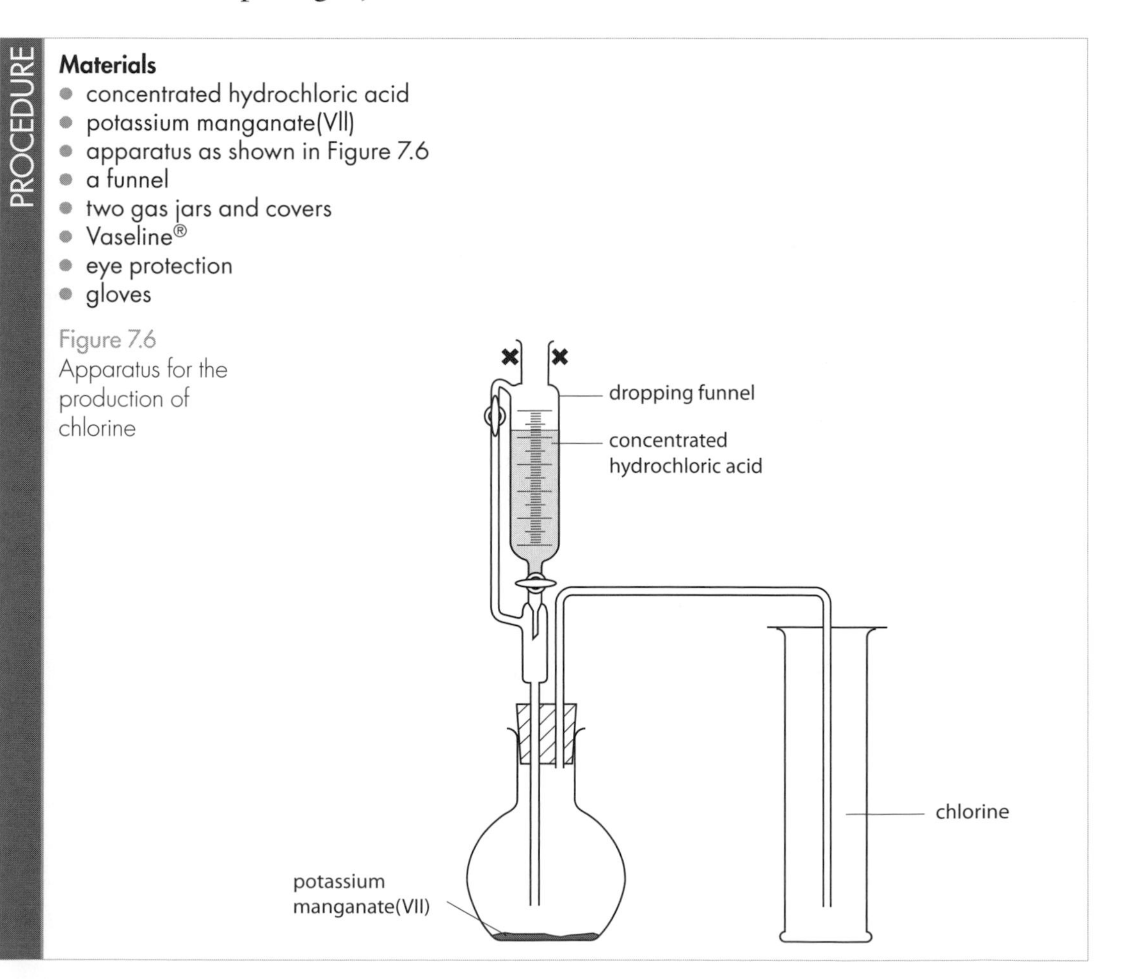

Figure 7.6 Apparatus for the production of chlorine

Safety

This practical must be carried out in a well ventilated fume cupboard. Chlorine gas is extremely dangerous if inhaled.

Potassium manganate(VII) is explosive when mixed with sulfuric acid. Ensure that you use hydrochloric acid.

Eye protection and gloves should be worn.

Method

1. Place a heaped spatula of potassium manganate(VII) into the bottom of the flask.
2. Carefully fit the dropping funnel and half-fill it with hydrochloric acid using a funnel. Remove the funnel.
3. Turn the tap on the dropping funnel to add concentrated hydrochloric acid (HCl) to the crystals of potassium manganate(VII), drop-wise. You will begin to see signs of yellowy-green chlorine gas forming in the flask and appearing in the bottom of the gas jar. Chlorine is denser than air and can be collected by downward delivery. Hold a white background behind the gas jar to see the green colour of the gas.
4. Put Vaseline on the lid of the gas jar and place it over the jar.

Discussion points

The hydrogen is 'taken away' from the hydrochloric acid by the oxidising agent potassium manganate(VII), leaving chlorine gas. The potassium manganate(VII) supplies oxygen which takes away the hydrogen to form water. The hydrochloric acid has lost hydrogen – we say that it has been oxidised.

$$16HCl + 2KMnO_4 \rightarrow 2KCl + 2MnCl_2 + 8H_2O + 5Cl_2$$

The chlorine generated can be used in an interesting reaction. The following demonstration is optional, but is useful to show students that it is not just oxygen that supports burning.

The reaction of sodium and chlorine – teacher demonstration

Students – especially those who have learnt the fire triangle – will assume that oxygen has to be present for something to burn. This demonstration challenges that assumption in a dramatic way. It helps to pave the way for the idea that we describe many reactions as 'oxidation' when they do not involve the reaction with oxygen.

PROCEDURE

Materials

- a gas jar filled with chlorine
- a dry, clean brick
- sodium, 3 mm cube
- a Bunsen burner or kitchen blow torch
- filter paper
- eye protection

Safety

This demonstration should be carried out in a fume cupboard. The demonstrator should wear goggles or a face shield; students should wear eye protection. The product should not be tasted.

Method

1 Clean any oil off the sodium using filter paper and place it on the clean dry brick. You need to ensure that there is enough room to place the gas jar over the top.
2 Heat the sodium using either the hottest part of a roaring Bunsen flame (the tip of the blue cone) or the kitchen blow torch until it catches fire.
3 Quickly remove the lid from the gas jar and place it over the sodium.
4 You will see a yellow sodium flame and clouds of white produced. If you leave the jar in place for a few minutes, the sodium chloride will settle out and there will be a clear circle of white, solid product – sodium chloride.

Possible questions and discussion points

Was the sodium burning? Yes it clearly was. Was there oxygen in the gas jar? No – it was full of chlorine. Depending on the level of the students, this could be presented as an addition to the fire triangle – either oxygen or chlorine are needed for something to burn – or as an example of oxidation (of sodium) by reaction with something other than oxygen. (In section 7.3 of this chapter, page 243, the burning of magnesium in carbon dioxide is described.)

■ Oxidising and reducing agents

Substances that do the oxidising are called oxidising agents. They are themselves reduced in the reactions. For example, in the thermit reaction discussed later in this chapter the iron oxide is the oxidising agent – it is the source of the oxygen which bonds to the aluminium.

Substances that do the reducing are called reducing agents. They are oxidised in the reactions. In the thermit reaction the aluminium is the reducing agent – it is removing the oxygen from the iron oxide and itself has been oxidised to aluminium oxide.

aluminium + iron oxide → aluminium oxide + iron

Students can be asked to identify the oxidising agents and the reducing agents in various reactions.

What happens to the electrons in redox reactions?

Chemists have a number of ways of categorising reactions and there are several models used to describe redox reactions. These are based on the following: adding or removing oxygen or hydrogen; the movement of electrons; or oxidation states. The model to be used is chosen according to which best suits the needs of the scientist at the time.

One of the redox models describes the movement of electrons. For example,

$$2Mg + O_2 \rightarrow 2MgO$$

can be split into:

$$Mg \rightarrow Mg^{2+} + 2e^-$$

and $O_2 + 4e^- \rightarrow 2O^{2-}$

In terms of electrons, the magnesium atoms lose two electrons each to become Mg^{2+} ions, and each oxygen gains two electrons to go from O_2 to $2O^{2-}$.

Mnemonics can be useful ways to help students to remember scientific concepts. In this instance, the mnemonic OIL RIG can help them remember that:

OXIDATION is LOSS of ELECTRONS

REDUCTION is GAIN of ELECTRONS

The separation of the oxidation and reduction reactions is seen most clearly during electrolysis reactions which are covered later in the chapter.

■ Application of ideas about redox reactions

In the following practical, chlorine and bromine solutions are added to potassium bromide and potassium iodide solutions. Students are required to observe any reactions and then explain what they have seen in terms of electron transfer.

PROCEDURE

Materials

- test tubes with bungs to fit
- teat pipettes
- chlorine water, 0.1% w/v
- bromine water, 0.3% w/v
- iodine solution ($I_2/KI_{(aq)}$)
- potassium bromide solution, 0.1 mol dm^{-3}
- potassium iodide solution, 0.1 mol dm^{-3}
- hydrocarbon, such as petroleum spirit (boiling range 120–160 °C) or cyclohexane, labelled 'hydrocarbon solvent'
- eye protection

Safety

Students should wear eye protection. Store all solutions in the fume cupboard.

Chlorine can cause severe damage to lungs and eyes.

Avoid inhaling any vapours.

Avoid any solutions coming into contact with skin.

Method

1. Put about 2 cm^3 of chlorine water in a test tube and add a small amount of hydrocarbon solvent. Shake the mixture and then allow it to settle.
2. Repeat the procedure using bromine water and iodine solution.
3. Note the distinctive colour of the hydrocarbon layer so that identification can be made easily in the following experiments.
4. Add about 1 cm^3 of chlorine water to an excess of a solution of potassium bromide. Then add about 1 cm^3 of chlorine water to an excess of a solution of potassium iodide.
5. Note any signs of reaction. Add a hydrocarbon solvent to each test tube, shake and allow the layers to settle. Examine the colours and use the results from part 3 to help identification.
6. Repeat the experiment using bromine water.

The students' tables of results should resemble Table 7.1.

Table 7.1 Results

	Potassium bromide solution	Potassium iodide solution
Chlorine water	Orange colour seen Orange colour seen when hydrocarbon added	Brown colour seen Purple colour seen when hydrocarbon added
Bromine water	No reaction	Brown colour seen Purple colour seen when hydrocarbon added

$$\text{chlorine} + \underset{\text{bromide}}{\text{potassium}} \rightarrow \underset{\text{chloride}}{\text{potassium}} + \text{bromine}$$

$$Cl_{2(aq)} + 2KBr_{(aq)} \rightarrow 2KCl_{(aq)} + Br_{2(aq)}$$

In terms of electron transfer:

$$Cl_{2(aq)} + 2e^- \rightarrow 2Cl^-_{(aq)} \quad \text{Gain of electrons so REDUCTION}$$

$$2Br^-_{(aq)} \rightarrow Br_{2(aq)} + 2e^- \quad \text{Loss of electrons so OXIDATION}$$

Chlorine molecules, Cl_2, accept the electrons from the bromide ions so chlorine is the oxidising agent and is itself reduced in the reaction.

Bromide ions, Br^-, give electrons to the chlorine molecules so they are the reducing agents and are themselves oxidised in the reaction.

Similarly in the other reactions:

chlorine + potassium iodide → potassium chloride + iodine

bromine + potassium iodide → potassium bromide + iodine

Cl_2 oxidises Br^- to Br_2
Cl_2 oxidises I^- to I_2
Br_2 oxidises I^- to I_2

Reducing agents give electrons and so become oxidised (OIL).
Oxidising agents accept electrons and become reduced (RIG).

For a video clip of the reaction of fluorine with solutions of the other halogens visit the Royal Society of Chemistry website: www.rsc.org

Depending upon the class, you can structure the discussion of these tests to best ensure students are thinking about the chemistry. For example, you could work through one example (such as the chlorine + potassium bromide example given previously) as a teacher-led whole-class discussion, and then set the students to work in pairs or small groups to work through the other examples. Finish the session with a plenary where you ensure everyone has written the correct equations and has identified the redox processes.

■ Balancing equations using electron transfers

When students are happy writing redox equations, they could be asked to balance equations of electron transfers. For example, the iron ions in iron(II) chloride solution are oxidised to iron(III) ions by passing chlorine through iron(II) chloride solution. We can look at the two parts of the reaction and write half equations to show what is happening.

Equation 1:

$$Fe^{2+} \rightarrow Fe^{3+} + e^-$$

iron(II) → iron(III)

This is loss of electrons so oxidation has taken place.

Equation 2:

$$Cl_2 + 2e^- \rightarrow 2Cl^-$$

This is gain of electrons so reduction has taken place.

To get a full equation we can add the two half equations together, but first we must make sure that the number of electrons in each half is the same. In this example we have only one electron in equation 1 but two electrons in equation 2. Equation 1 must be doubled so it becomes:

Equation 3:

$$2Fe^{2+} \rightarrow 2Fe^{3+} + 2e^-$$

Now we can add the two halves together (equations 2 + 3):

$$2Fe^{2+} + Cl_2 + 2e^- \rightarrow 2Fe^{3+} + 2Cl^- + 2e^-$$

The electrons can be cancelled out because they are the same on both sides.

$$2Fe^{2+} + Cl_2 \rightarrow 2Fe^{3+} + 2Cl^-$$

The equation is now balanced. Students could be challenged to balance other examples to ensure they have understood the principle.

Rusting

Rusting is a good example of a redox reaction that students will be familiar with. However, you may find that your students hold a number of misconceptions about the process, for example: rust acts as a protective layer so a metal nail will not be destroyed; rustiness exists in the air and attaches itself to a metal nail through some process; water and some other materials would produce rusting; rusting occurs because the water destroys a metal nail. It is important to shift students' thinking towards seeing rusting as a chemical change, in which iron is one reactant and iron compounds are the products.

Investigation of rusting

PROCEDURE

Materials

Each group of students will need:

- four test tubes fitted with corks
- eight iron nails
- a spatula
- a beaker, 100 cm^3
- anhydrous calcium chloride, small lumps
- cotton wool
- freshly boiled water/salt water
- oil
- distilled water
- eye protection

Safety

Eye protection should be worn.

Method

1. In test tube 1, partly cover two nails with distilled water and cork the tube (as a control).
2. In test tube 2, place a few pieces of anhydrous calcium chloride followed by a small plug of cotton wool and two nails. Cork the tube. The anhydrous calcium chloride absorbs water so these nails should be in a completely dry atmosphere.
3. In test tube 3, place two nails and cover them with water that has been boiled for several minutes to expel the dissolved air. Cover the surface of the water with some oil to form an air-proof layer. Cork the tube. These nails should be without oxygen.
4. In test tube 4, partly cover two nails with aqueous salt (sodium chloride) solution and cork the tube. See Figure 7.7 and Table 7.2 for a summary of the four test conditions.
5. Leave the tubes for several days.

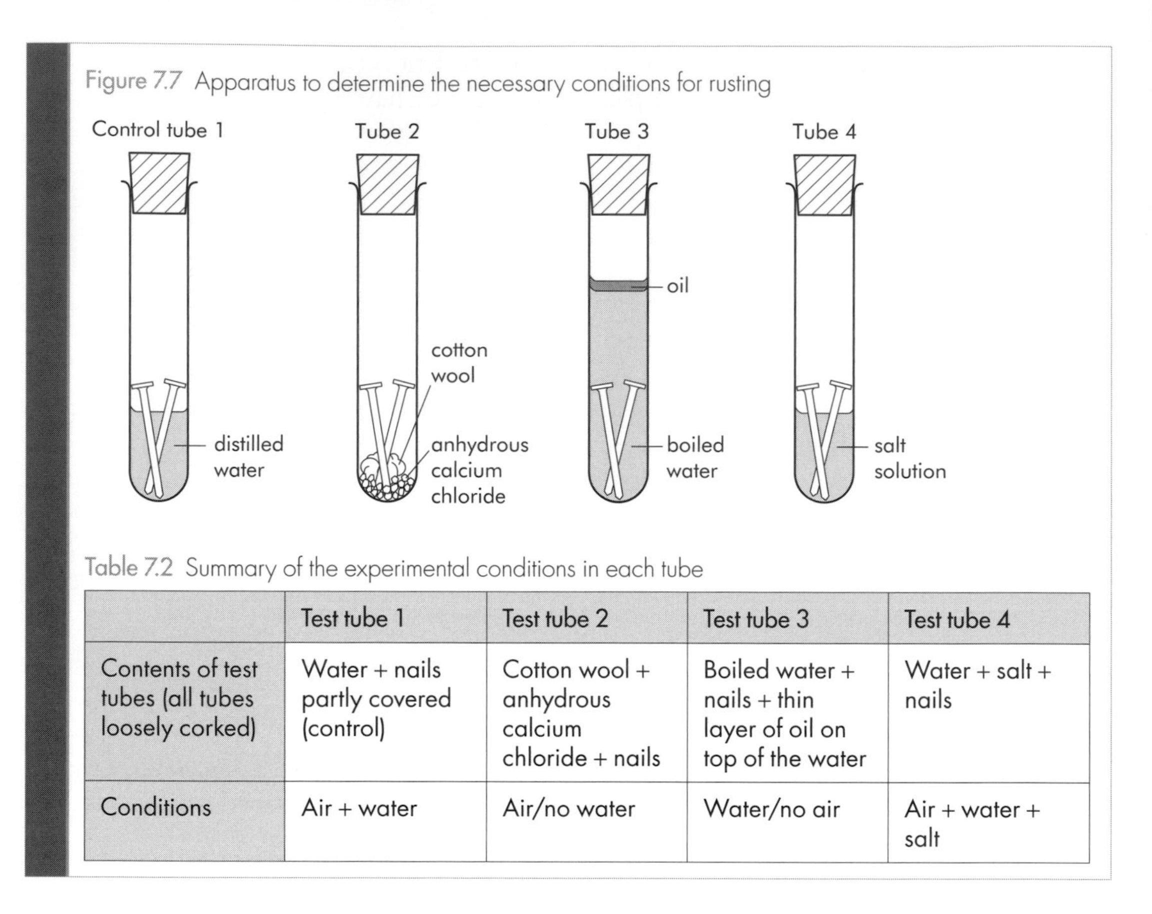

Figure 7.7 Apparatus to determine the necessary conditions for rusting

Table 7.2 Summary of the experimental conditions in each tube

	Test tube 1	Test tube 2	Test tube 3	Test tube 4
Contents of test tubes (all tubes loosely corked)	Water + nails partly covered (control)	Cotton wool + anhydrous calcium chloride + nails	Boiled water + nails + thin layer of oil on top of the water	Water + salt + nails
Conditions	Air + water	Air/no water	Water/no air	Air + water + salt

Students should observe that the nails in the salty water and exposed to the air rusted the most quickly followed by the ones with both oxygen (air) and water. The others should not rust – but in reality it will depend on how carefully the test tubes were set up. They should certainly rust far less. This allows students to see that water and air/oxygen are required for rusting and that salt speeds it up.

A couple of additional rusting experiments are given here.

■ How much air is used during rusting?

PROCEDURE

Materials

- a test tube
- a beaker (100 cm^3)
- a ruler
- iron wool

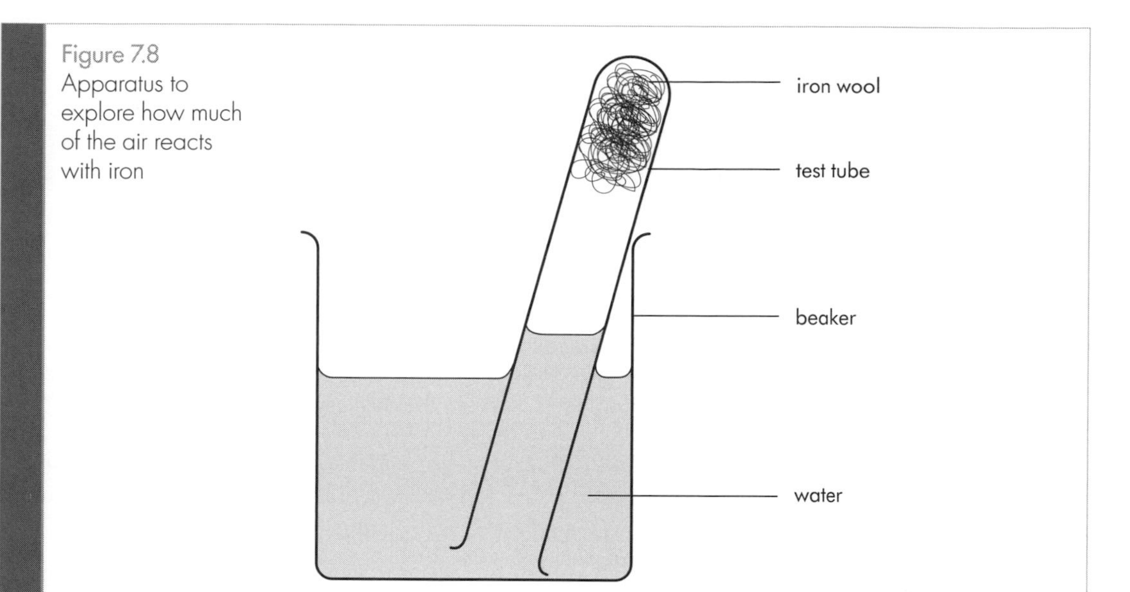

Figure 7.8 Apparatus to explore how much of the air reacts with iron

Method

1 Put about 3 cm depth of iron wool into the test tube and wet it with water. Tip away excess water.
2 Put about 20 cm^3 of water into the beaker. Invert the test tube and place it in the beaker of water as in Figure 7.8. Measure the length of the column of air with the ruler and mark the level on the tube using a grease pencil (china marker).
3 Leave the apparatus for at least a week. Measure the new length of the column of air, taking care not to lift the test tube out of the water.

Discuss the results with students, making sure that they understand that rusting is an oxidation reaction of iron with oxygen:

$$\text{iron} + \text{oxygen} \rightarrow \text{iron oxide}$$

(This is not the full story, however, as the formation of rust is a complex process.)

From their two measurements for the length of the column of air – before and after rusting takes place – students should be able to calculate the percentage of the air which has been removed by the rusting reaction. This should be about 20 %, which is approximately the percentage of oxygen in the air.

You could ask students how they could show that the reaction is complete. They may suggest leaving it for another week or so to see if any further air is used up. The iron is present in excess in this experiment, so it will not all rust.

Oxygen is needed for rusting

PROCEDURE

Materials
- two test tubes fitted with rubber bungs
- four nails
- distilled water
- cylinders of oxygen and nitrogen

Method
1. Boil some distilled water for a few minutes to expel the dissolved air.
2. Place two nails in each of two test tubes and half cover them with boiled water.
3. Pump nitrogen into one of the tubes from a cylinder through a delivery tube dipping into the water until it is thought all the air is displaced. Remove the delivery tube and quickly close the test tube with a bung that fits well.
4. Repeat the procedure for the other tube with oxygen gas and then leave the tubes for a week.

Results
The tube with the oxygen will have rusty nails, but the tube with the nitrogen will not.

■ Preventing rusting

This experiment, detailed on the Practical Chemistry website, can be set up in pairs or as a whole class and, by using corrosion indicator, the results can be seen within a lesson.
www.nuffieldfoundation.org/practical-chemistry/preventing-rusting

7.3 Reactivity series of metals

Previous knowledge and experience

Students are likely to have some knowledge of chemical reactions and may have met the idea of reactivity. Depending on the order in which you approach this topic, they might have met the concept of redox reactions. The reactions in the reactivity series are a good way to either introduce or reinforce redox.

Students will probably have studied acids and alkalis (covered in Chapter 6) but may need reminding. They may need to be taught how to test for hydrogen with a glowing splint.

Students' misconceptions

The misconceptions that are most often encountered are those relating to chemical reactions (Chapter 4) and writing equations (covered earlier in section 7.2). Work on the reactivity series provides an excellent opportunity for students to encounter a wide range of chemical reactions and gain a better understanding of exactly what a chemical reaction is. Encouraging students to come up with their own explanations for what is going on, prior to discussing the topic as a class, will help lead them to a better understanding and move their thinking forward.

A teaching sequence

A possible teaching sequence for the reactivity series is given here:

- The reactions of metals with acids
- The reactions of metals with water – for the group 1 metals Li, Na and K and group 2 metals Ca and Mg
- The reaction of magnesium with steam
- The reactions of metals with oxygen
- Reactions between a metal and another metal compound
- Using the reactivity series to make predictions about reactions

This is often split across different years with teaching about the idea of the reactivity series (the first four points above) coming earlier than reactions between a metal and another metal compound.

The advantage of starting with reactions of acids (which can often be done by students) before moving on to the more spectacular reactions of groups 1 and 2 with water or steam (which are usually teacher demonstrations) is that students are not shown the more exciting reactions before being asked to perform the less dramatic ones themselves.

Putting metals in order of reactivity with acids

PROCEDURE

Materials

- samples of the following metals: magnesium, zinc, iron, copper
- dilute hydrochloric acid ($0.4\,mol\,dm^{-3}$)
- test tubes
- splints
- eye protection

Safety

All students should wear eye protection.

Method

1. Pour 5 cm depth of acid into each of four test tubes.
2. Put a small amount of a different metal into each of the test tubes.
3. If the reaction is very slow, add another spatula end of the relevant metal powder. (Only allow students to add further quantities of metal if you are sure they will be responsible. If you cannot be sure of this, reserve this step for a demonstration in the plenary discussion.)

4 Note the evolution of hydrogen gas and test for the gas with a lighted splint.
5 Compare the different rates at which the bubbles form and show this on a diagram (such as Figure 7.9).
6 If this practical is done later in the teaching sequence, students could predict the positions of, and complete the diagram for, calcium and gold.

Figure 7.9 Expected results when adding different metals to dilute acid solution

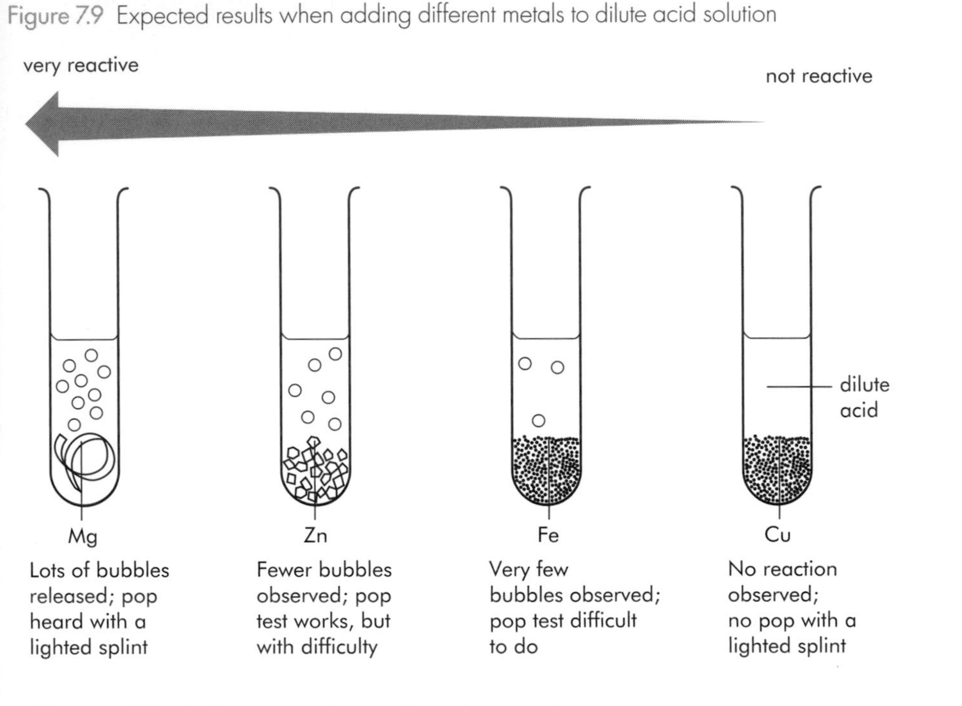

Table 7.3 Table of observations for metal–acid practical

Metal	Observations	Equations
Magnesium	Very vigorous with ribbon	$Mg + 2HCl \rightarrow MgCl_2 + H_2$
Zinc	Quite fast with granules once it gets going; much faster with powder	$Zn + 2HCl \rightarrow ZnCl_2 + H_2$
Iron	Slight reaction with iron foil or iron nail; steady reaction with powder	$Fe + 2HCl \rightarrow FeCl_2 + H_2$
Copper	No reaction	

Reactions with dilute sulfuric acid are similar. The reaction of copper with concentrated sulfuric acid can be demonstrated by the teacher in a fume cupboard. Goggles and chemical resistant gloves must be worn. Heat 1 g of copper turnings with 1 cm^3 sulfuric acid – sulfur dioxide will be evolved. To dispose of the mixture, place cooled test tubes into a bowl of cold water.

■ Putting metals in order of reactivity with water

Group 1 metals with water

For sodium and potassium, the reaction must be carried out as a teacher demonstration.

For calcium and lithium, depending on the age and ability of the class, the reaction may be a class experiment (see page 231).

PROCEDURE

Materials

- a large trough
- a Perspex cover
- a plastic safety screen
- a scalpel
- tweezers
- a white tile
- filter paper
- fresh, clean, small quantities of lithium, sodium and potassium in small sample bottles of paraffin oil
- universal indicator solution
- washing-up liquid

Safety

Eye protection must be worn by all students.

Lithium, sodium and potassium are very corrosive and flammable and should only be picked up with tweezers and never touched with a bare hand. Gloves should be worn.

Safety screens must be used and students should be 2–3 m from the experiments.

Teachers must wear eye protection and gloves, and a lab coat is advisable.

Method

1. Half-fill the trough with water.
2. Remove a sample of the lithium from the oil with tweezers and blot off the oil using filter paper.
3. Cut a rice-grain-sized piece.
4. Place the sample on the water in the trough.
5. Repeat using sodium and then potassium.
6. Make a list of similarities and differences noticed between the three metals.

Notes

- The metals will whizz around the surface reacting with the water. They can stick to the side of the trough and this can cause them to 'pop', spitting hot metal oxide out of the trough. To help prevent this, put one drop of washing up liquid into the water to reduce the surface tension. It is also a good idea to put a Perspex sheet over the top of the trough to prevent anything escaping from it.
- It is important to remove all oil from the sample.
- Visibility can be a problem for students, especially in a large class. Putting the trough on an old overhead projector helps to overcome this and a visualiser or other camera will also give an excellent view.
- Universal indicator can be added at the start – in which case you will see trails of purple through the green liquid – or at the end. For the best results, change the water between each metal as this allows a visual comparison of how quickly a basic solution is produced.
- To show students that the metal is shiny, it can be placed into a dry Petri dish, with the lid taped on, and it can then be passed around the class. The metals can also be shown to be conductors of electricity using a circuit tester. This can help students to accept that these strange substances are, indeed, metals, as they are not like any metals that they have met before.

- The gas can be collected from the lithium reaction. Take a test tube and place it over the top of the lithium. This will constrain the lithium in the trough, underneath the test tube. Collect the gas released and test it to show that it is hydrogen. When the hydrogen burns, look out for the red colour of the flame showing the presence of some lithium ions. Do not attempt this for sodium or potassium. The potassium will spontaneously catch fire anyway and the lilac flame characteristic of potassium will be seen.

Observations

Observations to be made could include:

- All the metals are kept under oil to prevent reaction with air and water.
- All the metals are soft – note any trend in softness.
- They are all shiny – this is typical of metals. Will they conduct electricity?
- They float on water.
- There is a visible trend in melting.
- They all fizz – gas is given off.
- Some of the metals produce coloured flames. Note the colours produced.
- All the solutions at the end turn purple with universal indicator.

Table 7.4 Comparing lithium, sodium and potassium

Property	Lithium	Sodium	Potassium
Cut with a knife	Fairly soft	Soft	Very soft
Floating on water	Floats	Floats	Floats
Shiny	Very shiny inside corroded outer coat	Very shiny inside	Very shiny inside; corrodes quickly
Heat given out to melt the sample	Does not melt as not enough heat given out	Sample melts into a ball	Sample melts into a ball very quickly
Fizzing – evolution of hydrogen gas	Sample does not skate about the surface and the gas given off does not self-ignite	Sample skates about the surface of the water, hydrogen is evolved but (probably) does not self-ignite	Sample skates around on the surface; enough heat is evolved to set fire to the hydrogen which burns with a lilac flame due to the potassium present
Universal indicator solution	Goes purple – alkaline	Goes purple – alkaline	Goes purple – alkaline
Reactivity	Least reactive	Reactive	Most reactive

The equation for the reaction between sodium and water is:

$$2Na_{(s)} + 2H_2O_{(l)} \rightarrow 2NaOH_{(aq)} + H_{2(g)}$$

Similar equations can be written for lithium and potassium.

For video clips of the reactions of rubidium and caesium see the Royal Society of Chemistry website: www.rsc.org

For a video tutorial in carrying out this demonstration see: www.rsc.org/Education/Teachers/Resources/Practitcal-Chemistry/Videos/index.asp

PROCEDURE

Reaction of calcium with water: class practical or teacher demonstration

Materials

- a test tube
- a large beaker
- calcium metal
- tweezers
- universal indicator solution
- splints

Safety

Calcium should only be handled with tweezers. All students should wear eye protection.

Method

1. Half fill a large beaker with water.
2. Invert a test tube full of water into the beaker.
3. Put a small piece of calcium into the water.
4. The calcium will sink and give off bubbles. Collect the gas given off in the test tube.
5. Remove the test tube and test the gas with a lighted splint.
6. Add universal indicator solution to the beaker and note the colour.

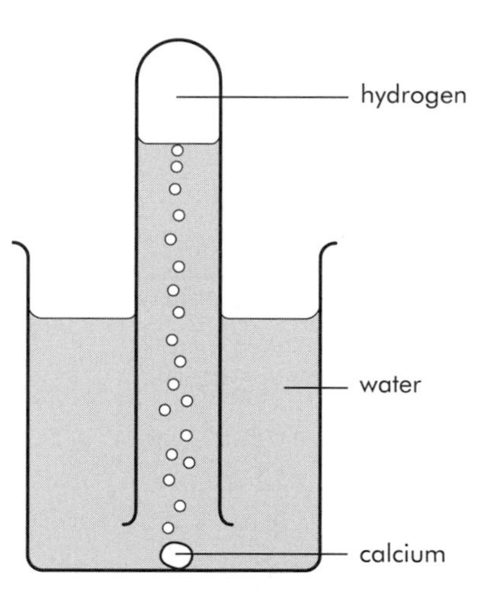

Figure 7.10 Apparatus for collecting hydrogen from the reaction of calcium with water

Results

Calcium is less reactive than lithium, sodium and potassium. The bubbles of gas that are given off displace the water in the test tube. When tested with a lighted splint, the gas gives a 'pop' and so it is hydrogen. Students can also note that the water becomes 'milky' and might be asked to think what this may mean in terms of solubility of the product formed. (The product of this reaction is less soluble than the products of the lithium, sodium and potassium reactions.) Universal indicator solution is added to the water to show that the product is alkaline. The equation for this reaction is:

$$Ca + 2H_2O \rightarrow Ca(OH)_2 + H_2$$

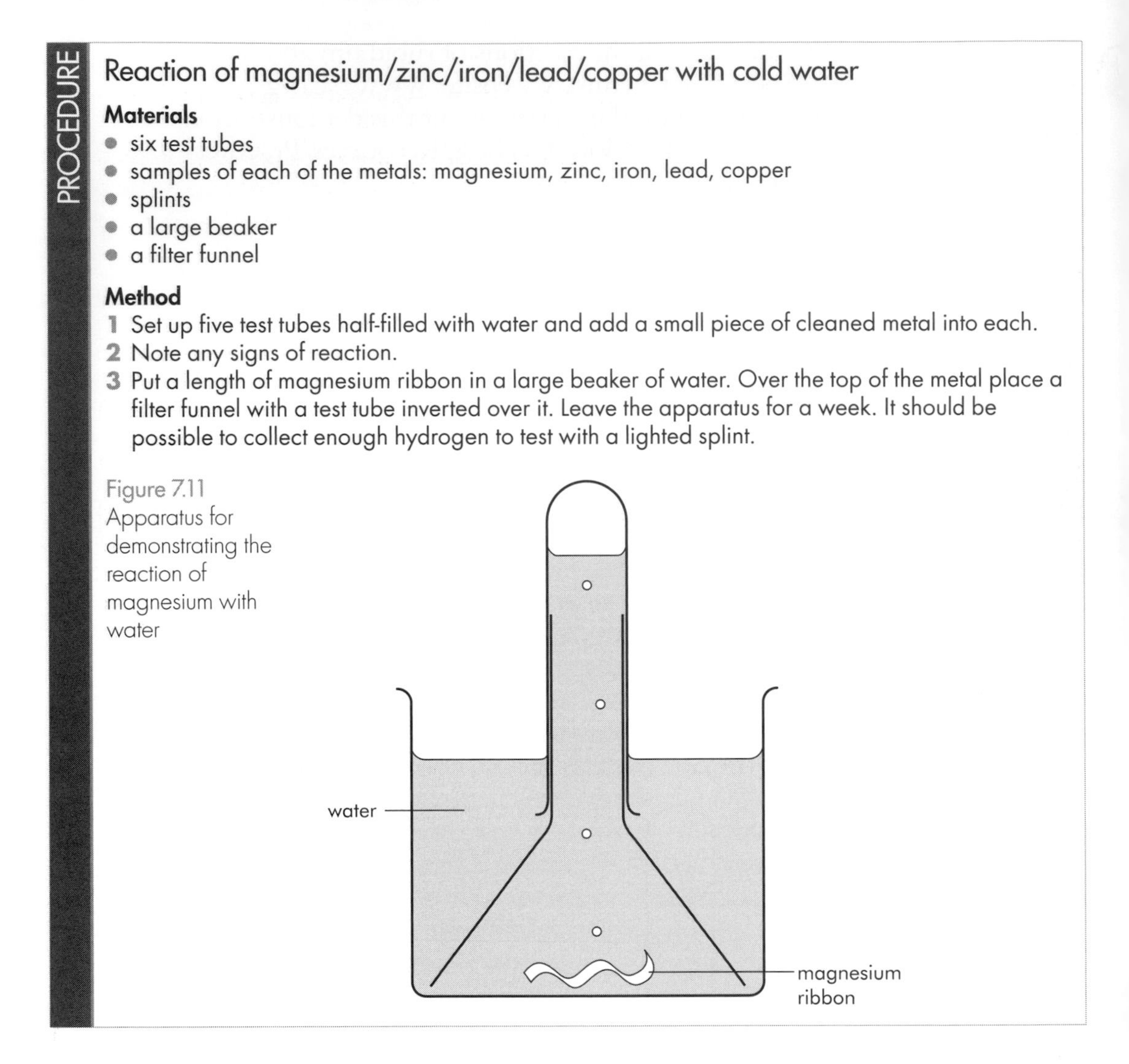

PROCEDURE

Reaction of magnesium/zinc/iron/lead/copper with cold water

Materials

- six test tubes
- samples of each of the metals: magnesium, zinc, iron, lead, copper
- splints
- a large beaker
- a filter funnel

Method

1. Set up five test tubes half-filled with water and add a small piece of cleaned metal into each.
2. Note any signs of reaction.
3. Put a length of magnesium ribbon in a large beaker of water. Over the top of the metal place a filter funnel with a test tube inverted over it. Leave the apparatus for a week. It should be possible to collect enough hydrogen to test with a lighted splint.

Figure 7.11 Apparatus for demonstrating the reaction of magnesium with water

Reaction of magnesium with steam – teacher demonstration

Although the reaction of magnesium with cold water is very slow, if the magnesium is heated and steam is passed over it, the reaction is much faster (see the discussion of reaction rates in Chapter 5).

PROCEDURE

Materials

- a hard glass boiling tube fitted with a bung and a small piece of glass delivery tube
- magnesium ribbon, about 20 cm
- mineral wool
- a Bunsen burner
- heatproof mats
- a clamp stand
- a safety screen
- splints

PROCEDURE

Safety

All students should wear eye protection. The reaction should be carried out behind a safety screen.

Magnesium will burn vigorously and will damage the tube – it may cause it to break. If the tube breaks, stop heating. Make sure the tube you use for this experiment is undamaged at the start.

Warn students not to look directly at the flame. Looking through partially opened fingers in front of the eyes or blue glass offers some protection.

Method

1 Soak the mineral wool with water and place it at the bottom of the boiling tube to a depth of about 3 cm.
2 Coil the magnesium around the outside of the boiling tube and then place it inside, ensuring good contact between the magnesium and the inside of the tube. Insert the bung as shown in Figure 7.12.
3 Clamp near the bung and tilt the tube slightly downwards so water does not flow onto the magnesium.
4 Heat the metal not the mineral wool until the reaction starts, then quickly flick the Bunsen to the mineral wool so that steam passes over the hot magnesium.
5 Apply a light to the end of the glass tubing.

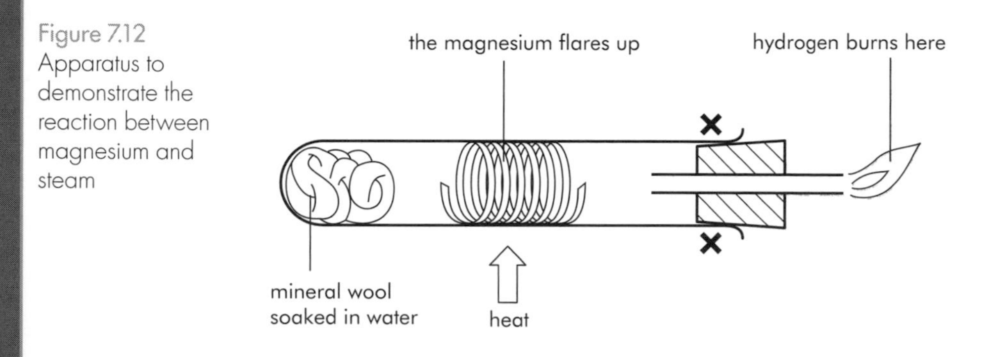

Figure 7.12 Apparatus to demonstrate the reaction between magnesium and steam

Results

The magnesium burns very brightly with a white flame and turns to a white powder. Hydrogen gas is given off in a large enough quantity to burn at the end of the delivery tube.

Note that the magnesium gets so hot that it may actually react with the boiling tube, forming magnesium silicide.

To dispose of the apparatus safely, place the cold boiling tube in a large bowl of water to destroy the magnesium silicide. This may ignite due to the production of silanes which combust in air.

Alternative version to collect the hydrogen

Using the apparatus shown in Figure 7.13, the hydrogen can be collected over water. Magnesium can be used for this reaction. Iron or zinc could also be used, but because they are less reactive than magnesium, less hydrogen is produced.

PROCEDURE

Materials
- a hard glass boiling tube
- a delivery tube and bung
- a 'Bunsen' valve
- a large beaker
- magnesium ribbon, zinc powder or iron powder (whichever is being tested)
- tests tubes with bungs to fit
- mineral wool
- a Bunsen burner
- a clamp stand
- heatproof mats

Safety

Eye protection should be worn by all students. The 'Bunsen' valve is used to prevent suck back.

If you stop heating then the delivery tube must be removed from the water or suck back will take place.

Method

1 Set up the apparatus as shown in Figure 7.13.

2 The details for conducting the experiment are exactly as in the previous experiment, except the gas is now collected over water.

Results

Zinc glows white hot and produces a powder which is yellow when hot and white when cold. A few test tubes of hydrogen will be produced.

Iron will show a slight change in colour during the reaction and some gas will be given off – it may be less than a test tube.

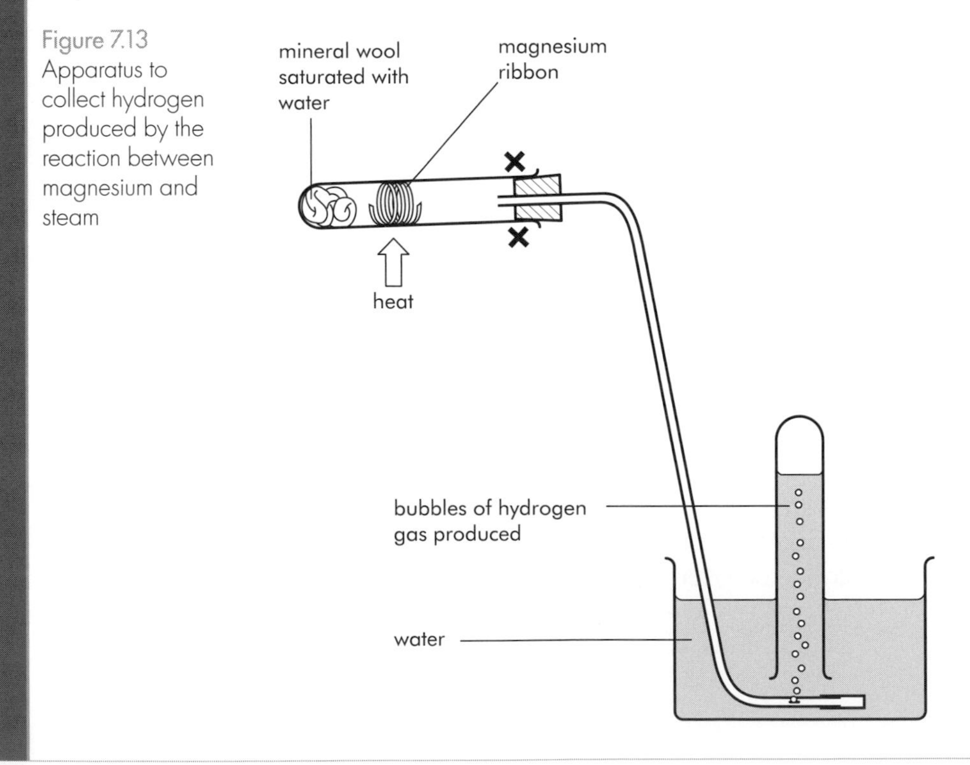

Figure 7.13 Apparatus to collect hydrogen produced by the reaction between magnesium and steam

Putting metals in order of reactivity with oxygen – teacher demonstration

PROCEDURE

Materials

- a combustion spoon to fit a gas jar
- gas jars full of oxygen and lids to fit (read Note below)
- tweezers
- filter paper
- a Bunsen burner
- universal indicator solution
- distilled water
- samples of the following elements: sodium, calcium, magnesium, iron wool, copper strip (see Notes on the metals below)

Safety

All students must wear eye protection.

Sodium must not be touched – use tweezers and remove all oil on the surface by pressing sample between filter paper.

Note on filling gas jars

Gas jars can be filled with oxygen by using a cylinder or by generating the gas in a reaction and collecting the gas over water. For details of the generation see: www.nuffieldfoundation.org/practical-chemistry/generating-collecting-and-testing-gases

Method

1. Fill a gas jar with oxygen (see Note above).
2. Take a sample of a metal and place it on a combustion spoon.
3. Hold it in the Bunsen flame until it has either caught fire or is red hot.
4. Place it quickly into the gas jar.
5. As soon as it has finished burning, remove the combustion spoon and replace the lid of the gas jar.
6. Note what you have seen.
7. Add a few cm^3 of distilled water and shake well, noting if the oxide dissolves or not.
8. Add a few drops of universal indicator solution and note the result.
9. Repeat with the other samples.

A table of results can be drawn up and patterns identified.

Notes on the metals to use

Sodium – usual precautions of working with sodium (do not touch, remove oil with filter paper and use a very small amount, such as a rice-grain-sized piece); a glass combustion spoon will give a cleaner result

Calcium – very difficult to ignite; may be better in metal tongs

Magnesium – look at the reaction through blue glass or with partially opened fingers in front of the eyes

Iron wool – plunge into oxygen when glowing

Copper strip – will not burn; just blackens

Results

Table 7.5 Table of expected results

Metal	Amount	Appearance of reaction	Appearance of product	Solubility in water	pH with UI solution
Sodium	Rice-grain size	Brilliant yellow flame	White marks on gas jar (must be clean spoon); product may remain in spoon	Dissolves completely	pH 11
Calcium	Small piece	Orange flame	White solid formed on the spoon or tongs	Dissolves but white suspension seen	pH 10
Magnesium	Ribbon, 5–10 cm	Brilliant white flame	White powder mainly on the spoon	Dissolves but forms a white suspension	pH 10
Iron	Iron wool	Bright glow but no flame	Surface of iron wool blackened	No change	pH 7
Copper	Foil strip	Glows	Black on surface	No change	pH 7

Silver and gold do not react with oxygen.

Defining and using the reactivity series

From these experiments it is possible to put the metals in order of reactivity depending on how reactive they are with water, steam, oxygen and acid. Some metals are not used in all the reactions because of safety reasons.

The reactivity series is as follows:

Potassium — very reactive
Sodium
Lithium
Calcium
Magnesium
Aluminium
Zinc
Iron
Lead
Copper
Silver
Gold — not reactive

Chemists can use the reactivity series to answer questions. Students can be encouraged to predict what they expect to happen in reactions and to explain what they observe.

1 Can metals be used to remove the oxygen from the oxides of other metals?
  The thermit reaction is an interesting demonstration of this (see page 238). A class experiment (page 239) illustrates the same point.
2 Can metals be displaced from solutions of their salts?
  There is a class experiment (page 240) and a teacher demonstration (page 242) to illustrate this point.
3 Can carbon be put into the reactivity series?
  Although carbon is not a metal, it is very useful to see if it can be fitted into the reactivity series of metals. A class experiment (page 242) and a teacher demonstration (page 243) demonstrate where it goes.
4 Can the reactivity series be used to help extract metals?
  Section 7.4 develops this idea.

■ 1 Can metals be used to remove the oxygen from the oxides of other metals?

The practical work described here concerns 'displacement' or 'competition' reactions. This language, and the terminology used here about the 'removal' of oxygen, reflects common ways that chemists think about, and discuss, certain types of chemical reactions. Students will meet this type of language in their textbooks and in other resources they refer to, but this might be a good context to emphasise the motto proposed in Chapter 4 – 'reacting with, not reacting to'. That is, these reactions concern the interactions at the submicroscopic level of two different substances: not one active chemical agent acting on another more passive substance.

The thermit reaction – teacher demonstration

The thermit reaction demonstrates the reaction between aluminium powder and iron(III) oxide. It must be carried out as a teacher demonstration

PROCEDURE

Materials

- safety screens
- a 1-litre beaker
- sand
- filter paper, 12 cm diameter
- a pipe clay triangle
- a tripod
- heatproof mats
- aluminium powder, 3 g
- iron(III) oxide, 9 g, thoroughly dried by heating in an evaporating dish over a Bunsen burner
- two pieces of paper
- magnesium powder and ribbon
- sandpaper
- a Bunsen burner with long enough pipe or a kitchen blow torch
- a magnet attached to a stick

Safety

- Safety screens should be used.
- The demonstrator should wear eye protection and a lab coat (it can become quite messy at the end). Students should be at least 4 m away and be wearing eye protection.
- The demonstrator should be able to get away from the reaction once it has started.
- There have been occasional explosions using mixtures similar to this. The quantities should not be exceeded.
- Some teachers have had accidents when performing this outside due to unexpected breezes catching the sparks.
- The bench should be clear of combustible materials and protected with mats or a sheet of hardboard.
- Students should not look directly at the burning magnesium but should cover their eyes with their fingers spread apart or look through blue glass.

Method

1. Fill a beaker one-third full of sand and add another one-third of water.
2. Fold two pieces of filter paper and place them in a pipe clay triangle on a tripod over the beaker (as shown in Figure 7.14).
3. Place the iron oxide and aluminium powders onto a piece of paper. Pour from one piece to another until a uniform mixture is produced.
4. Place this mixture into the filter paper.
5. Make a depression in the pile of powder and put a little magnesium powder in the depression. Sprinkle a little extra magnesium powder over the top of the pile.
6. Insert a piece of freshly cleaned magnesium ribbon (10 cm) through this pile to act as a fuse.
7. Light the magnesium ribbon with the tip of the blue cone of a Bunsen flame or with a kitchen blow torch. Stand back.
8. Observe the extremely vigorous reaction that takes place. The molten iron will fall through the filter paper and be cooled by the water.
9. Use a magnet to extract the lump of iron from the beaker.

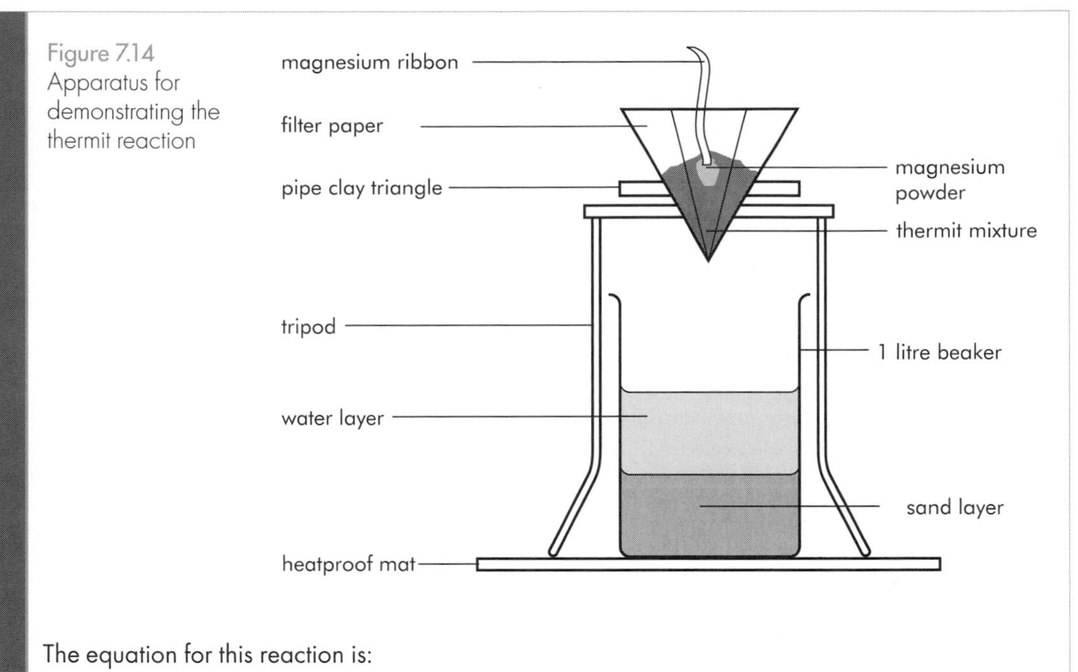

Figure 7.14 Apparatus for demonstrating the thermit reaction

The equation for this reaction is:

aluminium + iron oxide → aluminium oxide + iron

$2Al + Fe_2O_3 \rightarrow Al_2O_3 + 2Fe$

Iron oxide has lost oxygen so it has been reduced. The more reactive aluminium has removed the oxygen from the less reactive iron. This reaction is highly exothermic (Chapter 5) and is used in situ to produce the molten iron needed to weld railway lines together.

Class practical to investigate whether metals can be used to remove the oxygen from the oxides of other metals

PROCEDURE

Materials

- zinc oxide
- copper(II) oxide
- iron filings
- a tin lid
- test tubes
- a crucible or beer bottle top (see Note)
- dilute hydrochloric acid

Note

This experiment uses several crucibles which schools may find exceeds provision. Beer bottle tops are fine for this as long as the plastic inserts are removed before use. This will also tend to reduce the amount of material consumed.

Students can be given the chemicals pre-mixed in equal masses. Alternatively, if they are given them separately, they can be told to place the chemicals on a piece of paper and pass them back and forth to another piece of paper until they are well mixed.

Method

Before the experiment, ask students to predict the outcome of reacting iron filings with zinc oxide and copper(II) oxide and write word equations.

1 Mix together zinc oxide with an equal volume of iron filings. Place the mixture in a crucible or on a bottle top.
2 Heat the mixture gently at first and then more strongly.
3 Observe any signs of reaction, such as a glow spreading through the mixture or a change of colour.
4 Repeat the experiment with copper(II) oxide and iron filings.
5 After the reaction has ended and cooled down, transfer the product to a test tube and add some warm dilute hydrochloric acid to see if there are any signs of copper having been formed.

Results

The more reactive metal ends up bound with the oxygen which it removes from the less reactive metal, so only the copper(II) oxide and iron filings should react. A similar reaction is that between zinc and copper oxide. This is best done as a teacher demonstration and the details can be found at www.nuffieldfoundation.org/practical-chemistry/reaction-between-zinc-and-copper-oxide

■ 2 Can metals be displaced from solutions of their salts?

This series of reactions can be used in a number of slightly different ways. Students can predict what they expect to happen based on the reactivity series and observe to see if their predictions hold. This can be used as a reintroduction to the reactivity series, with students observing where reactions happen and using this to draw up a reactivity series of the metals they are using. It is also useful for reinforcing the idea of careful observation as some of the results are quite difficult to see.

A visualiser gives great results. Zinc, in particular, produces marvellous crystals which work really well on a large screen. It can also be useful to show the whole class interesting aspects of an individual group's work.

PROCEDURE

Materials

- a spotting tile with at least 16 depressions (or two smaller tiles)
- a dropping (teat) pipette
- a felt-tip pen or other means of labelling
- access to about 5 cm^3 of each of the following 0.1 mol dm^{-3} metal salt solutions: zinc sulfate, magnesium sulfate, copper(II) sulfate, lead(II) nitrate
- five samples, approximately 1 cm lengths or squares, of the following metals: zinc foil, magnesium ribbon, copper foil, lead foil

Safety

All students should wear eye protection.

Method

1 Using a dropping pipette, put a little of the zinc nitrate solution in four of the depressions in the spotting tile, using Figure 7.15 as a guide.
2 Label this row with the name of the solution.
3 Rinse the pipette well with water afterwards.

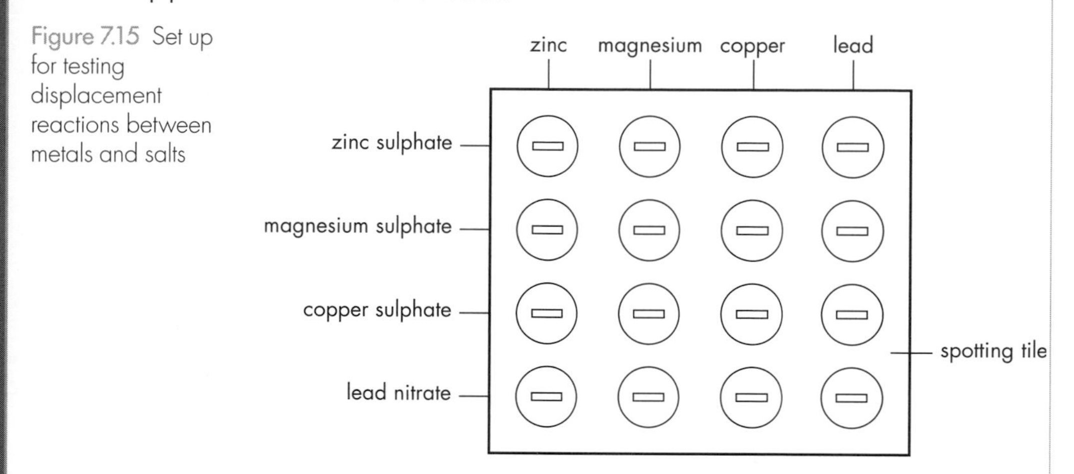

Figure 7.15 Set up for testing displacement reactions between metals and salts

4 Do this for each solution in turn, rinsing the pipette when you change solution.
5 Put a piece of each metal in each of the solutions, using Figure 7.15 as a guide.
6 Over the next few minutes observe which mixtures have reacted and which have not.

(Note that the magnesium will appear to react with the magnesium nitrate and perhaps the zinc with the zinc nitrate. For this reason, you may wish to direct students not to mix the metal with its own metal solution.)

The equation for the reaction between magnesium and copper nitrate is:

magnesium + copper nitrate → magnesium nitrate + copper

$$Mg + Cu(NO_3)_2 \rightarrow Mg(NO_3)_2 + Cu$$

This can be done as half equations to show oxidation and reduction:

$Mg \rightarrow Mg^{2+} + 2e^-$ Oxidation

$Cu^{2+} + 2e^- \rightarrow Cu$ Reduction

The more reactive metal removes the other ion from the less reactive metal, so displacing it from the solution.

Displacing silver from a silver salt – teacher demonstration

PROCEDURE

Materials
- 15 cm of thick copper wire, about 22 swg
- silver nitrate solution, 0.1 $mol dm^{-3}$
- a test tube, 100 × 16 mm, with a cork/bung

Safety
Wear eye protection and keep silver nitrate off the skin.

Method
1. Fill two-thirds of the test tube with silver nitrate solution.
2. Make a coil with the copper wire by winding it around a pencil but leave 5 cm at the end straight.
3. Place the coil in the silver nitrate solution, then bend the straight part over the top of the tube to prevent the coil from moving about.
4. Cork the tube and leave it undisturbed.

Results
Crystals of silver will start developing within a few minutes on the copper wire. After several hours there will be masses of crystals and the background solution will be blue due to the copper(II) ions in solution.

$$Cu_{(s)} + 2AgNO_{3(aq)} \rightarrow 2Ag_{(s)} + Cu(NO_3)_{2(aq)}$$

■ 3 Can carbon be put into the reactivity series?

Although carbon is not a metal, it is very useful to see if it can be fitted into the reactivity series of metals.

Competition between copper oxide and carbon

PROCEDURE

Materials
- a small, hard, glass test tube (ignition tubes)
- a dry teat pipette
- a spatula
- copper(II) oxide
- carbon (dry powdered wood charcoal)
- lime water

Safety
All students should wear eye protection.

Method
1. Put a spatula of copper(II) oxide into the ignition tube.
2. Carefully add a spatula of charcoal powder on top of the copper oxide without any mixing.
3. Heat the ignition tube strongly for about 5 minutes.
4. Optional: remove a sample of the gas above the heated solid and test it by bubbling it through a small amount of lime water. This can be done by squeezing an empty, dry teat pipette above the sample while it is being heated. The gas collected in this way can be squirted through about 1 cm^3 lime water.
5. Allow to cool and then look closely at where the powders meet in the tube.
   You should be able to see a line of copper clearly formed at the interface.

Competition between magnesium and carbon – teacher demonstration

PROCEDURE

Materials
- a gas jar and greased cover
- magnesium ribbon
- tongs or a combustion spoon
- a supply of carbon dioxide

Safety
All students should wear eye protection.

Method
1. Fill a gas jar with carbon dioxide and seal it with a greased cover.
2. Attach a piece of magnesium ribbon to a combustion spoon or hold it in tongs and ignite.
3. As soon as it ignites, plunge it quickly into the gas jar of carbon dioxide.
4. After it has stopped burning, examine the sides of the gas jar.

Results
A white solid – magnesium oxide – will be seen on the sides of the gas jar and black specks of carbon.

magnesium + carbon dioxide → magnesium oxide + carbon

This experiment is useful for challenging students' perceptions of what is burning, and of using carbon dioxide as a fire extinguisher. They will not tend to expect the result and may need convincing that this is really what happened. (Earlier in the chapter, section 7.2 page 218, an example of burning in chlorine was described.)

These experiments and similar ones allow us to place carbon below zinc and above iron in the reactivity series (page 236).

7.4 Extraction of metals

Metals are found in ores in the Earth's crust. Ores are rocks that contain minerals from which metals can be extracted. Most metals are too reactive to exist on their own in the Earth's crust. Some metals exist as metal oxides, such as iron oxide and aluminium oxide, some as carbonates, such as copper carbonate, and some are sulfides, including zinc sulfide. A few metals are not found combined with any other element because they are not reactive – they occur native. The best example of this is gold.

The experiments in the previous sections enabled common metals to be put in order of their reactivity. The reactivity series we developed can be used in this section to enable students to make predictions about extraction of metals and make observations in the practicals carried out.

Previous knowledge and experience

Students are likely to have come across ideas about extraction of metals in history lessons. For example, iron has been used since prehistoric times and the start of the Iron Age in 2500BC when it was first extracted from its ore. In geography lessons, students will have discussed the abundance of natural resources such as ores and may have studied the blast furnace.

In the course of this topic, students will have investigated the reactions of metals with air, oxygen, water and acids, so they will be familiar with the reactivity series that can be drawn up from such experiments (see page 237). They may have covered the idea that a metal or carbon can be used to remove oxygen from another metal. They may or may not have seen some electrolysis.

A teaching sequence

The way in which a metal is extracted from its ore depends upon the position of the metal in the reactivity series. It may take time to convince students that the production of a metal from its ore involves the reverse of the reactions that they have studied in the reactivity series sequence. In particular, a key idea here is the concept that the more reactive the metal, the harder it is to extract it from its ore. In other words, the more easily it loses electrons to become ions, the more difficult it is to reverse the process by adding electrons to those ions to form the metal.

A lot of energy is transferred from the chemical store when reactive metals react, and a lot of energy has to be transferred into the chemical store to extract them. Much less energy is transferred when unreactive metals react or when they are extracted from their compounds.

Students should look at the extraction of:

- a less reactive metal
- an unreactive metal
- a reactive metal.

Students can then be asked to use the reactivity series to predict how various metals might be extracted from their ores. The metals at the top of the list are very reactive and give up their electrons easily to form positive ions which are very stable. These metals lose electrons easily, therefore they are good reducing agents and they are themselves oxidised in the process. They include potassium, magnesium and zinc.

The metals at the bottom of the list are not very reactive and do not give up their electrons very easily to form positive ions. They are not good reducing agents. They include copper, silver and gold.

Although not a metal, it is extremely useful to have carbon in the list (just below zinc) as it is a good and cheap reducing agent.
The main methods for extracting metals are:

1 electrolysis (using electricity) – for the metals at the top of the reactivity series such as potassium, sodium, calcium, magnesium and aluminium
2 reduction by carbon or carbon monoxide (produced from carbon) – for the metals lower in the reactivity series
3 roasting the ores in air at a high temperature – for the least reactive metals
4 reacting the ore with a more reactive metal such as sodium – this is only used for a few metals, such as titanium, as it is extremely expensive.

Converting the metal ions in an ore to the metal is reduction, whatever the process. This involves gain of electrons.

$$M^{n+} + ne^- \rightarrow M_{(s)}$$

These ideas may be covered later in lower secondary or in upper secondary, but will need prior knowledge of the reactivity series.

Extraction of less reactive metals

Some of the less reactive metals can be extracted by reducing their ore with carbon. Iron is the most common example of this, but the following practical uses this method to extract lead.

■ A model 'blast furnace' to extract lead

PROCEDURE

Materials
- lead(II) oxide, PbO, 1 g
- carbon powder, 0.2 g
- a blowpipe, sterilised in Milton® fluid and rinsed
- a beer bottle top, plastic insert removed
- a Bunsen burner, tripod, gauze, heatproof mat

Safety
Pregnant women should be aware of the issues with lead compounds.
Demonstrate how to heat from above using a Bunsen burner before students try this.

Method

1 Mix together the lead oxide and carbon powder – the exact quantities are not vital.
2 Moisten the mixture slightly with a little water and mix into a paste to reduce the amount blown into the air.
3 Two people are needed for this procedure. The first person holds the Bunsen burner above the mixture and aims a blue flame onto the lead oxide and carbon paste. The second blows through the flame with the blowpipe into the lead oxide mixture. Beads of lead metal will be seen to appear.

If you prefer not to use lead or blow pipes then the experiment entitled 'Competition between copper oxide and carbon' from Section 7.3 (page 242) could be used instead or the simple extraction which follows here.

■ Extraction of iron on a match head

PROCEDURE

Materials

- non-safety/'strike anywhere' matches (these usually have pink ends)
- tongs (crucible tongs)
- a weighing boat (small white plastic ones are ideal)
- a spatula
- Petri dishes or watch glasses
- a Bunsen burner
- a heatproof mat
- a magnet (such as a bar magnet) – preferably covered in cling film
- iron(III) oxide powder
- sodium carbonate powder
- water

Method

It is easiest to provide small amounts of the powders in Petri dishes or watch glasses and the water in a small beaker. Several students can share the chemicals.

1 Dip the head of a match in water to moisten it.
2 Roll the damp match head first in sodium carbonate powder, then in iron(III) oxide powder.
3 Hold the match in a pair of tongs. Put the head of the match into a blue Bunsen flame (air hole open). The match will flare and burn. Do not allow the match to burn more than half-way along its length.
4 Allow the match to cool for about 30 seconds.
5 Use a spatula to crush the charred part of the match into a small plastic weighing boat.
6 Move a magnet around under the weighing boat. Some of the small particles will move around in the weighing boat following the track of the magnet. Do not dip the magnet into the particles directly, unless you have first wrapped the magnet in cling film – any pieces of iron will stick to the magnet and will be difficult to clean off.

The sodium carbonate can confuse students. It is not the source of the carbon (that is the wood of the match) but rather fuses (melts) easily, allowing the chemicals to come into contact with each other.

Large-scale extraction of iron

Full details of the blast furnace are readily available in chemistry textbooks. The essential details of the process are:

- The process takes place in the blast furnace.
- The raw materials – iron ore (iron oxide), coke and limestone – are put into the top of the blast furnace.
- The reducing agent is carbon monoxide formed by the following process:

Hot air is blown into the lower part of the furnace and the coke burns to form carbon dioxide.

$$C_{(s)} + O_{2(g)} \rightarrow CO_{2(g)} \text{ (this reaction is exothermic)}$$

The carbon dioxide moves up through the furnace and reacts with more coke.

$$CO_{2(g)} + C_{(s)} \rightarrow 2CO_{(g)}$$

This carbon monoxide reduces the iron oxide.

$$Fe_2O_{3(s)} + 3CO_{(g)} \rightarrow 2Fe_{(l)} + 3CO_{2(g)}$$

The molten iron, which is very impure, falls down the furnace. The main impurity in the iron ore is silica (sand) and this is now removed using the limestone by the following reactions:

$$CaCO_{3(s)} \rightarrow CaO_{(s)} + CO_{2(g)}$$

The CO_2 produced here is blown up the furnace and reacts to produce more $CO_{(g)}$.

$$CaO_{(s)} + SiO_{2(s)} \rightarrow CaSiO_{3(l)}$$

Silica, SiO_2, is an acidic oxide (non-metal oxide) and reacts with calcium oxide, which is a basic oxide (metal oxide), to form calcium silicate (a salt, known as slag).

The slag runs down to the bottom of the furnace and floats on top of the iron. Both the iron and the slag are run off from the bottom of the furnace. For a video clip about the extraction of iron see: www.rsc.org/Education/Teachers/Resources/Alchemy

Enhancement ideas

Students could find out about:

- purification of the cast iron produced from the blast furnace
- the production of steel and its uses
- the extraction of zinc: comparing it to the extraction of iron, noting the similarities and differences
- uses of zinc.

Extraction of unreactive metals

Extraction of copper

The main ore of copper is a sulfide. The ore is purified by crushing, froth flotation to remove impurities, drying and roasting to produce copper. The impure copper is purified by electrolysis.

Enhancement ideas

Students could research:

- the extraction of copper in greater detail
- the uses of copper
- the extraction of silver.

Purification of copper by electrolysis

When copper is extracted, in a process similar to that happening in the blast furnace, it is very impure. To purify the copper, it is used as the anode in a very large electrolysis cell containing copper sulfate solution as the electrolyte and a pure copper rod as the cathode. During the process, the anode dissolves and pure copper is deposited on the cathode. The impurities settle at the bottom of the cell. A simplified version of this process can be carried out as a class practical.

PROCEDURE

Materials

- a beaker
- a carbon rod
- a piece of copper foil
- two cables with crocodile clips at both ends
- a power pack, 6–8 V
- copper(II) sulfate solution, 1 mol dm^{-3}

Safety

Copper(II) sulfate is harmful if swallowed and may be irritating to the skin.

Do not touch the power pack with wet hands.

Wear eye protection when carrying out electrolysis.

Method

1. Pour about 2 cm depth of copper(II) sulfate solution into the beaker.
2. Put in place a carbon rod as the cathode (negative) and a piece of copper foil as the anode (positive) and connect, via crocodile clips and cables, to a power pack (Figure 7.16).
3. Pass a current through the circuit at a voltage of around 6–8 volts.
4. After about 3 minutes, switch off and remove the carbon rod; it should be coated with copper.
5. If you replace the cathode, reconnect the circuit and pass a current for about 15 minutes, eventually the anode will 'dissolve' away.

Figure 7.16 Purifying copper by electrolysis

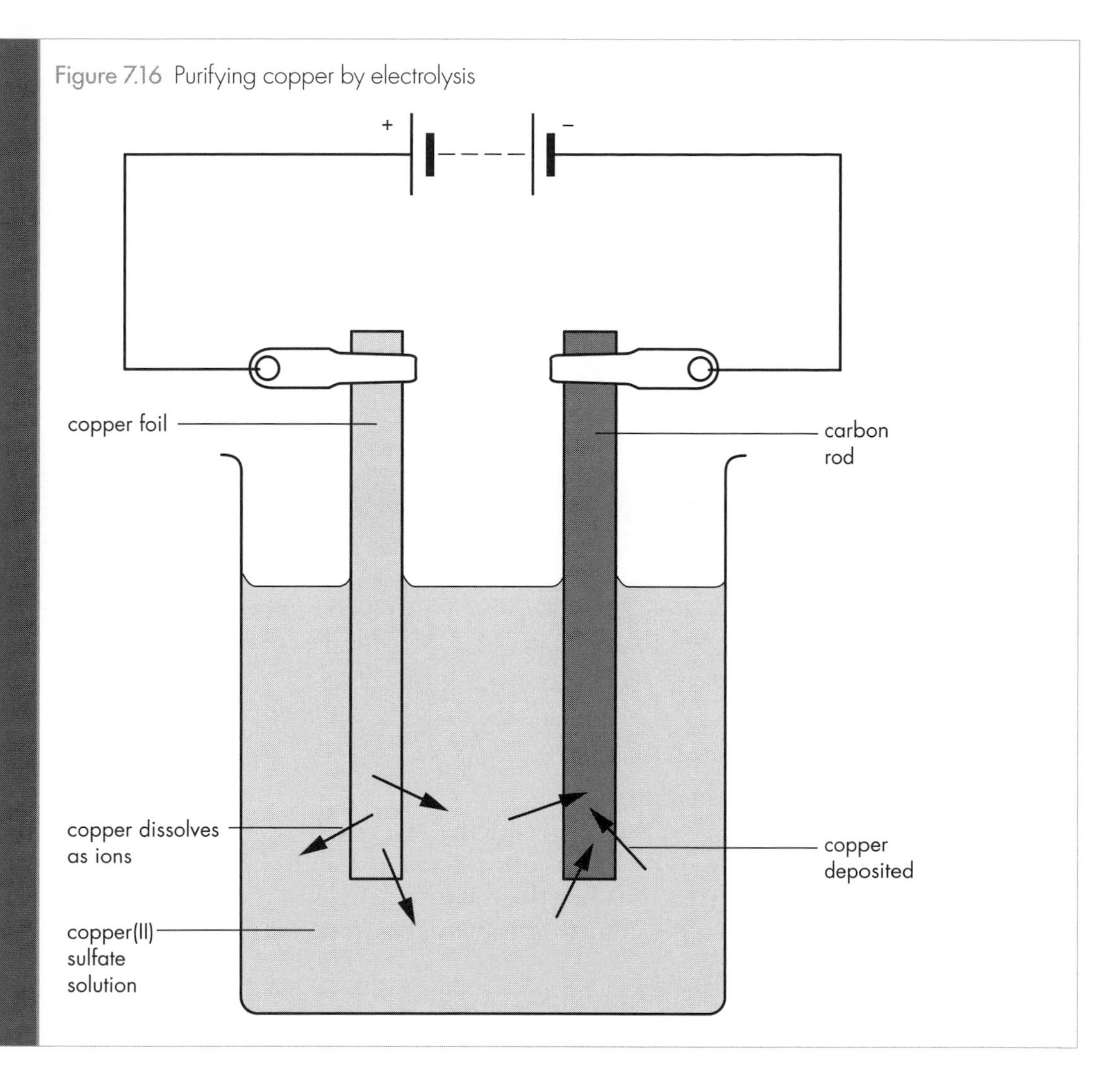

Electrolysis offers a good way of showing students that during a redox reaction, electrons are gained by one chemical and lost by another. As the two reactions happen at different places in a cell, it can be easier to see the distinction between the two. Electrolysis is discussed in greater depth in Chapter 8.

At the cathode:

$$Cu^{2+} + 2e^- \rightarrow Cu \text{ (reduction)}$$

At the anode:

$$Cu \rightarrow Cu^{2+} + 2e^- \text{ (oxidation)}$$

For a video clip on the purification of copper see: www.rsc.org/Education/Teachers/Resources/Alchemy

Extraction of reactive metals

Extraction of aluminium by electrolysis

Aluminium is generally extracted from its ore – bauxite (aluminium oxide, $Al_2O_3.2H_2O$) – by electrolysis. Bauxite has a high melting point so it is dissolved in molten cryolite (Na_3AlF_6) which gives a conducting solution that is molten at a lower temperature. Full details of the extraction and the electrolysis cell can be obtained from any standard chemistry textbook. The essential details are: electrode material – carbon graphite; ions present – Al^{3+} and O^{2-}.

Table 7.6 Electrolysis of aluminium

At the cathode (–)	At the anode (+)
$Al^{3+} + 3e^- \rightarrow Al_{(l)}$ The molten aluminium collects at the bottom of the electrolysis cell and is tapped off. This is reduction.	$2O^{2-} \rightarrow O_{2(g)} + 4e^-$ The oxygen formed reacts with the carbon anode to form carbon dioxide and so the anodes have to be replaced regularly. This is oxidation.

Enhancement ideas

As an extension activity, students can research:

- why, although aluminium is the most abundant metal in the Earth's crust, it is not the most widely used metal
- why and when aluminium was able to be extracted in larger quantities
- the uses of aluminium
- the environmental problems associated with the location of aluminium plants
- the process of anodising
- the extraction of sodium and calcium
- recycling aluminium.

For a video clip about the extraction of aluminium, together with some focus questions, see:
www.rsc.org/Education/Teachers/Resources/Alchemy

Other resources

Websites

For video clips of experiments which would be difficult to do in the classroom, such as the reactions of rubidium and caesium with water, see:
www.rsc.org

There are a number of 'how to' videos made by the Royal Society of Chemistry. These are designed to help teachers to do these experiments for themselves and include the thermit reaction and reactions of the alkali metals:
www.rsc.org

These are alternatively available on YouTube:
www.youtube.com/playlist?list=PL477054A3DD895431

The Practical Chemistry website has loads of ideas for practical activities for a wide range of age groups and covering many different topics:
www.nuffieldfoundation.org/practical-chemistry

The BBC Learning Zone is an excellent source of video clips which can be used to liven up lessons and put work in a wider context. There are several clips about chemical reactions and about the reactivity of metals and the implications in the real world.
www.bbc.co.uk/learningzone/clips

The book *Classic Chemistry Demonstrations* is published online by the Royal Society of Chemistry. It contains 100 demonstrations and is available free in pdf format:
www.rsc.org/Education/Teachers/Resources/Books/CCD.asp

The RSC-produced project *Alchemy?* comprises downloadable material covering 15 major topics, including aluminium extraction and copper refining:
www.rsc.org/Education/Teachers/Resources/Alchemy

Stuff and Substance is a resource from SEP with video clips and animations of chemical reactions. It is available on the National STEM centre website:
www.nationalstemcentre.org.uk/elibrary/collection/520/stuff-and-substance-teachers-resources

■ Resources aimed at students

Catalyst magazine is aimed at 14–16 year olds studying science and includes articles about a number of topics relating to redox and metal extraction. Several years of archived issues can be found here: www.nationalstemcentre.org.uk/elibrary/collection/567/catalyst
The magazine can also accessed from: www.catalyststudent.org.uk

For information on the purification of copper, see: www.bbc.co.uk/schools/gcsebitesize/science/add_aqa/ions/electrolysisrev3.shtml

And for electrolysis in general, see: www.bbc.co.uk/schools/gcsebitesize/science/add_aqa/ions/electrolysisrev1.shtml

Another electrolysis site with some simple animations can be found at:
www.s-cool.co.uk/gcse/chemistry/electrolysis

Electrolysis, electrolytes and galvanic cells

Georgios Tsaparlis

8.1 Electrolysis

Previous knowledge and experience

On commencing study of this topic, students should be comfortable with writing and balancing equations for chemical reactions (covered in greater depth in section 3.2 of Chapter 3). Knowledge of ions is also central. Concepts of basic electricity, such as electric current and potential difference (a concept many students struggle with), are essential. Redox reactions (covered in Chapter 7) enter the study of electrochemistry.

A teaching sequence

The teaching sequence I propose starts with a context: the production of the very common and useful metal aluminium by electrolysis. Context enters again at the end, with the use of Faraday's laws to calculate the amount of electricity and the cost of producing the aluminium that is needed for making a beverage can. This in turn is connected with the importance of recycling aluminium.

The teaching methodology suggested follows an active learning approach. It starts with observations and facts and then goes on to build particulate and structural mental models of electrolytes and the electrolysis phenomenon. Symbolic models in the form of chemical notation of ions and equations are then introduced.

Finally, although in this chapter I discuss galvanic cells after electrolysis, the reverse order is also possible.

The extraction of aluminium by electrolysis

Electrolysis is involved in some major industrial processes, including the extraction of a number of very reactive metals from their ores, such as potassium, sodium, calcium, magnesium and aluminium. In addition, very pure hydrogen can be produced from the electrolysis of water.

Aluminium is an extremely useful metal in our everyday lives – its alloys exhibit a high degree of durability and elasticity. Students may be familiar with the use of aluminium in the construction of spacecraft, aircraft and car engines, car wheels, external doors and windows. In addition, aluminium foil is commonly used in our homes.

Aluminium is the most abundant metallic element in the Earth's crust (8.3% by weight), but it is never found in free, metallic form. Aluminium is a stronger reducing agent than carbon, so direct reduction with carbon, as is used to produce iron, is not chemically possible (see section 7.3 of Chapter 7). For this reason, aluminium is extracted by electrolysis of alumina. Alumina is a white powder that is obtained by refining bauxite, a metal ore containing 40–60 % aluminium oxide or alumina, Al_2O_3. Because bauxite has a high melting point, it is dissolved in molten cryolite (Na_3AlF_6) which gives a conducting solution that is molten at a lower temperature. Al_2O_3 is reduced to the pure metal by electrolysis (see section 7.4).

The operational temperature of the reduction cells is around 950–980 °C. The economic cost of the process is considerable: the transferred electrical energy per kilogram of produced aluminium is 13–16 kWh (about 50 MJ). The two electrolysis half reactions are:

$$Al^{3+} + 3e^- \rightarrow Al$$

$$2O^{2-} \rightarrow O_2 + 4e^-$$

Figure 8.1 Bauxite

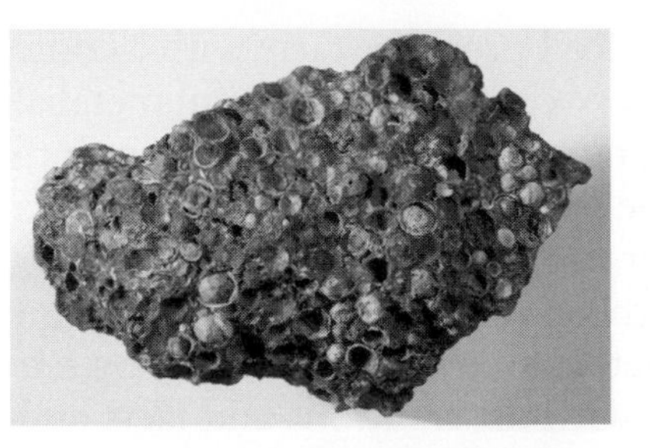

Electrical conduction

Students should be aware from everyday life about distinguishing materials as electrical conductors and insulators. An added feature of metallic conduction is that no chemical changes are taking place at the same time as current flows. From physics, students should know that the electric current in metallic conductors is the flow of electrons through the metal, which is caused by an electric potential difference (an electric field) across the conductor.

Electrolytes provide a new example of electrical conduction in which chemical changes (chemical reactions) are occurring at the two electrodes. In electrolytes, the electric charge carriers are not discrete electrons but positive and negative ions (cations and anions).

Educational research has shown that students are often confused about the nature of electric current both in metallic conductors and in electrolytes (assuming, for example, that current always involves drifting electrons, even in solution). Misconceptions have also been detected in identifying the anode and the cathode, its sign and its function in electrolytic cells. Students need to remember that an oxidation half reaction occurs always at the anode and a reduction half reaction occurs always at the cathode.

Electrolysis in terms of flow of charge by moving ions

To demonstrate electrolysis of a molten compound

PROCEDURE

A demonstration of the electrolysis of a molten ionic binary compound can be carried out by the teacher in a fume cupboard. Suitable compounds include lead bromide ($PbBr_2$, melting temperature 373 °C) or lead iodide (PbI_2, melting temperature 402 °C).

Materials

- a porcelain crucible
- two carbon (graphite) electrodes
- an external current supply
- an electric bulb or an ammeter
- lead bromide or lead iodide (solid state)
- a propane burner

Safety

Be aware that the process will release unpleasant vapours. Bromine vapour is very toxic and irritating to eyes and throats. Iodine vapour is an irritant, irritating the eyes and mucous membranes. Students should be kept at a safe distance from the procedure. (Note that at usual temperatures bromine is a liquid, and iodine is a solid.)

Method

1 Place the compound to be electrolysed (either $PbBr_2$ or PbI_2) inside a porcelain crucible.
2 Set up the electrolytic cell by inserting the two carbon electrodes into the crucible, making sure that they touch the compound.
3 Connect the two electrodes with the positive and the negative terminals of the power supply.
4 Complete the circuitry by connecting/inserting in series an electric bulb or an ammeter.
5 Turn on the power supply, adjusting the voltage in the power supply to a value higher than 2 V. Students should observe that nothing happens.
6 Then heat the outside of the crucible carefully with the propane burner until the compound melts. The lighting of the bulb or the reading on the ammeter indicates the passing of current.
7 Switch off the burner and leave the compound to solidify again. Students should note that the bulb will eventually go off.

Discussion point

Students should be able to explain what they have observed by making reference to the fact that the compound only conducted electricity when molten, and that this was because the electrical carriers (the ions) became mobile in the liquid state.

Students can also watch these experiments on the internet. As an alternative to $PbBr_2$ or PbI_2, sodium hydroxide (NaOH, melting temperature 318.4 °C) can be used, in which case the products are, in addition to sodium metal, oxygen gas and water ($4OH^- - 4e^- \rightarrow O_2 + 2H_2O$). This has the advantage of not producing the unpleasant vapours released by the two lead compounds. Sodium chloride (NaCl) is unsuitable for laboratory electrolysis as it has a melting temperature of 801 °C, so the flame of the burner is inadequate to melt it. (In the commercial Downs cell for the electrolysis of molten NaCl, a 3 : 2 mixture by mass of calcium chloride, $CaCl_2$, and NaCl is used. This has a melting temperature of 580 °C.)

Figure 8.2 shows an idealised cell for the electrolysis of molten sodium chloride. The dotted vertical line in the centre represents a diaphragm that keeps the chlorine gas produced at the anode from coming into contact and reacting chemically with the sodium metal generated at the cathode.

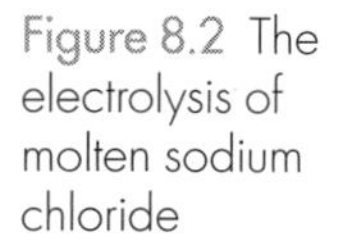

Figure 8.2 The electrolysis of molten sodium chloride

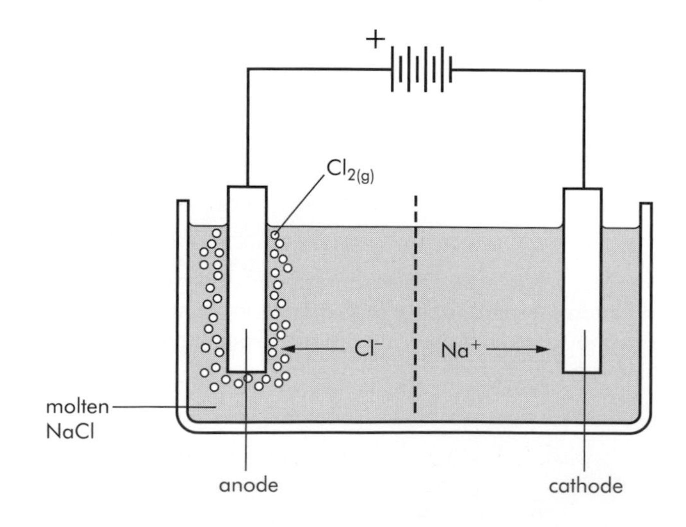

The electrode which is connected to the positive terminal of the power supply is termed the positive (+) electrode or anode, while the electrode which is connected to the negative terminal of the power supply is termed the negative (–) electrode or cathode. The positive and negative terminals of the supply are often coloured red and black (or blue), respectively.

■ The electrolysis of water

Figure 8.3 depicts a special electrolytic cell, called a Hoffmann voltameter, which demonstrates the famous electrolysis of water experiment. Note that pure water (a covalent compound – see Chapter 4) does not conduct electricity, so cannot be readily electrolysed. In the demonstration an electrolyte, such as sulfuric acid or a sodium hydroxide aqueous solution, is added to the water to allow electrolysis to occur at an observable rate. The oxygen and hydrogen evolved can be tested in the usual ways. The observations from this experiment provide data that support many chemical models, such as the composition of water and its H_2O formula.

Figure 8.3 The Hoffmann voltameter

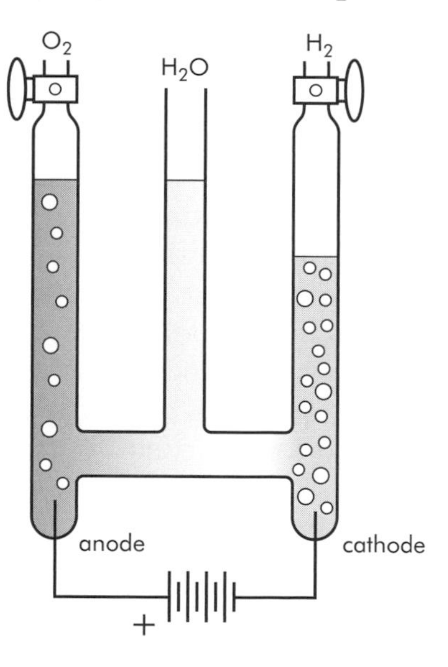

If sulfuric acid is used as the electrolyte in the electrolysis of water, the energetically favoured half reaction at the anode is provided by water itself and the hydrogen ions from the acid:

at the anode (+) $2H_2O_{(l)} \rightarrow O_{2(g)} + 4H^+_{(aq)} + 4e^-$

at the cathode (–) $2H^+_{(aq)} + 2e^- \rightarrow H_{2(g)}$

Note that the hydronium ion, H_3O^+, is the formula preferred by many chemists for the solvated (hydrated) hydrogen ion, $H^+_{(aq)}$.

Because, however, there is no experimental evidence supporting the existence of H_3O^+ in aqueous solutions, it is advisable to stick with $H^+_{(aq)}$. (See also section 6.4 of Chapter 6.)

If sodium hydroxide is used as the electrolyte, the energetically favoured half reactions are provided by water itself and the hydroxide ions from the base:

at the anode (+): $4OH^-_{(aq)} \rightarrow O_{2(g)} + 2H_2O_{(l)} + 4e^-$

at the cathode (–): $2H_2O_{(l)} + 2e^- \rightarrow H_{2(g)} + 2OH^-_{(aq)}$

In both cases, the overall reaction corresponds to the decomposition of water to its component elements:

$$2H_2O_{(l)} \rightarrow 2H_{2(g)} + O_{2(g)}$$

Students need to understand that not every liquid can be electrolysed. Pure water, pure sulfuric acid, pure ether and an aqueous sugar solution cannot be electrolysed, hence they are not electrical conductors. However, an aqueous solution of sulfuric acid, molten sodium chloride and an aqueous solution of sodium chloride can undergo electrolysis, hence they are electrical conductors.

A question that some students ask is why, while water is not an electrical conductor, it is dangerous to expose damp hands or in general a damp body, to mains electricity, for example in a bathroom. The explanation is that dry skin has a resistance of 100 000 ohms or more, while wet skin may have a resistance of only 1000 ohms. Such a low resistance of the wet skin allows current to pass into the body more easily and give a greater electric shock. Note also that pure water should be contrasted with natural water (such as tap water), which has various substances dissolved in it, among which are electrolytes. Hence natural water has some electric conductivity, which depends on the concentration of the water in various electrolytes. Electrical conductivity measurements of natural waters are used to predict the salinity, major solute concentrations, and total dissolved solids concentrations of natural waters.

Teachers need to be aware of two different uses of the term 'electrolyte'. In the strict sense, an electrolyte is a *liquid that can undergo electrolysis*. This can be a single substance, as in the case of a molten salt, or a solution. The most typical electrolytes are the aqueous solutions of salts (in general of ionic compounds), of acids, and of bases. By extension, we also call electrolytes the pure substances (solid, liquid, or gaseous) that, when dissolved in water, provide liquid electrolytes. Some biological substances (such as DNA or polypeptides) and synthetic polymers (such as polystyrene sulfonate) contain multiple charged functional groups and their dissolution leads to electrolyte solutions; these are termed polyelectrolytes.

We cannot tell an electrolyte from a non-electrolyte (in the 'melted' or in the solution phase) by just looking at them. There must then be something intrinsic in an electrolyte that makes possible the conduction of electric current in an electrolysis experiment. Although we cannot see the mechanism of conduction, we can speculate about it – in other words, we will construct a model.

■ The electrolytic dissociation model

It is known that each pure liquid substance has a characteristic boiling temperature (boiling point) that identifies it (as discussed more fully in Chapter 2). When a solution is created by dissolving a solid substance (the solute) into this liquid substance (the solvent), the phenomenon of boiling point elevation is demonstrated. For a given solvent and for solutes that are non-electrolytes (for example sugar with water as solvent), the boiling point elevation depends only on the concentration of the solution, but it is independent of the solute. In the case of solid solutes that are electrolytes (for example sodium chloride or calcium chloride with water as solvent), boiling point elevation is also observed, but its magnitude is much higher than in the case of a non-electrolyte solution of the same concentration – grossly a double value in the case of sodium chloride and a triple value in the case of calcium chloride. The chemical formulae tell us that sodium chloride 'consists of two species; one species of sodium and one species of chlorine', while calcium chloride 'consists of three species; one species of calcium and two species of chlorine'.

These findings, along with others, led the Swedish physicist and chemist Svante Arrhenius (1859–1927) to propose his *Theory of Electrolytic Dissociation* in 1884. The main features of the model are:

An electrolyte such as a salt when dissolved in water, breaks up into the particles of the species of which it consists: for instance, sodium chloride breaks up into particles of sodium and particles of chlorine, while calcium chloride breaks up into particles of calcium and particles of chlorine.

It is important to bear in mind that some students may hold the common misconception that 'a compound is a mixture of its constituent elements' when they interpret the idea of 'particles of sodium' and 'particles of chlorine'. The particles here are actually the corresponding ions, so they are different from the particles of the elemental 'sodium' (Na ions plus delocalised electrons in a solid lattice) and of elemental chlorine (diatomic Cl_2 molecules in the gas state), respectively. (This is discussed in greater depth in Chapters 1 and 4.) This issue should be made clear to students.

To accommodate the capability of the electrolyte to conduct electricity, Arrhenius assumed that these particles are charged. They were what the English physicist and chemist Michael Faraday had named 'ions' since 1834. There are two types of ions, one type (cation) carrying a positive electric charge and the other type (anion) carrying a negative electric charge. In the form of a balanced chemical equation, we show electrolytic dissociation as follows:

$$NaCl_{(s)} \xrightarrow{+H_2O_{(l)}} Na^{+}_{(aq)} + Cl^{-}_{(aq)}$$

$$CaCl_{2(s)} \xrightarrow{+H_2O_{(l)}} Ca^{2+}_{(aq)} + 2Cl^{-}_{(aq)}$$

In its modern form, the model assumes that solid electrolytes, such as the salts, are composed of ions which are held together by electrostatic forces of attraction. In other words, the ions do not form during the dissolution process, but pre-exist in the solid state (Chapter 3). Remind students that sodium chloride and calcium chloride are ionic compounds. When such an electrolyte is dissolved in a solvent, these forces are overcome and the electrolyte undergoes dissociation into ions. In the solution, the ions are moving about, each surrounded and followed in its movement by a number of solvent molecules, and so we say that in the solution the ions are solvated (hydrated in the case of aqueous solutions).

Taking into account the fact that aqueous solutions of certain covalent compounds (notably bases that are not hydroxides, and acids) are electrolytes (while the pure covalent compounds, such as liquid sulfuric acid, gaseous hydrogen chloride, or gaseous ammonia, are not), the model accepts that during the dissolution process, the molecules of the solute are split into ions by reacting chemically with water. This process is called ionisation. It has been observed that electrolytes do not all ionise to the same extent. Some are almost completely ionised (strong electrolytes, for example sulfuric acid and hydrochloric acid) while others are feebly ionised (weak electrolytes, for example acetic acid and the base ammonia).

■ Electrolytes conduct because of the presence of mobile ions

If one sets up the circuit for electrolysis and closes it by turning on the switch of the power supply, the electric field across the circuit sets electrical charges in motion throughout the circuit. For example, from the negative terminal of the source, electrons move along the metallic conductor to the negative electrode of the electrolytic cell. (Recall that, by convention, the positive electrode of the electrolytic cell is the electrode which is connected with the positive terminal of the power supply, and similarly for the negative electrode of the electrolytic cell.)

It is accepted that inside the electrolyte the role of current carrier is taken up by the ions. In electrolysis of a molten ionic compound, ions move to the oppositely charged electrode: cations move to the negative electrode, where they discharge their charge; anions move to the positive electrode, where they also discharge their charge. Cations take electrons from the negative electrode (this corresponds to reduction); anions release electrons to the positive electrode (this corresponds to oxidation) (Chapter 7). The electrons that are released at the positive electrode flow along the external metallic conductor that connects the positive electrode with the positive outlet of the current source. To sum up, the closed circuitry functions by the electric charge being pushed by the electric field due to the external power source (by the potential difference between the two source teminals).

Schematically, one 'observes' a 'moving up' of the electrons from the anode and a 'moving down' of electrons onto the cathode. These two processes result in chemical changes taking place at each electrode and this is a main feature of electrolytic conduction.

The chemical changes that occur at each electrode are written as chemical equations including electrons; the corresponding chemical reaction is termed a half reaction and the equation a half equation. Summing up the two chemical half equations and balancing so that the electrons cancel out, one gets a total equation, which represents the overall reaction that has occurred in the electrolysis. With the electrolysis of molten NaCl as an example, we have:

At the negative electrode: $Na^+{}_{(aq)} + e^- \rightarrow Na_{(s)}$

At the positive electrode: $2Cl^-{}_{(aq)} - 2e^- \rightarrow Cl_{2(g)}$

or, equivalently:

$2Cl^-{}_{(aq)} \rightarrow Cl_{2(g)} + 2e^-$

Overall equation: $2Na^+{}_{(aq)} + 2Cl^-{}_{(aq)} \rightarrow 2Na_{(s)} + Cl_{2(g)}$

■ Some common student misconceptions

It should be emphasised that in electrolytic cells, it is the two half reactions that actually take place, while the total reaction represents a 'mathematical' overall result. (The same is also true for galvanic cells, covered in section 8.2.)

Students need also to understand the basic elements of half reactions:

- The chemical equation of a half reaction includes one or more electrons on its left-hand side (representing a reduction) or on its right-hand side (representing an oxidation).

- This chemical equation must be balanced both in terms of the species involved and in terms of electric charge.
- A half reaction can never occur alone. It must be accompanied by a second half reaction that occurs at the other electrode.
- One half reaction is always a reduction and the other is always an oxidation.
- The overall reaction in electrolysis cannot occur on its own; it cannot occur naturally – it is a non-spontaneous reaction. The transfer of energy through electric current is responsible for forcing the half reactions. (The opposite is the case in galvanic cells.)

The concept of transfer of electrical charge by the ions in the liquid electrolyte appears difficult for students to understand, looking like transport of electrons, a familiar concept from prior physics lessons. In general, many students encounter difficulties with the concept of ions and ionic equations in aqueous solutions, so special attention should be given to these issues. The writing and balancing of ionic half equations is also tricky for many students.

Electrolysis of aqueous sodium chloride

The result of the electrolysis of molten sodium chloride (Figure 8.2, page 256) is the production of the chemical elements sodium (a soft, very reactive solid at normal conditions of temperature and pressure) and chlorine (a yellowish, reactive and poisonous gas).

In the case of electrolysis of an aqueous solution of NaCl (Figure 8.4), at the negative electrode the product is hydrogen gas, because for energetic reasons it is much easier to reduce water to hydrogen, than sodium ions to sodium metal (as discussed later in this section). At the positive electrode, in principle the products could be oxygen gas resulting from the oxidation of water or chlorine gas resulting from the oxidation of chloride ions. Although energetically it is slightly easier for water to be oxidised, in practice chlorine is produced, except at dilute solutions. The explanation has to do with the very slow rate of the corresponding half reaction: this is due to the fact that the half reaction for the oxidation of water has a much larger overpotential (or overvoltage) than that of the oxidation of chloride ions; this knowledge is, however, beyond secondary level.

At the negative electrode: $2H_2O_{(l)} + 2e^- \rightarrow H_{2(g)} + 2OH^-_{(aq)}$

At the positive electrode: $2Cl^-_{(aq)} - 2e^- \rightarrow Cl_{2(g)}$

or, equivalently:

$2Cl^-_{(aq)} \rightarrow Cl_{2(g)} + 2e^-$

Overall equation: $2H_2O_{(l)} + 2Cl^-_{(aq)} \rightarrow H_{2(g)} + 2OH^-_{(aq)} + Cl_{2(g)}$

Note that the sodium ions move to the negative electrode but they are not discharged/not reduced. The resulting surplus of positive charge (following the corresponding loss of negative charge due to the discharge/oxidation of the chloride ions at the positive electrode) is balanced by the production of an equal number of hydroxide ions. The production of these hydroxide ions makes the solution alkaline, so the presence of phenolphthalein indicator will give the initially colourless solution a shade of red.

Figure 8.4 The electrolysis of aqueous NaCl

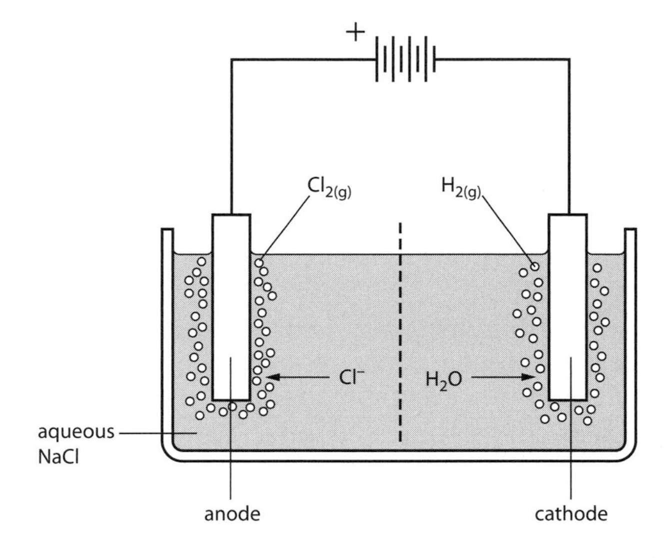

■ The reactivity series

When an aqueous solution of NaCl is electrolysed as above, metallic sodium is not deposited on the cathode because sodium ions (Na^+) have a very low tendency to gain electrons and turn into metal. For this reason the formation of hydrogen is favoured instead.

Recall from Chapter 7 that the more reactive a metal is, that is, the more easily it loses electrons to become ions, the more difficult it is to reverse the process by adding electrons to those ions to form the metal. Figure 8.5 overleaf shows a chart that lists ions in order of difficulty of being discharged during electrolysis of aqueous solutions. The higher an ion is in the chart, the more difficult it is to discharge the ion during electrolysis, and vice versa. (Students should be reminded that discharging results in reduction for a cation, but in oxidation for an anion.) Thus, Pb^{2+} is easier to discharge than Zn^{2+}, H^+ is easier to discharge than Na^+, and OH^- is easier to discharge than Cl^-. These series of ions are forms of the activity or reactivity series (as covered in Chapter 7), and are part of the so-called electrochemical series. Note that in the polyatomic anions OH^-, NO_3^- and SO_4^{2-}, oxygen is oxidised to oxygen gas (O_2).

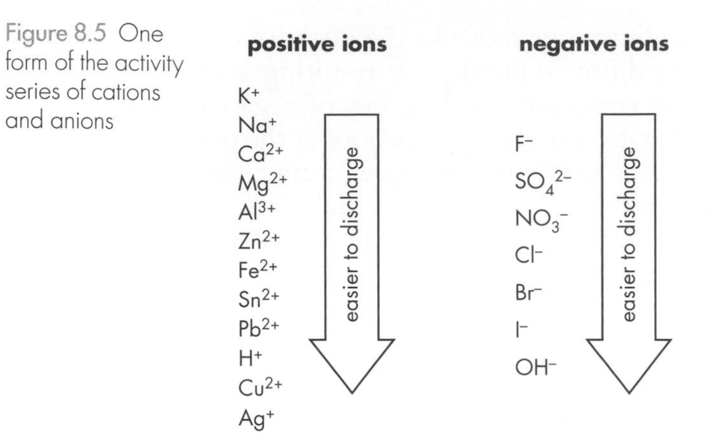

Figure 8.5 One form of the activity series of cations and anions

The solvent (water) can sometimes itself be the reactant in an electrode half reaction. This is the case in neutral or alkaline solutions where, instead of H^+, H_2O is the source of hydrogen (water is reduced to hydrogen). On the other hand, in neutral or acidic solutions, instead of OH^-, H_2O is the source of oxygen (water is oxidised to oxygen).

Electrolysing copper sulfate solution

In groups, some students can conduct an experiment to electrolyse an aqueous solution of copper sulfate ($CuSO_4$) using carbon (graphite) electrodes (Figure 8.6). They should observe that the cathode gets plated with copper, bubbles of oxygen gas (O_2) are released at the anode and the blue colour of the solution slowly fades. Other groups can electrolyse copper sulfate solution using copper rods as electrodes. They should observe that the cathode gets plated with copper, while the blue colour of the solution remains unchanged.

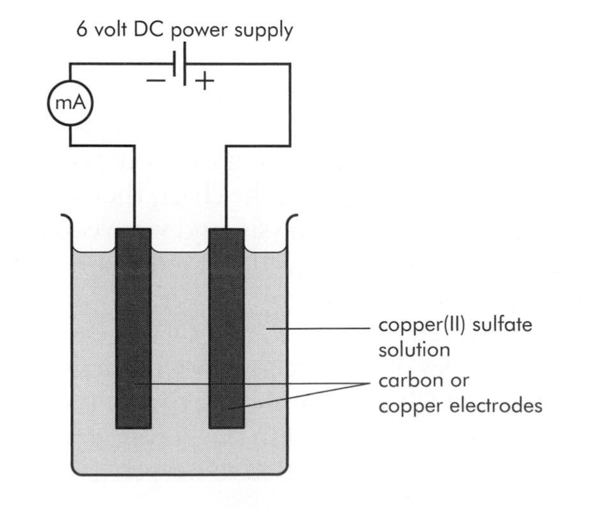

Figure 8.6 Apparatus for the electrolysis of copper sulfate solution

PROCEDURE

Materials

Each group will need:

- a large beaker (250 or 400 cm^3)
- two carbon (graphite) electrodes or two metallic copper electrodes
- 1 M $CuSO_4$ aqueous solution
- an external power supply (6 V DC)
- a milliammeter
- a stopclock
- a balance

Method

1. Pour some copper sulfate solution into the beaker.
2. Immerse the two electrodes (that are properly fixed on stands) into the solution.
3. Connect the two electrodes to the positive and the negative terminals of the power supply.
4. Adjust the voltage of the power supply to a value higher than 2 V, so that a constant current (measured in mA) is achieved.

In both practicals, students could also record the mass of deposited copper by weighing carefully the mass of the cathode before and after the electrolysis. The mass of the carbon anode will remain unchanged, while that of the copper anode will decrease. The value of current in milliamperes (mA) and the duration of the electrolysis in minutes should also be recorded.

The half reactions in the two electrodes and the overall reaction are as follows:

At the positive carbon electrode: $2H_2O_{(l)} \rightarrow O_{2(g)} + 4H^+_{(aq)} + 4e^-$

At the positive copper electrode: $Cu_{(s)} \rightarrow Cu^{2+}_{(aq)} + 2e^-$

At the negative carbon or copper electrode: $Cu^{2+}_{(aq)} + 2e^- \rightarrow Cu_{(s)}$

Overall equation with carbon electrodes: $2H_2O_{(l)} + 2Cu^{2+}_{(aq)} \rightarrow O_{2(g)} + 4H^+_{(aq)} + 2Cu_{(s)}$

Overall equation with copper electrodes: no overall chemical equation. An amount of metallic copper dissolves at the anode and at the same time an equivalent amount of metallic copper is deposited at the cathode. The overall change is a transfer of metallic copper from one electrode to the other!

Students should conclude that in electrolysis, the metal of an electrode itself can undergo a chemical change. They should also be asked to account for the fading or the stability of the blue colour of the $CuSO_4$ solution after the electrolysis for the two experiments.

Faraday's laws of electrolysis

Electrochemistry is one area where the ideas can be developed in quantitative terms. This is not normally undertaken by most secondary students but could be the basis of extension work, illustrating the use of mathematics in chemistry. Faraday's two laws of electrolysis offer the means, in the form of mathematical equations, for numerical calculations. Michael Faraday proposed his laws in 1934, while he was working at the Royal Institution in London.

■ Faraday's first law

The mass of a substance altered at an electrode during electrolysis is directly proportional to the quantity of electricity (electrical charge, typically measured in coulomb, C) *transferred at that electrode* (or equivalently the amount of electrical charge transferred through the electrolytic cell).

■ Faraday's second law

For a given quantity of electricity (electrical charge), the mass of an elemental material altered at an electrode is directly proportional to the element's equivalent weight.

As a consequence of the second law it follows that whenever the same quantity of charge is passing in an electrolysis practical, the amount of the product at each electrode, measured in moles, is the same when measured for similarly charged ionic species (for instance for Na^+, K^+ and Cl^-, or for Ca^{2+}, Zn^{2+} and Cu^{2+}).

The second law is explained in a straightforward manner by using the submicroscopic model of electrolytes to interpret the electrolysis phenomenon. In the case of copper deposited on the cathode in the electrolysis of copper sulfate solution, this model is expressed symbolically by the equation of the corresponding reduction half reaction:

$$Cu^{2+}_{(aq)} + 2e^- \rightarrow Cu_{(s)}$$

From the stoichiometry of this reaction, it follows that for the deposition of 1 mole Cu, 2F of charge must be passed through the electrolytic cell; equivalently, the passing of 1F results in the deposition of ½ mole Cu. (The mass of ½ mole Cu, that is 65.34/2 = 32.67 g, is copper's '*equivalent weight*' in this case.)

The numerical data that were collected by the students in the copper sulfate electrolysis practicals can be used now as an application of Faraday's laws.

Taking into account that the total charge of 1 mole of electrons, which is termed the faraday (1 F), is found by multiplying the Avogadro constant, N_A, by the charge of one electron (the elementary unit of electrical charge, q_e), it follows that:

$$\text{charge passed, } q = (\text{no. of moles of species} \div \text{its absolute charge}) \times 1\,\text{F}$$

where 1 F = 96 485 C

The above equation expresses Faraday's laws and allows the students to calculate the moles of copper, and hence the mass of copper, that was deposited on the cathode. This mass should be compared with the measured mass, providing thus an indirect verification of the laws and demonstrating the viability of the way chemists model the process of electrolysis.

Worked example
Duration of electrolysis: 60 minutes (3600 s)
Constant current: 100 mA
Students should recall from their physics lessons that current is a measure of the rate of flow of electrical charge:

$$\text{current} = \frac{\text{electric charge}}{\text{time}}$$

$$i = \frac{q}{t}$$

from which we get:

$$q = it$$

With these values, $q = it = 100 \times 10^{-3}\,\text{A} \times 3600\,\text{s} = 360\,\text{C}$
Hence:

$$\text{number of moles of Cu} = 2 \times \frac{360\,\text{C}}{1\,\text{F}} = 2 \times \frac{360\,\text{C}}{96\,485\,\text{C}} = 7.46 \times 10^{-3}$$

This corresponds to a mass of:

$$7.46 \times 10^{-3}\,\text{mol Cu} \times 65.34\,\text{g/mol} = 0.487\,\text{g}$$

Electroplating

Plating is the process by which a metal is deposited on a conductive surface. Plating has been undertaken since ancient times to decorate objects by giving them a silver or gold finish. It is also used in modern technology, for instance for corrosion inhibition, for the production of electronic circuitry and in nanotechnology.

In electroplating, the metallic object which is to be electroplated (that is, to be covered with a thin film of another metal) is made the cathode in an electrolytic cell. A solution with the ionic form of the metal is used as the electrolyte. Finally, as anode, either the metal which is being plated is used (soluble anode) or an insoluble anode made of carbon, platinum, titanium, lead or steel.

Aluminium recycling

The study of electrolysis can be completed by assigning the students a project on recycling.

Figure 8.7 Aluminium recycling symbols

Ask students to weigh an empty aluminium can. Using the proper chemical equations (page 254) and Faraday's laws (page 266), students should be able to calculate the energy (in units of kWh) that needs to be transferred electrically for the production of this mass of aluminium. Make the class aware that 1 kWh of (electrical) energy is sufficient to light one 60 W electric bulb for about 16 hours. This will lead students to realise the importance of recycling aluminium. Aluminium is 100% recyclable without any loss of its natural qualities. Recycling involves melting the scrap, a process that requires only 5% of the energy transferred to produce aluminium from ore, though a significant part (up to 15% of the input material) is lost as dross (ash-like oxide).

Enhancement ideas

Other applications of electrolysis that students could research include sodium production by electrowinning (the Downs process) (a useful website on this process is detailed in the 'Other resources' section at the end of the chapter).

8.2 Electrochemical power production – galvanic cells

Previous knowledge and experience

In the previous section electrolytes, electrolysis and electrolytic cells were studied, and the concepts and models were introduced that are also necessary for building up an understanding of galvanic cells. In this section, I will discuss galvanic cells and the accompanying transfer of energy electrically from the chemical store, and I will explore their many modern applications, including fuel cells. The concepts of electric current and potential difference are essential here, too. Students must be comfortable with writing and balancing equations for half reactions and reactions. Knowledge of ions is also central.

A teaching sequence

In the previous section we saw that electric current can drive a non-spontaneous chemical reaction in an electrolytic cell. The opposite is also true: a spontaneous chemical reaction can produce electricity. This is the principle on which galvanic cells operate.

'Galvanic' cells are named after Luigi Galvani (1737–1798), an Italian physician and physicist who discovered that the muscles of a dead frog's legs twitched when struck by a spark. Galvani interpreted the phenomenon as a special type of electricity intrinsic to animals. The Italian physicist Alessandro Volta (1745–1827) refuted Galvani's model, claiming that the mere contact between different metals 'set the electric fluid in motion'.

Electrolytic cells and galvanic cells together constitute electrochemical cells. An electrochemical cell, therefore, is a device capable of either driving chemical reactions through the introduction of electrical power or for deriving electrical power from chemical reactions. In everyday life we use galvanic cells in the form of 'batteries'. I suggest starting teaching with the study of climate change and the need for 'green cars'. These are connected with fuel cells that are covered later in the section.

A report on climate change

Written by chief executives across the business community, the CBI's report *Climate change: everyone's business* (see the 'Other resources' section at the end of the chapter) argued that a much greater sense of urgency is required if the UK is to meet its targets for reducing greenhouse gas emissions, and that the issue at hand needs an immediate and direct response. This report aspired to bring British businessmen to the world's forefront in the battle against climate change. Large British companies have suggested new products and services that aim to reduce carbon dioxide emissions. The use of renewable power sources, is currently increasing.

Similarly, the global car industry is spending vast sums of money developing 'green cars'. For over 100 years cars have been contributing (along with other major causes such as industry) to the pollution of our planet. The cars being developed today are more environmentally friendly and should one day replace the fleet of older polluting cars.

This could be a research topic for students to tackle in groups. Different groups in the class could be given a different theme or question to explore and then each group could report back to a class conference (using posters or audio–visual presentations). There are opportunities for differentiation in how groups are set up (for example, less confident students could be supported by more-able ones), how themes are assigned to groups and the level of scaffolding provided to different groups to guide their work. Possible topics for research could include:

1 The technologies that aim to reduce fuel consumption, including more effective engines, lighter cars and features such as the 'Stop and Start' system, which automatically stops and starts the engine of a car when the car stops for a short time (for example at traffic lights). This feature is said to reduce fuel consumption by about 10%.

2 Hybrid cars: How do they work? What percentage of the car market do their sales comprise?
3 Biodiesel: What is it and how is it made? What are the advantages and disadvantages of the production and use of biodiesel?
4 The use of hydrogen as fuel
5 Totally electric cars

With the exception of biodiesel, all the above solutions are dependent on the consumption of oil – even the production of hydrogen. Protecting the environment, however, requires that in the not-so-remote future, cars should be 100% independent of oil. The development and use of alternative renewable power sources, such as those using the power of the Sun and the wind, is inevitable and urgent.

Electrochemical power

Today, electrical power is the most useful and widespread means of transferring energy. Its production is based to a great extent on the transfer of energy from burning hydrocarbons – by transferring energy from a chemical store in an engine. The chemical store consists of the hydrocarbon and the oxygen from the air. This method, however, suffers from two large drawbacks:

- thermodynamics dictates that efficiency of this energy transfer is low (for instance, for a steam engine that operates in temperatures between 350 and 100 °C, the efficiency is about 40% [calculated using the formula: $\frac{(T_1 - T_2)}{T_1} = \frac{(623 - 373)}{623} = 0.40$])
- the production of air-polluting gases.

In contrast, electrochemical power sources are based on the direct transfer of energy during a chemical reaction through a flow of electrical current. This has an efficiency which, in practice, is around 90%.

In addition, electrochemistry allows the carrying out of a combustion reaction at normal temperatures (cool combustion). The most useful reaction to use for the production of power by electrochemical means is the combustion of hydrogen (discussed later in the section, see page 273) – not only for energetic reasons but also because the product (water) is harmless for man and the environment:

$$(-)\ H_{2(g)} \rightarrow 2H^+_{(aq)} + 2e^-$$

$$(+)\ O_{2(g)} + 4H^+_{(aq)} + 4e^- \rightarrow 2H_2O_{(l)}$$

Overall redox equation:

$$2H_{2(g)} + O_{2(g)} \rightarrow 2H_2O$$

■ Producing electricity with a lemon

Lemon juice is an electrolyte (an aqueous solution containing citric acid and many other substances). Students can be shown that by inserting two dissimilar electrodes into the lemon, electricity is produced. The voltage output depends on the metals and the concentration of the electolyte. Figure 8.8 shows a battery of two lemons connected in series.

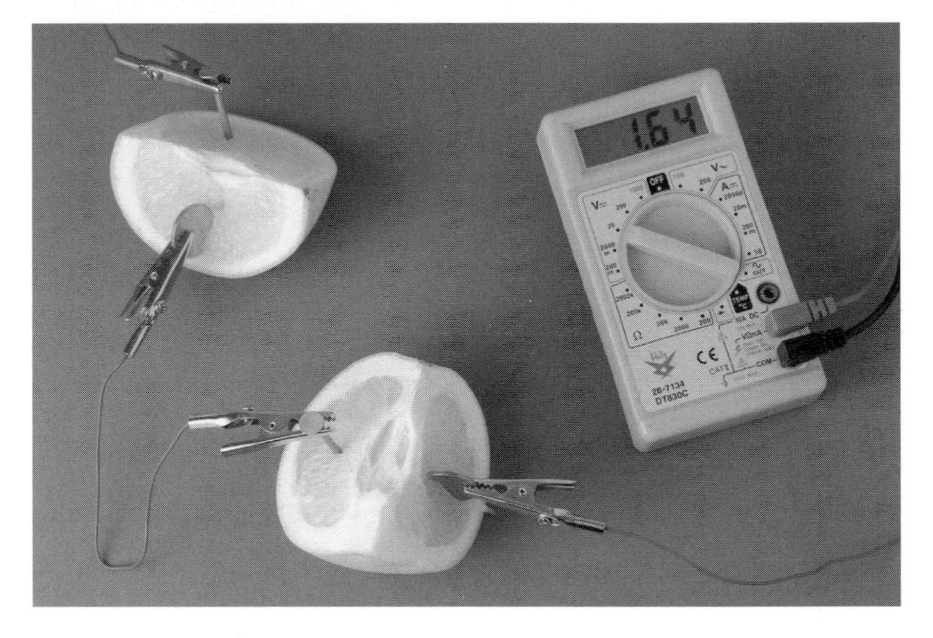

Figure 8.8 The lemon battery

■ The Volta cell

In his creation of the first electric battery in 1799, Alessandro Volta laid the foundation of electrochemistry. His battery consisted of a zinc and a copper electrode dipped into an aqueous solution of sulfuric acid and connected externally by copper wires. The procedure for setting up and running this experiment is similar to the earlier practical of the electrolysis of aqueous copper sulfate solution (page 265), but without the external power supply. If a voltmeter is connected in parallel to the external circuit, it can be used to measure the potential difference. The half reactions and the overall reaction are as follows:

At the negative electrode (anode): $Zn_{(s)} \rightarrow Zn^{2+}_{(aq)} + 2e^-$ (oxidation)

At the positive electrode (cathode): $2H^+_{(aq)} + 2e^- \rightarrow H_{2(g)}$ (reduction)

Overall redox equation: $Zn_{(s)} + 2H^+_{(aq)} \rightarrow Zn^{2+}_{(aq)} + H_{2(g)}$

A device in which two half reactions take place at the interface of two electrodes with a common electrolyte or two separate electrolytes is called a galvanic cell.

As is the case with electrolysis, in galvanic cells it is the two half reactions that have actually taken place, while the total reaction represents an overall result. Zinc has a higher tendency to be oxidised than hydrogen, hence electrons have a larger tendency to leave Zn (which is oxidised) and be taken up by $H^+_{(aq)}$ which is reduced (see the activity series of cations in Figure 8.5 page 264). The overall reaction is a redox reaction.

In a galvanic cell, there is a reversal of the anode and the cathode in relation to electrolysis. In this case, the anode is the negative electrode, while the cathode is the positive electrode. The reason is that here electrons are freed on the anode, from where they 'move up' in the external metallic circuit and then 'move down' onto the cathode where they are taken up by chemical species in the electrolyte. In common with electrolysis, an oxidation half reaction occurs at the anode and a reduction half reaction occurs at the cathode. Figure 8.9 shows an electrolytic cell which draws electric current from a galvanic cell. Electron flow, electrode polarity, anodes and cathodes, and the half reactions (oxidations and reactions) that take place are shown.

Table 8.1 summarises the main features and functions of electrochemical cells.

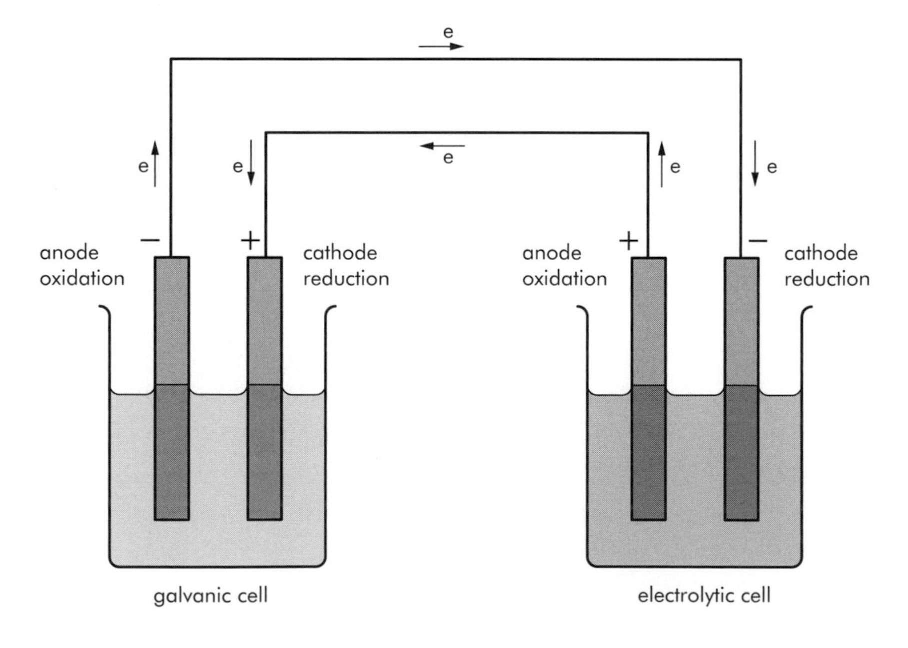

Figure 8.9 An electrolytic cell (right) connected to a galvanic cell (left). The galvanic cell supplies electric power to the electrolytic cell.

Table 8.1 A summary of electrochemical cells

Context	Sign and role of anode	Sign and role of cathode
Electrolysis – external current used to drive a non-spontaneous chemical reaction in the cell	**Positive** – through the external metallic circuit, it drives away current (electrons) to the positive terminal of an external power supply. These electrons are freed on the electrode and come from an oxidation half reaction that occurs at the electrode–electrolyte interface.	**Negative** – through the external metallic circuit, it attracts current (electrons) from the negative terminal of an external power supply. These electrons cause a reduction half reaction to occur at the electrode–electrolyte interface.
Galvanic cell – current generated by an internal spontaneous chemical reaction in the cell	**Negative** – through the external metallic circuit, it drives away current (electrons). These electrons are freed on the electrode and come from an oxidation half reaction that occurs at the electrode–electrolyte interface.	**Positive** – through the external metallic circuit, it attracts current (electrons) that are driven away from the other (the negative) electrode. These electrons cause a reduction half reaction to occur at the electrode–electrolyte interface.

Fuel cells

A galvanic cell in which a 'combustion reaction' takes place is called a fuel cell. Students could be encouraged to carry out some internet research about fuel cells. The most important fuel cell is the one in which the combustion of hydrogen takes place (the reaction between hydrogen and oxygen). Such a cell is supplied with hydrogen (the fuel) and oxygen. A fuel cell provides an efficient electrochemical device for the production of electrical power (Figure 8.10). Fuel cells have been used for a long time to provide power to spacecraft.

Students can understand why a hydrogen–oxygen fuel cell does not form polluting waste products. Note that apart from water being the only product substance, the cell functions at normal

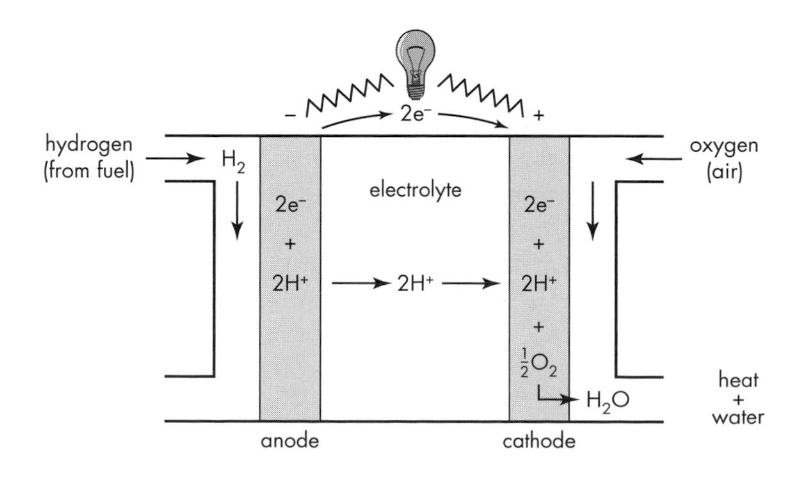

Figure 8.10 Schematic diagram of the experimental set-up for a PEM (proton exchange membrane or polymer electrolyte membrane) hydrogen fuel cell

temperatures so there are no secondary gas pollutants produced. (In comparison, at the high temperatures of internal combustion engines, reaction between the atmospheric nitrogen and oxygen leads to the formation of nitric oxides.) In principle, fuel cells that use standard fuels, such as oil, gasoline or natural gas, can be used in cars instead of the conventional thermal engines, but there are still technological difficulties that need to be overcome. Students could research these difficulties and present their findings.

Batteries

One or more electrochemical cells connected in series constitute an electrical 'battery'. Primary electrochemical (galvanic) cells are ready to produce current immediately and do not need to be charged in the way secondary cells (described below) do. In disposable cells, the chemical half reactions are not easily reversible, so the cells cannot be reliably recharged. Common disposable cells include the zinc–carbon cells and the alkaline cells. Secondary electrochemical cells contain the active materials in the discharged state, so they must be charged before use. The oldest form of rechargeable cell is the lead–acid battery.

Non-rechargeable cells are disposed of when their life is completed. Recycling of the cells protects the environment from pollution and is very important; many modern shops have a signed bin near their entrance for the disposal of used batteries. Rechargeable cells can be used and reused for a long time by recharging them by means of a proper supplied charger or recharger. This is an electric power transformer which is connected to the house mains supply and converts mains alternating current (AC) into direct current (DC) to recharge the discharged cells. During the recharging cycle, the cell functions as an electrolytic cell and the reverse half reactions take place.

Students can search the internet to find out about the composition of the dry cells that are in common use in electric and electronic devices. For practical reasons, in dry cells the electrolyte is in the form of a thick 'jam'. Given the composition of a dry cell, there is a certain voltage output (usually 1.5 V). By connecting several cells in series, we get a voltage output which is a multiple of the output of a single cell. Each dry cell has a given capacity that depends on its size (in other words, on the quantity of the reactant chemicals). A dry cell with the same composition gives the same voltage output, but can provide current for longer if it is larger in size. The life of the cells depends also on the time they are not used, because they lose capacity over time. Keeping cells in a fridge prolongs their life. Finding out why cells discharge when not in use and why keeping them at a lower temperature helps prolong their life could be suitable questions for group discussion work and/or some internet research for some classes.

■ Dry cells – the standard zinc–carbon dry cell

Georges Leclanché invented the Leclanché cell in 1866, which formed the basis for the construction of a dry cell. The well known standard dry cell (the zinc–carbon dry cell) consists of a cylindrical vessel made of zinc, which constitutes the negative electrode (anode). Into this vessel is placed a paste made of solid manganese dioxide (MnO_2), zinc chloride ($ZnCl_2$), ammonium chloride (NH_4Cl) and carbon (C). In the centre of this paste a carbon (graphite) rod is embedded which constitutes the positive electrode (cathode). $ZnCl_2$ and NH_4Cl are the electrolytes. The half reactions and the overall reaction are as follows:

$$(-)\ Zn_{(s)} \rightarrow Zn^{2+}_{(aq)} + 2e^-$$

$$(+)\ 2MnO_{2(s)} + 2NH_4{}^+{}_{(aq)} + 2e^- \rightarrow Mn_2O_{3(s)} + 2NH_{3(aq)} + H_2O$$

$$\text{(overall)}\ Zn_{(s)} + 2MnO_{2(s)} + 2NH_4Cl_{(aq)} \rightarrow ZnCl_{2(aq)} + Mn_2O_{3(s)} + 2NH_{3(aq)} + H_2O$$

This dry cell provides a voltage of 1.5 V which remains constant during its function. When the cell is nearing the end of its useful life, the terminal voltage falls below 1.5 V (because of increasing internal resistance) and the cell needs to be replaced.

■ Lithium batteries

Lithium batteries are used in modern electronic devices, such as mobile phones. They are rechargeable and have certain advantages, for instance they can be recharged at any time.

■ Lead–acid batteries

The lead–acid battery, which is used mainly in cars, was invented by Gaston Planté in 1859. It contains a liquid electrolyte which is a aqueous sulfuric acid solution with density 1.290 g/cm^3 (about 4.5 M H_2SO_4 in concentration). The electrodes are two lead rods or plates that have each undergone a special treatment. The negative electrode consists of a lead plate covered with spongy lead. The positive electrode consists of lead(II) oxide deposited on the lead rod or plate. Each cell provides a 2 V output, so six cells in series give a 12 V battery.

The half reactions taking place at the the anode (–) and the cathode (+) when the cell is functioning as a galvanic cell (supplying electrical current), and the overall redox reaction, are as follows:

$$(-)\ Pb_{(s)} + HSO_4{}^-{}_{(aq)} \rightarrow PbSO_{4(s)} + H^+{}_{(aq)} + 2e^- \text{ (oxidation)}$$

$$(+)\ PbO_{2(s)} + 3H^+{}_{(aq)} + HSO_4{}^-{}_{(aq)} + 2e^- \rightarrow PbSO_{4(s)} + 2H_2O_{(l)} \text{ (reduction)}$$

$$\text{(overall) } Pb_{(s)} + PbO_{2(s)} + 2HSO_4{}^-{}_{(aq)} + 2H^+{}_{(aq)} \rightarrow 2PbSO_{4(s)} + 2H_2O_{(l)}$$

The lead–acid cell is rechargeable. When the car engine is working, an electric motor transfers energy from the car engine as electricity, which charges the battery. In this way, the battery functions as an electrolytic cell and the reverse half reactions occur.

In the lead–acid battery, the positive and negative terminals (coloured red and black or blue, respectively) are fixed. However, they change their role as anode or cathode depending on whether the battery is charging or discharging. In both cases, a reduction half reaction occurs at the cathode and an oxidation half reaction at the anode.

Further research into the lead–acid battery is suitable for a homework study task that can be assigned to students. An integral part of this task could be a private visit of the students, accompanied by one parent or guardian, to a car electrician, who would provide a wealth of relevant useful information.

Enhancement ideas

■ Thermochemical calculations

Chapter 5 introduced basic thermodynamic ideas largely in qualitative terms. Electrochemistry offers the opportunity to develop these ideas in quantitative terms, reinforcing the application of mathematics in chemistry. For example, students could use thermochemical data (such as values of standard enthalpies of formation, available from any chemical data book or the internet) to calculate the standard enthalpy change during chemical reactions such as the hydrogen–oxygen reaction used in fuel cells. This is not, however, a normal task for most secondary students, so it could be used for extension work only.

■ Representing reactions with the reaction profile

These values can also be used for the construction of the energy (enthalpy) level diagram for the reaction (based on the general form shown in Figure 8.11). Research suggests that learning is supported when students are asked to represent information in different forms. Using mathematical and graphical representations of particular reactions gives students an opportunity to learn how chemistry uses different methods of presenting information about chemical reactions.

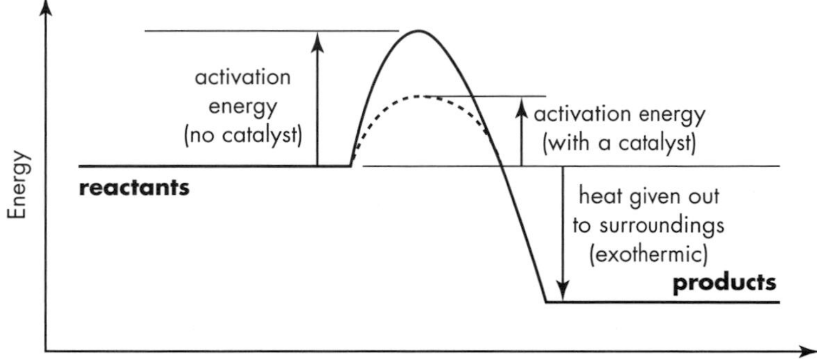

Figure 8.11 An energy (enthalpy) level diagram for a reaction (a reaction profile)

Animations

Animations are helpful in visualising chemical processes on the molecular level. Computer animations and simulations are most effective when coupled with actual demonstrations or working in the laboratory with electrochemical cells. For instance, Yang et al. (see 'Other resources' below) depicted how a battery-operated flashlight works using a computer animation representing the internal system of the flashlight. The project identified students' misconceptions about electricity and electrochemistry, and was reported to be successful in achieving conceptual change in students.

Other resources

Books and journals

De Jong and Treagust have reviewed the teaching and learning of electrochemistry. They consider that 'electrochemistry is ranked by teachers and students as one of the most difficult curriculum domains taught and learnt in secondary school chemistry'. See:

De Jong, O. and Treagust, D. (2003). Teaching and learning of electrochemistry. In Gilbert, J.K., de Jong, O., Justi, R., Treagust, D.F. and van Driel, J.H. (eds) *Chemical education: towards research-based practice* (Chapter 14). New York: Springer.

For another review of education research on electrochemistry, see:
Tsaparlis, G. (2007). Teaching and learning physical chemistry – review of educational research. In Ellison, M.D. and Schoolcraft, T.A. (eds) *Advances in teaching physical chemistry* (Chapter 7). Washington DC: American Chemical Society (distributed by Oxford University Press).

The following article is about lithium batteries:
Treprow, R.S. (2003). Lithium batteries: a practical application of chemical principles. *Journal of Chemical Education*, **80**, 1015–1020.

For the use of a computer animation representing the internal system of a battery-operated flashlight, see:
Yang, E.-M., Greenbowe, T. J. and Andre, T. (2004). The effective use of an interactive software program to reduce students' misconceptions about batteries. *Journal of Chemical Education*, **81**, 587–595.

Websites

Details about the *Bayer process* for alumina production, as well as the *Hall-Heroult process* for aluminum production, can be found in the *Electrochemistry Encyclopedia* accessible at:
http://electrochem.cwru.edu/encycl/art-a01-al-prod.htm

For more information about aluminium recycling see:
http://en.wikipedia.org/wiki/Aluminium_recycling

For information about sodium production by electrowinning (the Downs process):
http://corrosion-doctors.org/Electrowinning/Sodium.htm

There is a useful video about how fuel cells work on the HowStuffWorks website:
http://auto.howstuffworks.com/fuel-efficiency/alternative-fuels/fuel-cell.htm

For more information about lead–acid batteries:
http://en.wikipedia.org/wiki/Lead%E2%80%93acid_battery.
batteryuniversity.com

The CBI's report *Climate change: everypone's business* is accessible at:
www.cbi.org.uk/media/1058204/climatereport2007full.pdf

9 Inorganic chemical analysis

Kim Chwee Daniel Tan

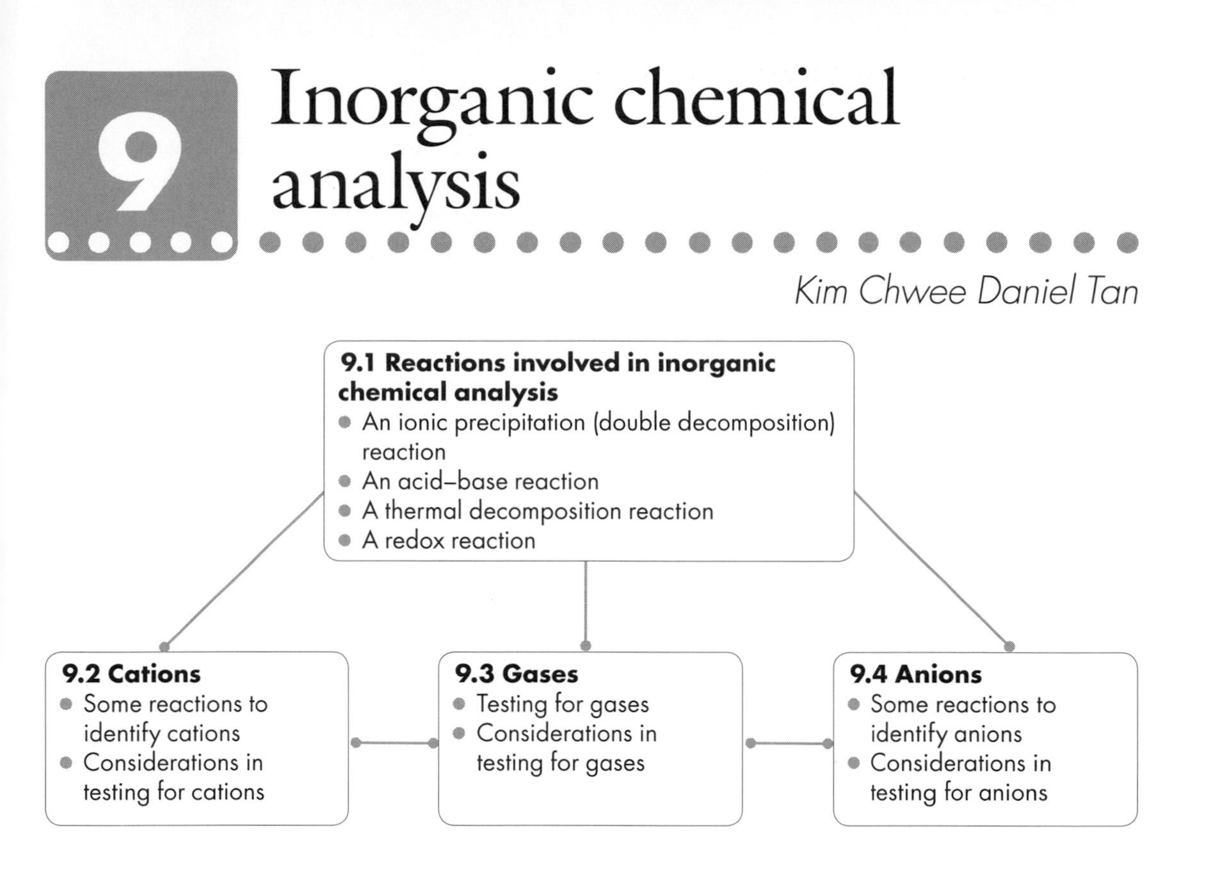

Introduction

Chemical analysis is used to identify and determine the amount of chemical substances (Chapter 1) present in a sample. It is often used for forensic, medical and environmental purposes, and for quality control in the food and chemical industries. Inorganic chemical analysis in secondary school involves the identification of cations such as zinc and lead(II), anions such as sulfate(VI) and iodide, and gases such as carbon dioxide and ammonia. It provides a context for the learning of the topics 'Acids, bases and salts' (covered in depth in Chapter 6) and 'Redox' (Chapter 7), as the reactions in these two topics are relevant to the identification of cations, anions and gases. Chemical analysis also enables students to develop their process skills, such as manipulation, observation and the making of inferences.

Learning difficulties in this topic

Research has shown that secondary students find inorganic chemical analysis difficult. This is mainly because they do not understand the purpose of the procedures and lack mastery of the skills required. Many students do not know how to carry out the procedures in the correct manner, what to observe and what valid inferences to make. They may also fail to see the links between the chemistry that they learn in the classroom and what they do in inorganic chemical analysis practical work. In addition, students' working memory may be overloaded when they carry out chemical analysis tests, as they need to read instructions and execute them, prepare additional tests (such as tests for gases), observe what happens, record and interpret their results, as well as be mindful of the time left to complete their practical work and reports. The resulting overload may leave little space for thought and learning. Hence, their actions are reduced to simply following instructions and their aim becomes one of 'getting the right answer'.

Choosing a route

The key concepts in inorganic chemical analysis are that:

- cations, anions and gases are identified by their properties and reactions
- positive and/or negative test results give clues to the identity of the unknown cations/anions/gases.

This chapter is divided into four sections. The first section introduces the main reactions involved in inorganic chemical analysis: ionic precipitation (double decomposition), acid–base, thermal decomposition and reduction–oxidation (redox). This is followed by sections on cations, gases and anions. It is probably easier to begin inorganic chemical analysis with cations (section 9.2) because many of the reactions involve only the formation of precipitates (for example when sodium hydroxide and/or ammonia solutions are added to solutions containing the cations). As the tests for anions usually involve testing for gases, it is suggested that you cover section 9.3 on gases next. Alternatively, you can proceed with the section on anions (section 9.4) and include the appropriate tests for gases where necessary.

The sections on cations, gases and anions start with introductions to the properties and reactions of the various ions and gases. Students can then familiarise themselves with the substances and see the reactions which the substances undergo. The reactions which result in the formation of gases are also included so that students can understand why they need to prepare to test for gases when they carry out certain procedures. Next, the manipulative, observational and inferential skills required to carry out the chemical analysis will be discussed, along with the strategies needed to identify the ions and gases. Research findings on students' difficulties in inorganic chemical analysis will be highlighted and suggestions are provided on how to address these difficulties.

9.1 Reactions involved in inorganic chemical analysis

Previous knowledge and experience

Students may have some experience of ionic precipitation, acid–base, thermal decomposition and redox reactions from lower secondary science. If they have already learnt the topics 'Acids, bases and salts' (covered here in Chapter 6) and 'Redox reactions' (Chapter 7) before starting on inorganic chemical analysis, you may want to skip this section and go directly to sections 9.2 Cations, 9.3 Gases and 9.4 Anions. All the relevant properties of substances and reactions that they have already learnt should be highlighted to students so that appropriate links can be made to what they do in inorganic chemical analysis.

Main reactions in inorganic chemical analysis

Students will need to be familiar with the main reactions involved in inorganic chemical analysis. These are ionic precipitation, acid–base, thermal decomposition and redox reactions. It is hoped that the formation and colour of products in the activities below will capture students' attention and interest and motivate them to learn the reactions involved.

■ An ionic precipitation (double decomposition) reaction

The following procedure outlines a common ionic precipitation reaction in which two soluble compounds react to form an insoluble compound (precipitate) and a soluble compound.

PROCEDURE

Materials
- test tubes
- a test-tube rack
- a test-tube holder
- a Bunsen burner
- potassium iodide solution, 0.04 mol dm^{-3}
- lead(II) nitrate(V) solution, 0.02 mol dm^{-3}

Safety

- All students should wear eye protection.
- Do not pour excess reagent back into the stock bottles. If too much is taken, dispose of the excess.
- Flush with plenty of water if the reagents come into contact with skin or eyes.
- Never point the test tube being heated at anybody as the contents may shoot out during heating.
- Be aware of the hot glassware.

Method

1. Add enough lead(II) nitrate(v) solution to fill the curved base of a test tube.
2. Add potassium iodide solution to another test tube to a height of 1 cm.
3. Observe the appearance of the solutions before adding the potassium iodide solution to the lead(II) nitrate solution. Observe the resulting mixture.
4. Heat the mixture until no further change is observed. Allow the mixture to cool.

Results

A yellow precipitate is formed when the two colourless solutions are added together. This is an example of a ionic precipitation reaction in which two soluble compounds react to form an insoluble compound (lead iodide) and a soluble compound (potassium nitrate):

$$Pb(NO_3)_{2(aq)} + 2KI_{(aq)} \rightarrow PbI_{2(s)} + 2KNO_{3(aq)}$$
$$Pb^{2+}_{(aq)} + 2I^-_{(aq)} \rightarrow PbI_{2(s)}$$

Discussion points

Questions such as 'Why does lead iodide form a precipitate?' or 'How does a precipitate form?' can lead students to link the low solubility of lead(II) iodide to the formation of the precipitate. They should also be able to visualise how the precipitate is formed – the lead(II) and iodide ions meet and attract each other strongly such that they cannot be 'pulled apart' by water molecules; they then grow in size as more ions come together (refer to Figure 4.7 in Chapter 4). The use of animations which can be found on the internet can be helpful for this purpose.

Ionic equations can also help make the reaction explicit as they focus only on the reacting species. Lead(II) iodide is soluble in hot water and recrystallises upon cooling; this example can be used to illustrate the effect of heat on the solubility of some compounds (the students can see the precipitate disappearing as the mixture is heated and crystals appearing upon cooling) and its application in the crystallisation process.

An acid–base reaction

PROCEDURE

Materials
- two test tubes
- some marble chips
- lime water
- hydrochloric acid, 1 mol dm^{-3}
- litmus paper – red and blue
- a bung with a hole in connected to a delivery tube

Safety
- All students should wear eye protection.
- Do not pour excess reagent back into the stock bottles. If too much is taken, dispose of the excess.
- Flush with plenty of water if the reagents come into contact with skin or eyes.

Method
1. Add marble chips to a test tube to a height of 1 cm.
2. Add enough lime water to fill the curved base of another test tube.
3. Add hydrochloric acid to the test tube containing the marble chips to a height of 2 cm.
4. Hold a moist blue litmus paper and a moist red litmus paper at the mouth of this test tube, taking care not to touch the wall of the test tube. Observe what happens.
5. Next, bubble the gas produced into the lime water using the delivery tube (Figure 9.1).
6. Add more marble chips and hydrochloric acid if insufficient gas is generated.

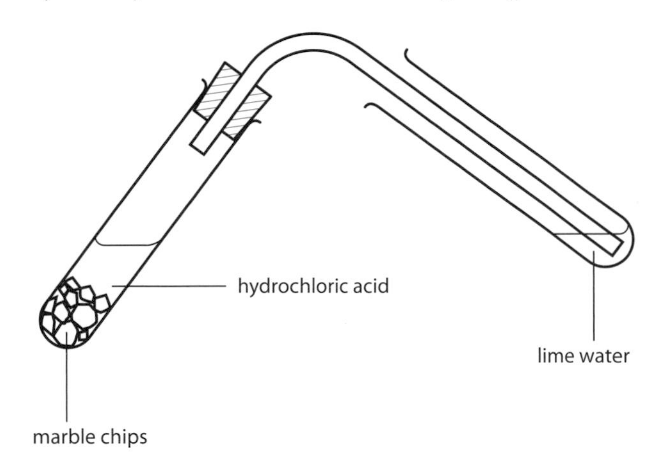

Figure 9.1 Apparatus for the reaction between marble chips and hydrochloric acid

Results

A colourless gas, carbon dioxide, is produced when the hydrochloric acid is added to the marble chips:

$$CaCO_{3(s)} + 2HCl_{(aq)} \rightarrow CaCl_{2(aq)} + H_2O_{(l)} + CO_{2(g)}$$

The carbon dioxide turns moist blue litmus red and reacts with lime water to produce a white precipitate. On further bubbling, the white precipitate disappears and a colourless solution is formed. You may want to ask students how the amount of lime water present in the test tube affects the test – a smaller amount of lime water will give a quicker result, as a smaller amount of carbon dioxide is required to react with it to give a white precipitate. The evolution of a gas when an acid or alkali is added to the unknown sample gives clues to the identity of the ions present in the sample. The reactions involved in the test for carbon dioxide will be discussed in section 9.3 Gases.

A thermal decomposition reaction

PROCEDURE

Materials
- two test tubes
- a test-tube holder
- copper(II) carbonate
- lime water
- a bung with a hole in connected to a delivery tube
- a Bunsen burner

Safety
- All students should wear eye protection.
- Do not pour excess reagent back into the stock bottles. If too much is taken, dispose of the excess.
- Flush with plenty of water if the reagents come into contact with skin or eyes.
- Never point the test tube being heated at anybody as the contents may shoot out during heating.
- On completion of the test for carbon dioxide when copper(II) carbonate is heated, continue heating while removing the delivery tube from the lime water to prevent the lime water from being 'sucked back' into the hot test tube (due to the decrease in pressure in the tube on cooling leading to greater external pressure pushing liquid into the still hot tube).
- Be aware of hot glassware.

Method
1. Put copper(II) carbonate into a test tube to a height of 1 cm.
2. Add enough lime water to fill the curved base of another test tube.
3. Heat the copper(II) carbonate strongly until there is no further change.
4. Bubble the gas produced into the lime water using the delivery tube.

Results

The green copper(II) carbonate will turn black upon heating. The carbon dioxide which is liberated will react with lime water to produce a white precipitate. Thermal decomposition is a reaction in which a compound is broken up by heating (Chapter 3). For example, in this reaction:

$$CuCO_{3(s)} \rightarrow CuO_{(s)} + CO_{2(g)}$$

The evolution of a gas when the unknown sample is heated and the colour changes which occur during the heating process can help identify the ions present.

A redox reaction

PROCEDURE

Materials
- a test tube
- sodium sulfate(VI)
- potassium dichromate(VI) solution, 1 mol dm^{-3} acidified with an equal volume of sulfuric(VI) acid, 1 mol dm^{-3}

Safety
- All students should wear eye protection.
- Do not pour excess reagent back into the stock bottles. If too much is taken, dispose of the excess.
- Flush with plenty of water if the reagents come into contact with skin or eye.
- Do the tests in a fume cupboard or in a well ventilated area.
- Avoid breathing in the gas evolved.

Method

1 Put enough sodium sulfate(VI) to half-fill the curved base of a test tube.
2 Add acidified potassium dichromate(VI) to the test tube to a height of 1 cm.
3 Observe any reactions.

Results

In this redox reaction, the orange potassium dichromate(VI) will turn green when it reacts with sodium sulfate(IV) in the presence of sulfuric(VI) acid. A redox reaction is one in which the oxidation states of atoms in substances are changed, indicating the presence of an oxidising or reducing agent (Chapter 7) in the unknown sample. For example, in this reaction:

$$K_2Cr_2O_{7(aq)} + 4H_2SO_{4(aq)} + 3Na_2SO_{3(s)} \rightarrow K_2SO_{4(aq)} + Cr_2(SO_4)_{3(aq)} + 3Na_2SO_{4(aq)} + 4H_2O_{(l)}$$

In the above reaction, the oxidation state of chromium changes from +6 to +3, and the oxidation state of sulfur changes from +4 to +6. (The idea of oxidation states is introduced in Chapter 3.)

Discussion points

Students could be asked to consider the following questions:

- Why does potassium dichromate(VI) need to be acidified? (to provide the hydrogen ions necessary for the redox reaction)
- What acid should be used? (sulfuric(VI) acid as it will not be oxidised by the potassium dichromate(VI))
- How much acid should be added? (usually equal volumes of potassium dichromate(VI) and 1 mol dm^{-3} sulfuric(VI) acid to provide excess hydrogen ions)

Note that the focus of this activity is to allow students to experience the colour change resulting from the redox reaction. Sulfur dioxide will be evolved when the acid reacts directly with the sulfate(IV) but this will be discussed in section 9.3.

9.2 Cations

Previous knowledge and experience

Students may have learnt about ionic precipitation reactions and the formation and properties of ammonia in the topic on 'Acids, bases and salts'. They also need to draw upon knowledge of amphoteric compounds (those that react with both acids and alkalis) to understand the further reactions of the insoluble hydroxides of zinc, aluminium and lead with excess sodium hydroxide solution to form soluble salts. The reactions of the insoluble hydroxides of copper and zinc with excess ammonia solution are similar in that soluble salts are also formed.

A teaching sequence

In basic inorganic chemical analysis, the cations that students may be required to identify are aluminium, ammonium, calcium, copper(II), iron(II), iron(III), lead(II) and zinc. The reagents used are usually sodium hydroxide and ammonia solutions. Whether a precipitate (an insoluble hydroxide) is formed, the colour of the precipitate and further reaction, if any, of the precipitate with excess sodium hydroxide and/or ammonia solution, all give clues to the identity of the cations present. The liberation of ammonia when sodium hydroxide solution is added to an unknown compound and gently heated indicates the presence of the ammonium ion. Flame tests are also used to identify certain cations, for example, sodium, potassium and copper(II), as they give distinctive coloured flames when heated in a non-luminous flame.

Some reactions to identify cations

PROCEDURE

Materials

- sodium hydroxide solution, 1 mol dm^{-3}
- ammonia solution, 1 mol dm^{-3}
- iron(III) chloride, 0.2 mol dm^{-3}
- zinc chloride, 0.2 mol dm^{-3}
- solid ammonium chloride
- solid copper(II) chloride
- solid potassium chloride
- solid sodium chloride
- solid calcium chloride
- solid barium chloride
- solid zinc chloride
- red and blue litmus paper
- deionised water in a wash bottle
- eye protection
- test tubes
- a test-tube rack
- a test-tube holder
- a 50–100 ml beaker
- a watch glass or evaporating dish
- wooden splints
- a Bunsen burner
- a lighter

Safety

- All students should wear eye protection.
- Do not pour excess reagent back into the stock bottles. If too much is taken, dispose of the excess.
- Flush with plenty of water if the reagents come into contact with skin or eyes.

Method 1

1 Add enough iron(III) chloride solution to fill the curved base of a test tube.
2 Add two drops of sodium hydroxide solution to the iron(III) chloride solution.
3 Add more sodium hydroxide solution until the test tube is half filled.
4 Repeat steps 1 to 3, this time adding ammonia solution instead of sodium hydroxide solution.

Results 1

A brown precipitate is formed when the two drops of sodium hydroxide or ammonia solution are added to the iron chloride solution. An ionic precipitation reaction has occurred forming insoluble iron(III) hydroxide which does not react with excess sodium hydroxide or ammonia solution. Ammonia reacts with water to produce ammonium ions and hydroxide ions which react with the iron(III) ions. The brown precipitate indicates the presence of iron(III) ions.

$$FeCl_{3(aq)} + 3NaOH_{(aq)} \rightarrow Fe(OH)_{3(s)} + 3NaCl_{(aq)}$$
$$FeCl_{3(aq)} + 3NH_{3(aq)} + 3H_2O_{(l)} \rightarrow Fe(OH)_{3(s)} + 3NH_4Cl_{(aq)}$$
$$Fe^{3+}{}_{(aq)} + 3OH^-{}_{(aq)} \rightarrow Fe(OH)_{3(s)}$$

Iron(II) ions form a green precipitate which does not react with excess sodium hydroxide solution or ammonia solution.

$$Fe^{2+}{}_{(aq)} + 2OH^-{}_{(aq)} \rightarrow Fe(OH)_{2(s)}$$

Calcium ions react with sodium hydroxide solution to give a white precipitate which does not react with excess sodium hydroxide solution.

$$Ca^{2+}{}_{(aq)} + 2OH^-{}_{(aq)} \rightarrow Ca(OH)_{2(s)}$$

However, calcium ions may give no precipitate or a very small amount of white precipitate with ammonia solution. This is because calcium hydroxide is slightly soluble in water (forming lime water) compared to, for example, iron(III) hydroxide which is almost insoluble. Ammonia solution is a weak alkali, forming a smaller amount of hydroxide ions compared to the same concentration of sodium hydroxide solution, a strong alkali. (Chapter 6 covers acids and alkalis in greater depth.) Thus, the amount of calcium hydroxide formed with dilute ammonia solution is likely to be soluble, so no precipitate may be formed unless a concentrated ammonia solution is used.

Method 2

1 Add enough zinc chloride solution to fill the curved base of a test tube.
2 Add two drops of sodium hydroxide solution to the zinc chloride solution.
3 Add more sodium hydroxide solution (1 cm height portions at a time) until the test tube is half filled, shaking the tube after each portion of the sodium hydroxide solution is added (Figure 9.2).

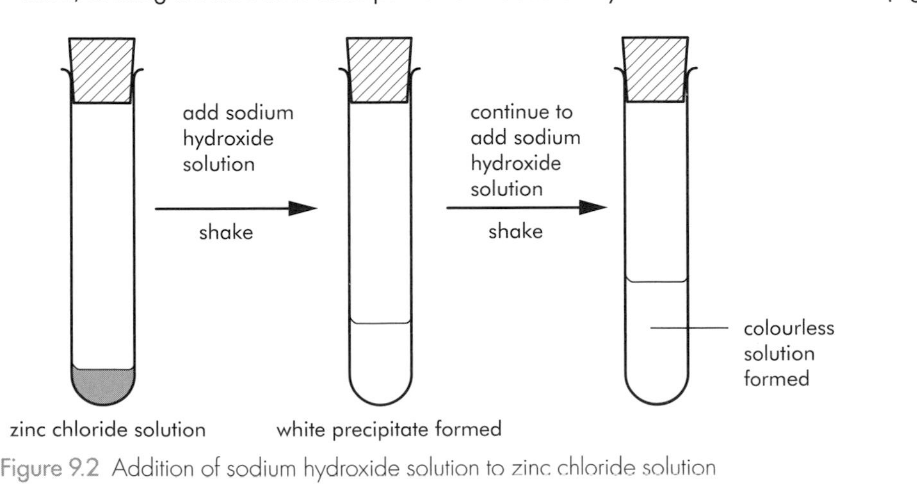

Figure 9.2 Addition of sodium hydroxide solution to zinc chloride solution

Results 2

A white precipitate is formed when the two drops of sodium hydroxide solution are added to the zinc chloride solution. An ionic precipitation reaction has occurred forming insoluble zinc hydroxide.

$$ZnCl_{2(aq)} + 2NaOH_{(aq)} \rightarrow Zn(OH)_{2(s)} + 2NaCl_{(aq)}$$
$$Zn^{2+}{}_{(aq)} + 2OH^{-}{}_{(aq)} \rightarrow Zn(OH)_{2(s)}$$

The white zinc hydroxide reacts with excess sodium hydroxide solution to form a complex salt (zincate) which dissolves to give a colourless solution.

$$Zn(OH)_{2(s)} + 2NaOH_{(aq)} \rightarrow Na_2Zn(OH)_{4(aq)}$$
$$Zn(OH)_{2(s)} + 2OH^{-}{}_{(aq)} \rightarrow Zn(OH)_4{}^{2-}{}_{(aq)}$$

In general, secondary students do not need to know the above equations; they only need to know that zinc hydroxide reacts further with excess sodium hydroxide to give a colourless solution.

Aluminium and lead(II) ions react in a similar manner with sodium hydroxide solution, forming white precipitates which react with excess sodium hydroxide solution to form complex salts (aluminate and plumbate), which dissolve to give colourless solutions.

$$Al^{3+}{}_{(aq)} + 3OH^{-}{}_{(aq)} \rightarrow Al(OH)_{3(s)}$$
$$Al(OH)_{3(s)} + OH^{-}{}_{(aq)} \rightarrow Al(OH)_4{}^{-}{}_{(aq)}$$

$$Pb^{2+}{}_{(aq)} + 2OH^{-}{}_{(aq)} \rightarrow Pb(OH)_{2(s)}$$
$$Pb(OH)_{2(s)} + 2OH^{-}{}_{(aq)} \rightarrow Pb(OH)_4{}^{2-}{}_{(aq)}$$

Research has shown that students tend to regard ionic precipitation reactions involving sodium hydroxide solutions as displacement reactions. This is most likely because they learn that sodium is a reactive metal and will displace less reactive metal ions from their compounds, and because they do not differentiate clearly between metals and metal ions. Thus, you may want to compare the two reactions to clarify that an ionic precipitation reaction involves metal ions and not metals, and also that it does not involve a redox reaction:

ionic precipitation reaction: $Zn^{2+}{}_{(aq)} + 2OH^{-}{}_{(aq)} \rightarrow Zn(OH)_{2(s)}$

displacement (redox reaction): $Zn_{(s)} + Cu^{2+}{}_{(aq)} \rightarrow Zn^{2+}{}_{(aq)} + Cu_{(s)}$

The example of a displacement reaction between sodium metal and a zinc solution is not used because the sodium metal is more likely to react with water molecules, as there will be more water molecules than zinc ions in solution. In addition, sodium reacts with water to produce sodium hydroxide which will react with zinc ions, giving rise to another competing reaction.

Method 3

1. Add enough zinc chloride solution to fill the curved base of a test tube.
2. Add two drops of ammonia solution to the zinc chloride solution.
3. Add more ammonia solution (1 cm height portions at a time) until the test tube is half filled, shaking the test tube after each portion of the ammonia solution is added.

Results 3

A white precipitate is formed when the two drops of ammonia solution are added to the zinc chloride solution. An ionic precipitation reaction has occurred forming insoluble zinc hydroxide.

$$ZnCl_{2(aq)} + 2NH_{3(aq)} + 2H_2O_{(aq)} \rightarrow Zn(OH)_{2(s)} + 2NH_4Cl_{(aq)}$$
$$Zn^{2+}{}_{(aq)} + 2OH^{-}{}_{(aq)} \rightarrow Zn(OH)_{2(s)}$$

This is similar to the reaction of zinc ions and sodium hydroxide. However, when excess ammonia solution is added, the zinc hydroxide will react with the ammonia molecules (rather than the small amount of hydroxide ions present in ammonia solution) to form a tetraamminezinc hydroxide which is soluble.

$$Zn(OH)_{2(s)} + 4NH_{3(aq)} \rightarrow [Zn(NH_3)_4]^{2+}{}_{(aq)} + 2OH^-{}_{(aq)}$$

Copper(II) ions react in a similar way with ammonia solution. A light blue precipitate is formed which reacts with excess aqueous ammonia to give a deep blue solution.

$$Cu^{2+}{}_{(aq)} + 2OH^-{}_{(aq)} \rightarrow Cu(OH)_{2(s)}$$
$$Cu(OH)_{2(s)} + 4NH_{3(aq)} \rightarrow [Cu(NH_3)_4]^{2+}{}_{(aq)} + 2OH^-{}_{(aq)}$$

Students at secondary level do not need to know the equations for complex ion formation but they do need to understand that the precipitate will react with excess ammonia solution to form a soluble compound.

Method 4

1. Add enough ammonium chloride to half-fill the curved base of a test tube.
2. Add sodium hydroxide solution to the test tube to a height of 1 cm.
3. Hold a moist blue litmus paper and a moist red litmus paper at the mouth of the test tube, taking care not to touch the wall of the test tube, and observe what happens.
4. Heat the test tube gently.

Safety

- Students should wear eye protection, especially when heating substances.
- Never point the test tube being heated at anybody as the contents may shoot out during heating.
- Be aware of the hot glassware.

Results 4

A colourless and pungent gas is formed which turns the moist red litmus paper blue.

$$NH_4Cl_{(s)} + NaOH_{(aq)} \rightarrow NH_{3(g)} + H_2O_{(l)} + NaCl_{(aq)}$$

Sodium hydroxide will react with ammonium salts to produce ammonia gas. Thus, students need to test for ammonia gas when they add sodium hydroxide solution to an unknown sample, especially if the mixture is heated, even if the testing for ammonia is not explicitly stated in the instructions. It is, however, uninformative to test for ammonia gas or the ammonium cation when ammonia solution is added to unknown compounds. You may want the students to explain why the test is unnecessary in this situation, as research shows that some students do carry out this unnecessary procedure.

Method 5

1. Soak a few clean wooden splints in a beaker half-filled with deionised water.
2. Place a tiny amount of solid copper(II) chloride on the watch glass or evaporating dish.
3. Remove a wooden splint and press the top end of the splint that has been soaked in water onto the solid copper(II) chloride.
4. Wave the end of the splint containing the copper(II) chloride in the non-luminous flame and observe the colour of the flame.
5. Do not hold the splint in the flame for long as it will catch fire.
6. Repeat the experiment with solid potassium chloride, sodium chloride, calcium chloride, barium chloride and zinc chloride.

Safety
Students should wear eye protection, especially when heating substances.

Results 5

Table 9.1 Results

Ions	Colour
Copper(II)	green
Potassium	purple (or lilac)
Sodium	orange
Calcium	brick red
Barium	light green
Zinc	greenish white

Using wooden splints is a cheaper and more convenient alternative to using platinum or nichrome wire. Chlorides are used as they tend to vaporise at a lower temperature. The high temperature of the non-luminous flame causes electrons in the vaporised metal ions to jump to orbitals with higher energy levels. When cooled, these electrons tend to fall back to the lower energy levels and release energy as light of a specific colour.

Considerations in testing for cations

While this chapter discusses a topic that is largely practical in nature, it is important that students engage in thinking about the processes used in analysis. Table 9.2 summarises the rationale behind different aspects of the tests described here. A useful activity with some classes might be to provide them with a version of this table with the second column blank, and ask students to work in small groups to suggest sensible reasons for the details of the procedures. This will ensure they understand why certain practice procedures are adopted and makes it more likely that they will remember to follow those procedures in their own analytical work.

Table 9.2 Reasons behind the procedures involved in the identification of cations

Procedure	Reasons
Adding enough unknown solution to half-fill the curved base of a test tube	To ensure that reaction can occur quickly (especially when adding reagents to excess) and to minimise wastage of reagents
Use of moist (rather than dry) litmus paper	To allow the gas to dissociate in the moist litmus paper and react with the dyes present
Litmus paper should not touch the wall of the test tube	To prevent reaction between the litmus paper and any residual reagent on the wall of the test tube
Adding sodium hydroxide or ammonia solution to a solution of the unknown solid rather than directly to the unknown solid	To ensure that the precipitate is the result of reaction between the two solutions. If a solid is used, it may be difficult to differentiate between undissolved solid and any precipitate formed. Thus, any unknown solid should be made into a solution before adding sodium hydroxide or ammonia solution to it.
Adding two drops of sodium hydroxide or ammonia solution initially to the unknown solution	To prevent excess sodium hydroxide or ammonia solution from being added so that any precipitate formed can be observed
Adding 1 cm height portions at a time and shaking the test tube after each portion of sodium hydroxide or ammonia solution is added in excess to the unknown solution	To prevent formation of layers of substances and to ensure that the substances are thoroughly mixed, facilitating any reaction

When carrying out the procedures, students need to take note of the following:

- whether a precipitate is formed when a small amount (two drops) of sodium hydroxide or ammonia solution is added
  - If a precipitate is obtained, record its colour; for example, 'A white precipitate is obtained.'
  - If there seems to be no reaction, then write 'No visible reaction'.
- what happens to the precipitate when adding a reagent until in excess or adding a reagent until there is no more change occurs (which means the same thing)
  - If the precipitate disappears, record it and the colour of the solution formed, for example 'The white precipitate disappears and a colourless solution is obtained.'
  - If there seems to be no reaction, then write, for example, 'There is no visible reaction, the white precipitate remains.'

You may want the students to practise adding two drops of reagent to the unknown solution followed by addition of the reagent to excess, making sure that they shake the test tube after each addition of reagent. They need to be able to carry out the procedures 'automatically' so that they can focus more on making sense of the results rather than being overwhelmed by the 'doing' of the practical work. When they are familiar with the techniques involved, they can be given a few unknown solids/solutions and instructed to design their own experiments to identify the cations present.

9.3 Gases

Previous knowledge and experience

The gases that students need to identify in secondary inorganic chemical analysis are usually oxygen, hydrogen, carbon dioxide, ammonia, sulfur dioxide, chlorine and nitrogen dioxide. Students may already have experience of the tests for oxygen, hydrogen and carbon dioxide from earlier lessons in secondary science, and ammonia from the previous section on cations. Besides the properties and reactions of gases, reactions producing the gases are also emphasised so that students know why they need to prepare to test for gases when they carry out certain procedures.

Testing for gases

PROCEDURE

Materials

- test tubes
- a test-tube rack
- a test-tube holder
- a delivery tube
- a dropper
- a Bunsen burner
- a lighter
- wooden splints
- red and blue litmus paper
- strips of filter paper
- deionised water in a wash bottle
- lime water (saturated calcium hydroxide solution)
- potassium dichromate(VI) solution, 0.1 mol dm^{-3}, acidified with an equal volume of sulfuric(VI) acid, 1 mol dm^{-3}
- solid sodium sulfate(IV)
- solid copper(II) nitrate(V)
- solid manganese(IV) oxide
- hydrochloric acid, 1 mol dm^{-3}
- concentrated hydrochloric acid
- a mixture of potassium iodide solution, 0.1 mol dm^{-3} and 5% starch solution

Safety

- All students should wear eye protection.
- Wear gloves when handling concentrated hydrochloric acid.
- Do the tests in a fume cupboard or in a well ventilated area.
- Avoid breathing in the gases evolved, those with asthma must take particular care. (However, at the same time, it would be an advantage for students to briefly experience the characteristic odour of certain gases such as sulfur dioxide or nitrogen dioxide.)
- Flush with plenty of water if acid comes into contact with skin or eyes.
- Be aware of the hot glassware.

Method 1 – testing for sulfur dioxide

1. Add enough sodium sulfate(IV) to half-fill the curved base of a test tube.
2. Dip a strip of filter paper into acidified potassium dichromate(VI).
3. Add enough lime water to fill the curved base of another test tube.
4. Add hydrochloric acid to the test tube containing sodium sulfate(IV) to a height of 2 cm.
5. Hold a moist blue litmus paper and a moist red litmus paper at the mouth of the test tube, taking care not to touch the wall of the test tube, and observe what happens.
6. Next, hold the strip of filter paper with acidified potassium dichromate(IV) at the mouth of the test tube, taking care not to touch the wall of the test tube.
7. After testing the gas with acidified potassium dichromate(VI), bubble the gas produced into the lime water using the delivery tube until there is no further change.
8. Repeat steps 1 and 4 to generate more sulfur dioxide, if necessary. Though it may be convenient to add more sodium sulfate(IV) at the start, try to limit the amount of sulfur dioxide released into the surroundings.

Results 1

A colourless, pungent gas (similar to that produced by fireworks or sparklers) is produced which turns moist blue litmus red.

$$Na_2SO_{3(s)} + 2HCl_{(aq)} \rightarrow 2NaCl_{(aq)} + H_2O_{(l)} + SO_{2(g)}$$

Sulfur dioxide is a reducing agent and it turns orange acidified potassium dichromate(VI) green.

$$K_2Cr_2O_{7(aq)} + 4H_2SO_{4(aq)} + 3SO_{2(g)} \rightarrow K_2SO_{4(aq)} + Cr_2(SO_4)_{3(aq)} + H_2O_{(l)}$$

Sulfur dioxide reacts with calcium hydroxide (lime water) to produce a white precipitate, calcium sulfate(IV).

$$SO_{2(g)} + Ca(OH)_{2(aq)} \rightarrow CaSO_{3(s)} + H_2O_{(l)}$$

On further bubbling into lime water, the white precipitate disappears and a colourless solution is formed.

$$CaSO_{3(s)} + H_2O_{(l)} + SO_{2(g)} \rightarrow Ca(HSO_3)_{2(aq)}$$

Thus, sulfur dioxide reacts with lime water in a very similar way to the way carbon dioxide reacts with it (in other words, a white precipitate forms which disappears on prolonged bubbling of carbon dioxide into lime water).

$$CO_{2(g)} + Ca(OH)_{2(aq)} \rightarrow CaCO_{3(s)} + H_2O_{(l)}$$
$$CaCO_{3(s)} + H_2O_{(l)} + CO_{2(g)} \rightarrow Ca(HCO_3)_{2(aq)}$$

Thus, it is advisable to carry out a second test using acidified potassium dichromate(VI) in addition to a lime water test, to differentiate between sulfur dioxide and carbon dioxide (which does not have any effect on acidified potassium dichromate(VI)). The odour of the gas gives another indication of its identity – sulfur dioxide has a characteristic odour and students can get whiffs of it when it escapes into the surroundings. However, on safety grounds, students must not breathe in any gas directly from the test tube.

Method 2 – testing for oxygen and nitrogen dioxide

1 Add enough copper(II) nitrate(V) to fill the curved base of a test tube.
2 Heat strongly.
3 Hold a moist blue litmus paper and a moist red litmus paper at the mouth of the test tube, taking care not to touch the wall of the test tube, and observe what happens.
4 Insert a glowing splint into the test tube.

Results 2

The blue copper(II) nitrate(V) will turn black upon heating (due to the formation of copper(II) oxide). A brown pungent gas, nitrogen dioxide, is evolved which turns moist blue litmus red.

$$2Cu(NO_3)_{2(s)} \rightarrow 2CuO_{(s)} + 4NO_{2(g)} + O_{2(g)}$$

Oxygen will also be evolved. The increased amount of oxygen in the test tube will result in the glowing splint burning more vigorously, causing it to relight. A blue solid turning black on heating indicates the presence of copper(II) ions and the liberation of nitrogen dioxide and oxygen indicates the presence of the nitrate(V) ion. Students need to be able to interpret the observations to identify the ions present.

Method 3 – testing for chlorine

1 Add enough manganese(IV) oxide to fill the curved base of a test tube.
2 Dip a strip of filter paper into a mixture of potassium iodide solution (0.1 mol dm^{-3}) and 5% starch solution.
3 Use a dropper to add three drops of concentrated hydrochloric acid to the test tube.
4 Hold a moist blue litmus paper and a moist red litmus paper at the mouth of the test tube, taking care not to touch the wall of the test tube, and observe what happens.
5 Next, hold the strip of filter paper with potassium iodide and starch solutions at the mouth of the test tube, taking care not to touch the wall of the test tube.

Results 3

A yellowish-green pungent gas is evolved which turns moist blue litmus red and then bleaches it (turns it white).

$$4HCl_{(aq)} + MnO_{2(s)} \rightarrow MnCl_{2(aq)} + 2H_2O_{(l)} + Cl_{2(g)}$$

Chlorine is an acidic gas which is also an oxidising agent – it oxidises the dye in litmus paper, turning it white. The gas also reacts with potassium iodide, liberating iodine which reacts with starch solution, turning it blue.

$$Cl_{2(g)} + 2KI_{(aq)} \rightarrow I_{2(aq)} + 2KCl_{(aq)}$$

Considerations in testing for gases

Research has shown that students find inorganic chemical analysis very tedious, especially testing for gases. They tend to test for all the gases listed in the chemistry syllabus without first considering which gases can be liberated from the given reagents and the procedures; time and effort are wasted testing for gases which cannot be liberated by the reactions involved and students may run out of gas before completing their battery of tests. Some rules of thumb are given here which might help to point students in the right direction when it comes to testing for gases.

1 Gases which might be liberated when a dilute acid is added to an unknown solution/solid include:
   - hydrogen
   - sulfur dioxide
   - carbon dioxide.
2 If the unknown solid looks metallic, students need to prepare to test for hydrogen as an acid–metal reaction liberates hydrogen, for example:

   $$Zn_{(s)} + 2HCl_{(aq)} \rightarrow ZnCl_{2(aq)} + H_{2(g)}$$

   Before adding the acid to the unknown sample, students need to light the Bunsen burner and get a wooden splint ready. This is because when the acid is added to the metallic sample, a gas may be liberated instantly and escape into the surroundings before preparations to test it are completed. To test for hydrogen, students should place a lighted splint about 1 cm into the mouth of the test tube. The hydrogen liberated should be able to mix with the oxygen in the air in that part of the test tube to give an audible explosion (a 'pop' sound) when ignited.

   $$2H_{2(g)} + O_{2(g)} \rightarrow 2H_2O_{(l)}$$

3 If the unknown solid does not look metallic, or an unknown solution is given, students need to test for carbon dioxide and sulfur dioxide as only carbonates and sulfate(IV) will react with acids to produce a gas. The choice of acid to use is important. Sulfuric(VI) acid and hydrochloric acid may react with the unknown sample to produce insoluble sulfates(VI) and chlorides, respectively, which will prevent the hydrogen ions from reacting further with the carbonate or sulfate(IV). For example, marble chips will react with sulfuric(VI) acid to form insoluble calcium sulfate(VI):

   $$CaCO_{3(s)} + H_2SO_{4(aq)} \rightarrow CaSO_{4(s)} + H_2O_{(l)} + CO_{2(g)}$$

   Thus, initially, there will be effervescence, but the reaction will soon stop when the layer of calcium sulfate(VI) coats the marble chips, preventing further reaction between the marble and the acid (Figure 9.3). If the carbonate and sulfate(IV) are finely powdered, then sufficient gas may be evolved to enable it to be identified. All nitrate(V) salts are soluble, so dilute nitric(V) acid can be used; however, do note that as nitric(V) acid is an oxidising reagent, there may be a redox reaction if the unknown sample is a reducing agent.

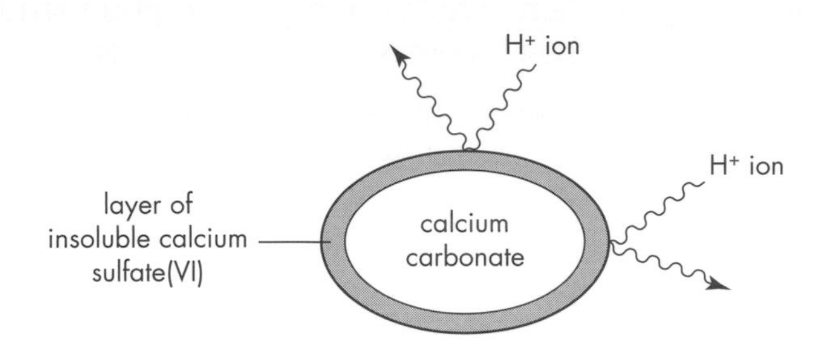

Figure 9.3 Layer of insoluble calcium sulfate(VI) preventing further reaction between the marble chip and acid

4 Before adding the acid, students need to prepare the moist litmus papers (both sulfur dioxide and carbon dioxide turn moist blue litmus red, but the change is faster with sulfur dioxide), strips of filter paper dipped into acidified potassium dichromate(VI) (sulfur dioxide turns it from orange to green) and lime water (both carbon dioxide and sulfur dioxide give a white precipitate with lime water).

As previously mentioned in section 9.2 on cations, students need to be able to carry out the procedures involved 'automatically' and, to achieve this, they need to practise them; a practical session or two can solely be devoted to mastering the required manipulative skills. When the students are proficient in carrying out the procedures, you may want them to propose which gases they need to prepare to test for before they add dilute acid to the unknown sample, the sequence of tests and the possible results expected. They can then be given an unknown sample to try out their proposals and thus determine their viability.

5 Concentrated sulfuric(VI) acid is added to the unknown sample with either an oxidising or reducing reagent and the gases that secondary students can be asked to identify are:

- chlorine
- nitrogen dioxide
- sulfur dioxide.

When manganese(IV) oxide and concentrated sulfuric acid are added to a chloride, chlorine gas is produced (this is similar to the reaction of concentrated hydrochloric acid and manganese(IV) oxide). For example:

$$4NaCl_{(s)} + 2H_2SO_{4(aq)} + MnO_{2(s)} \rightarrow MnCl_{2(aq)} + 2H_2O_{(l)} + Cl_{2(g)} + 2Na_2SO_{4(aq)}$$

If nitrate(V) is present in the unknown sample and concentrated sulfuric acid is added together with copper turnings, nitrogen dioxide will be produced. For example:

$$4NaNO_{3(s)} + 2H_2SO_{4(aq)} + Cu_{(s)} \rightarrow 2Na_2SO_{4(aq)} + u(NO_3)_{2(aq)} + 2NO_{2(g)} + 2H_2O_{(l)}$$

In the above reactions, the concentrated sulfuric(VI) acid will react with the chloride and nitrate(V) to produce hydrochloric and nitric(V) acid, respectively, which will undergo further redox reactions. Again students can be asked to propose a sequence or flowchart of tests to identify the three gases when concentrated sulfuric(VI) acid is used.

6 An unknown sample can liberate these gases upon heating:
- carbon dioxide
- sulfur dioxide
- nitrogen dioxide
- oxygen
- ammonia (if the unknown sample is heated with sodium hydroxide solution)
- chlorine (if the unknown sample is heated with an oxidising agent such as manganese(IV) oxide).

Heating an unknown sample and testing for gases is difficult as there is usually no visible sign that a gas is evolved, unlike the effervescence produced when an acid reacts with carbonate or sulfate(IV). Thus, students need to practise the individual tests for gases until they are proficient. They also need to be able to carry out a battery of tests to determine the identity of the gas(es) evolved, if any, when an unknown sample is heated. Students can be asked to propose a sequence or a flowchart to test the gases which can be liberated when a substance is heated. Generally, the appearance of the gas, if any, is noted, followed by the use of moist litmus paper to determine whether the gas is acidic, basic or neutral, and whether it bleaches the litmus paper. Subsequent tests would depend on results of the litmus test. When students have mastered the necessary skills and produced a suitable sequence, they can be given unknown samples to heat.

Table 9.3 overleaf provides a summary of the reactions which liberate gases.

Table 9.3 A summary of the reactions which liberate gases

Gas	Reactions which produce the gas
Hydrogen	acid + metal → salt + hydrogen
Oxygen	• Thermal decomposition of nitrate(V) • Decomposition of hydrogen peroxide
Carbon dioxide	• acid + carbonate/hydrogencarbonate → salt + water + carbon dioxide • Thermal decomposition of carbonate/hydrogencarbonate
Sulfur dioxide	• acid + sulfate(IV)/hydrogensulfate(IV) → salt + water + sulfur dioxide • Thermal decomposition of sulfate(IV) • Reduction of concentrated sulfuric(VI) acid
Chlorine	Oxidation of chlorides
Nitrogen dioxide	• Thermal decomposition of nitrate(V) • Reduction of nitrate(V)
Ammonia	ammonium salt + alkali → salt + water + ammonia

9.4 Anions

Previous knowledge and experience

The anions that are involved in secondary inorganic chemical analysis are, in general, carbonate, chloride, iodide, nitrate(V), sulfate(IV) and sulfate(VI). Students may be familiar with the carbonate ion from lower secondary science; the reactions of the remaining anions have been introduced in the tests for gases or ionic precipitation (iodide). Silver nitrate(V) solution, barium chloride/nitrate(V) solution, lead(II) nitrate(V) solution and dilute acids are generally used to test for anions.

Some reactions to identify anions

PROCEDURE

Materials
- test tubes
- a test-tube rack
- a test-tube holder
- a Bunsen burner
- a lighter
- red and blue litmus paper
- deionised water in a wash bottle

- sodium chloride solution, 0.2 mol dm^{-3}
- sodium nitrate(V) solution, 0.2 mol dm^{-3}
- copper(II) sulfate(VI) solution, 0.2 mol dm^{-3}
- barium nitrate(V) solution, 0.2 mol dm^{-3}
- silver nitrate(V) solution, 0.2 mol dm^{-3}
- nitric(V) acid, 1 mol dm^{-3}
- ammonia solution, 1 mol dm^{-3}
- sodium hydroxide solution, 1 mol dm^{-3}
- aluminium powder

Safety

- All students should wear eye protection.
- Do not pour excess reagent back into the stock bottles. If too much is taken, dispose of the excess.
- Flush with plenty of water if the reagents come into contact with skin or eyes.
- Never point the test tube being heated at anybody as the contents may shoot out during heating.
- Be aware of hot glassware.

Method 1

1. Add enough copper(II) sulfate(VI) solution to half-fill the curved base of a test tube.
2. Add barium nitrate(V) solution to the test tube containing copper(II) sulfate(VI) to a height of 2 cm.
3. Add nitric(V) acid to the test tube to a height of 2 cm.

Results 1

A white precipitate is formed when barium nitrate(V) is added to copper(II) sulfate(VI) and there is no visible reaction when nitric(V) acid is added. Sulfate(VI) does not react with nitric(V) acid.

$$Ba(NO_3)_{2(aq)} + CuSO_{4(aq)} \rightarrow BaSO_{4(s)} + Cu(NO_3)_{2(aq)}$$
$$Ba^{2+}{}_{(aq)} + SO_4{}^{2-}{}_{(aq)} \rightarrow BaSO_{4(s)}$$

Students may have difficulty determining the colour of the precipitate because of the blue liquid present, so you may want them to suggest how they can determine the colour of the precipitate more easily (decant the blue liquid or filter off the precipitate).

You may also want to ask students the following questions to help them make sense of the procedures involved in the identification of anions:

- Why is nitric(V) acid added?
- Could hydrochloric acid or sulfuric(VI) acid be used instead of nitric(V) acid?
- Could the nitric(V) acid be added to the unknown solution before barium nitrate(V) solution?

Generally, tests for anions involve the formation of precipitates in ionic precipitation reactions with silver nitrate(V), barium nitrate(V)/chloride or lead(II) nitrate and whether the anions react with acid; if they do, a gas is usually evolved and has to be identified. Barium ions form precipitates with carbonate, sulfate(IV) and sulfate(VI) ions, while silver ions form precipitates with chloride, iodide, carbonate and sulfate(IV) ions; silver sulfate(VI) is sparingly soluble and a precipitate may not be formed if the concentrations of the reagents used are low. Lead(II) nitrate(V) is used to determine the presence of iodide and chloride ions. Lead(II) iodide and lead(II) chloride precipitates are soluble in hot water (when the mixture is heated) but will recrystallise when cooled (as discussed in section 9.1). The equations for some of these reactions are given overleaf.

$Ba^{2+}{}_{(aq)} + CO_3{}^{2-}{}_{(aq)} \rightarrow BaCO_{3(s)}$

$Ba^{2+}{}_{(aq)} + SO_3{}^{2-}{}_{(aq)} \rightarrow BaSO_{3(s)}$

$Ba^{2+}{}_{(aq)} + SO_4{}^{2-}{}_{(aq)} \rightarrow BaSO_{4(s)}$

$2Ag^{+}{}_{(aq)} + CO_3{}^{2-}{}_{(aq)} \rightarrow Ag_2CO_{3(s)}$

$2Ag^{+}{}_{(aq)} + SO_3{}^{2-}{}_{(aq)} \rightarrow Ag_2SO_{3(s)}$

$Ag^{+}{}_{(aq)} + Cl^{-}{}_{(aq)} \rightarrow AgCl_{(s)}$

$Ag^{+}{}_{(aq)} + I^{-}{}_{(aq)} \rightarrow AgI_{(s)}$

$Pb^{2+}{}_{(aq)} + 2Cl^{-}{}_{(aq)} \rightarrow PbCl_{2(s)}$

$Pb^{2+}{}_{(aq)} + 2I^{-}{}_{(aq)} \rightarrow PbI_{2(s)}$

Except for silver iodide which is pale yellow and lead(II) iodide which is yellow, all the other precipitates are white and need to be differentiated. To determine whether carbonate and sulfate(IV) are present, acid is added to the precipitate. If present, carbonate and sulfate(IV) will react to form carbon dioxide and sulfur dioxide, respectively, and the gas has to be identified. To avoid introducing an additional anion to the test, nitric(V) acid is added when barium nitrate(V), silver nitrate(V) and lead(II) nitrate(V) are used, and hydrochloric acid is added when barium chloride is used. Sulfuric(VI) acid has to be used with caution as it has an additional disadvantage of forming precipitates with calcium, lead(II) and barium ions. Figure 9.4 describes how an anion may be identified using barium nitrate(V) or chloride solution.

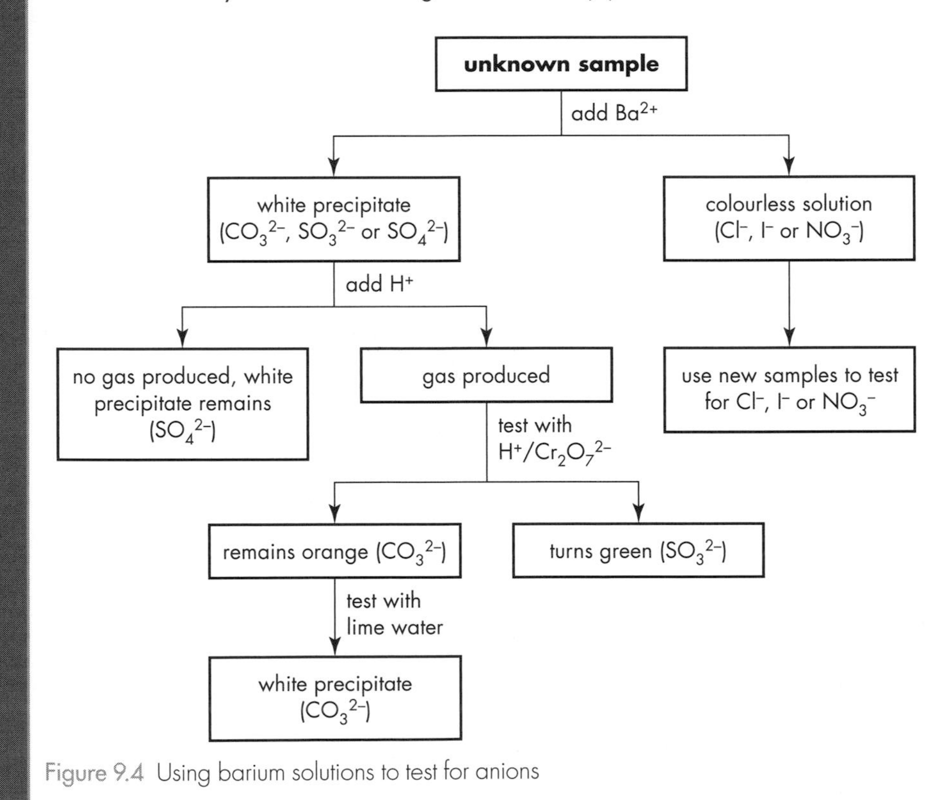

Figure 9.4 Using barium solutions to test for anions

You may want students to draw similar flowcharts for the use of silver nitrate(V) and lead(II) nitrate(V) solutions followed by the addition of acid. You may also want students to draw flowcharts for the addition of acid followed by silver, barium and lead(II) solutions and compare them with the flowcharts in which the acid was added after the silver, barium and lead(II) solutions. This is to help them to understand the impact of the procedures on the results obtained so that they may better understand what they do in the inorganic chemical analysis practical work; research has shown that students often do not understand the purpose of the procedures. If students have difficulty, you may want them to carry out the relevant experiments.

Method 2

1 Add enough sodium chloride solution to half fill the curved base of a test tube.
2 Add silver nitrate(V) solution to the test tube containing sodium chloride to a height of 2 cm.
3 Pour half of the contents of the test tube into another test tube.
4 To one test tube containing the sodium chloride/silver nitrate(V) mixture, add nitric(V) acid to a height of 2 cm.
5 To the other test tube containing the sodium chloride/silver nitrate(V) mixture, add ammonia solution until there is no further change.

Results 2

A white precipitate is formed when silver nitrate(V) is added to sodium chloride:

$$Ag^+_{(aq)} + Cl^-_{(aq)} \rightarrow AgCl_{(s)}$$

Silver chloride does not react with nitric(V) acid but will react with excess ammonia to form diamminesilver ions which are soluble to give a colourless solution. This reaction is similar to that of zinc and copper(II) hydroxides with ammonia solution.

$$AgCl_{(s)} + 2NH_{3(aq)} \rightarrow [Ag(NH_3)_2]^{2+}_{(aq)} + Cl^-_{(aq)}$$

Method 3

1 Add sodium nitrate(V) solution to a test tube to a height of 1 cm.
2 Add sodium hydroxide solution to the test tube containing sodium nitrate(V) to a height of 2 cm.
3 Add a spatula of aluminium powder to the mixture.
4 Heat gently.
5 Hold a moist blue litmus paper and a moist red litmus paper at the mouth of the test tube, taking care not to touch the wall of the test tube, and observe what happens.

Results 3

A pungent gas is evolved which turns red litmus blue. This indicates that ammonia is formed. The nitrate(V) ion is reduced by the aluminium to form an ammonium ion which will react with the sodium hydroxide to form ammonia. This is a test for nitrate(V).

$$3NaNO_{3(aq)} + 8Al_{(s)} + 5NaOH_{(aq)} + 18H_2O_{(l)} \rightarrow 3NH_{3(g)} + 8Na[Al(OH)_4]_{(aq)}$$

Considerations in testing for anions

To identify the anion present in an unknown sample, silver, barium or lead(II) solutions are generally added to determine whether a precipitate forms and whether the precipitate reacts with acids (carbonate or sulfate(IV)). The unknown sample can also be given in a solid form and students can be asked to heat the substance. When adding acid or when heating, students must be prepared to test for the appropriate gases as discussed in section 9.3.

Extension activities

Quantitative and qualitative analyses can be combined together to present guided inquiry tasks to students. For example, students can be given a sample containing two salts such as calcium carbonate and calcium chloride. They are told that they need to identify the common cation and two anions, as well as calculate the percentages by mass of the two salts. The students will be required to test for gases, anions and cations, and need to propose methods to determine the percentages by mass. One possible way is to identify the ions first and then to react a known mass of the sample completely with a known (excess) volume and concentration of dilute hydrochloric or nitric(V) acid. A sample of the acid left behind after the reaction can then be titrated with a known volume of dilute sodium hydroxide solution. The amount of acid reacted and the amount of carbonate reacted can be calculated, and hence the percentage by mass of the two salts.

Other resources

A study of students' perceptions of learning inorganic chemical analysis can be found in:
Tan, K.C.D., Goh, N.K., Chia, L.S., & Treagust, D.F. (2001). Secondary students' perceptions about learning qualitative analysis in inorganic chemistry. *Research in Science & Technological Education*. **19**(2): 223–234.

More information on the difficulties that students have in inorganic chemical analysis, and a diagnostic instrument to determine their understanding of the topic can be found in:
Tan, K.C.D., Goh, N.K., Chia, L.S., & Treagust, D.F. (2002). Development and application of a two-tier diagnostic instrument to assess high school students' understanding of inorganic chemistry qualitative analysis. *Journal of Research in Science Teaching*. **39**(4): 283–301.

The following book provides more information on how to facilitate students' learning in practical work:
Woolnough, B. & Allsop, T. (1985). *Practical work in science*. Cambridge: Cambridge University Press.

More information on chemical analysis can be found by visiting:
http://en.wikipedia.org/wiki/Analytical_chemistry

10 Organic chemistry and the chemistry of natural products

Vanessa Kind

10.1 Introduction

This chapter aims to provide activities that will help students develop a secure understanding of the basic principles of organic chemistry and natural products. Organic chemistry is the study of compounds containing carbon, usually combined with hydrogen, oxygen and other elements such as nitrogen and sulfur. 'Natural products' are chemicals such as alcohol, vinegar, glucose (and other sugars) and esters, a group of compounds found in ripe fruit. Examples of reactions that make some of these substances are provided.

Starting points – origins of organic chemistry

Organic chemistry refers to compounds containing the chemical element carbon, and usually a combination of hydrogen and oxygen. 'Organic' describes the fact that many of these compounds are found in plants and animals. For example, glucose (chemical formula $C_6H_{12}O_6$) is a simple sugar made in photosynthesis that helps to keep most cells alive, as well as being the basic 'building block' for more complicated sugars such as sucrose, starch and cellulose.

The eighteenth-century Swedish chemist Carl Wilhelm Scheele (pronounced 'Shaylur') was probably the first person in Europe to isolate organic compounds from plant and animal sources. He showed that milk turns sour because of the compound lactic acid (formal, or systematic, name 2-hydroxypropanoic acid, $CH_3CHOHCOOH$). At that time, chemists believed organic compounds could only be made in reactions in cells, by means of a 'life force'. This had to change after 1828, when the German chemist Friedrich Wohler made urea, the compound excreted in animal urine, by heating a solution of the inorganic compound ammonium cyanate:

$$\underset{\text{ammonium cyanate}}{NH_4CNO} \rightarrow \underset{\text{urea}}{CO(NH_2)_2}$$

This prompted the understanding that organic compounds could be synthesised in laboratories, leading to the development of 'biochemistry' as the specialist branch of chemistry investigating reactions in living cells and life processes.

Choosing a route

In this chapter we first look at the simplest organic compounds, the hydrocarbons (section 10.2), before moving on to consider an important class of substances that students will be familiar with: polymers (section 10.3). The simplest of these are made from polymerising unsaturated hydrocarbons and others can be understood as being based upon the same principles, but starting with substituted monomers. The final section (10.4) looks at a range of natural products – materials which are again likely to be familiar to students, but are here considered as chemical substances, thus exemplifying the way in which chemistry studies substances which are found in the materials around us (Chapter 1).

Previous knowledge and experience

Relevant information from prior experiences means students are likely to know that:

- alcohol, vinegar and sugars exist
- photosynthesis produces carbohydrates (sugars) in green plants
- crude oil and gas are the remains of long-dead organisms
- some compounds in fruit and perfumes have very sweet smells
- food includes carbohydrates, fats, proteins, vitamins and fibre
- plastics are made from crude oil.

They may also be aware of properties of some compounds, for example, that:

- alcohol is produced by fermentation of sugar using yeast
- vinegar is ethanoic acid, so behaves as an acid
- acidic chemicals in milk and wine change the taste, for example make milk go sour
- fuels such as petrol and butane burn with a yellow, smoky flame.

Students' difficulties with organic chemistry

Students' misconceptions about organic chemistry are less well documented than those for other parts of the subject. Aspects of the topic they are likely to find difficult are described.

■ Students may find the whole idea of organic chemistry difficult

The notion that there is a 'special' area of chemistry called organic chemistry may be problematic initially. The terminology does not help. It is useful to help students understand the distinction between organic chemistry and other aspects of the subject, namely 'inorganic' as the chemistry of the chemical elements in the periodic table and 'physical' as the laws governing how chemicals react, such as energetics, rates and equilibria. Some chemists may regard these as rather false distinctions, but in order to help students learn the subject, they are helpful. Finding out what students might understand by the term organic chemistry would be useful, as this will generate insight into their thinking that can be addressed. Describing the origins of organic chemistry (see the previous page) will help to set the context. Establishing and drawing on students' prior knowledge will help them realise they have relevant knowledge about the topic.

■ Students may struggle with nomenclature

Learning how to name organic compounds is a challenge for some students. The names of organic compounds are governed by strict international regulations. The names and rules can get quite complicated. For example, the formal name for lactic acid, 2-hydroxypropanoic acid, does not exactly run easily off the tongue, but it tells a chemist exactly what the structure of the molecule is (see the section on naming, page 309), in a way that the colloquial name 'lactic acid' does not. Helping students learn the basics of nomenclature by showing that the same prefixes and suffixes occur in many series of organic compounds is necessary. However, systematic names are not used consistently, often because of their complexity, so colloquial ones are often substituted. Examples include fatty acids: stearic acid (pronounced 'steeric' or 'stairic') has the systematic name octadecanoic acid; and sugars – glucose is 6-(hydroxymethyl)oxane 2,3,4,5-tetrol.

10.2 Hydrocarbons

Organic chemistry stems from the ability of each carbon atom to form four bonds. Although other chemical elements also form four bonds per atom, bonds involving carbon atoms have relatively high bond enthalpy values, so require input of large amounts of energy to break them. Also, carbon atoms bond readily with each other, a property called 'catenation'. So carbon–carbon bonds are commonplace. These factors lead to ranges of compounds based on chains of carbon atoms, sometimes up to thousands of atoms long. Hydrocarbons are the simplest type of organic molecules, comprising hydrogen and carbon only.

There are three types of hydrocarbon – alkanes, alkenes and alkynes. At upper secondary level knowledge of only the first four alkanes and first three alkenes is usually required. These are discussed in more detail as we go through the chapter. The alkynes have a carbon–carbon triple bond in their molecules. No knowledge of alkynes is expected at this level.

Modelling alkanes and alkenes

This activity (page 308) is a good one to introduce nomenclature and some basic principles of organic chemistry. Tables 10.1 and 10.2 show the names, molecular formulae and structural formulae of the first five alkanes and the first four alkenes, so go slightly beyond the typical requirements for upper secondary chemistry.

Table 10.1 Names, molecular formulae, boiling temperatures and structural formulae of some alkanes

Number of carbon atoms	Prefix	Alkanes			
		Suffix	Molecular formula	Structural formula(e)	Boiling temperature (°C)
1	Meth-	-ane	CH_4	H | H—C—H | H	–162
2	Eth-		C_2H_6	H H | | H—C—C—H | | H H	–89
3	Prop-		C_3H_8	H H H | | | H—C—C—C—H | | | H H H	–42
4	But-		C_4H_{10}	H H H H | | | | H—C—C—C—C—H | | | | H H H H	–1
5	Pent-		C_5H_{12}	H H H H H | | | | | H—C—C—C—C—C—H | | | | | H H H H H	36

Table 10.2 Names, molecular formulae, boiling temperatures and structural formulae of some alkenes

Number of carbon atoms	Prefix	Alkenes			
		Suffix	Molecular formula	Structural formula(e)	Boiling temperature (°C)
1	Meth-		–	–	–
2	Eth-	-ene	C_2H_4	H H C=C H H	–104
3	Prop-		C_3H_6	H H H C=C–C–H H H	–47
4	But-		C_4H_8	H H H H C=C–C–C–H H H H	4
5	Pent-		C_5H_{10}	H H H H H C=C–C–C–C–H H H H H	30

PROCEDURE

Materials

- Molymod® sets including up to 5 carbon atoms and 12 hydrogen atoms
- copies of Table 10.1 and Table 10.2

Method

Ask students to make models of the alkanes shown in Table 10.1 and the alkenes shown in Table 10.2. This may seem a straightforward, descriptive activity, but see the questions below and the discussion points opposite for aspects of the subject that this draws out.

Questions to ask

- How many bonds does each carbon atom make?
- What do you notice about the shape of the bonds around each carbon atom?
- How many bonds does each hydrogen atom make?
- What does the prefix tell you? What does the suffix tell you?
- What combination of carbon and hydrogen atoms is added to make the next chemical in the list of alkanes/alkenes?
- Write down a formula that could apply to all alkanes.
- Write down a formula that could apply to all alkenes.
- In what ways are the formulae of the alkanes and alkenes the same?
- In what ways are the formulae of the alkanes and alkenes different?
- How many carbon atoms could there be in one alkane or alkene molecule?
- What chemicals do we know (we use them every day!) that have very large numbers of carbon atoms in one molecule?

Discussion points

- Note that:
  - the prefix (meth-, eth-, prop-, etc.) corresponds to the number of carbon atoms
  - the suffix (-ane or -ene) stays constant.
- The naming of organic compounds is similar to people's first names and surnames – the surname, or suffix, shows the 'family' of compounds to which the chemical belongs. The prefix changes according to the number of carbon atoms. An organic chemical family is called a homologous series.
- The same prefixes are found throughout organic chemistry.
- Most organic compounds are members of homologous series. Each family has its own general formula. The general formula for alkanes is C_nH_{2n+2} and for alkenes is C_nH_{2n}. Note that:
  - A combination of one carbon and two hydrogen atoms ($-CH_2-$) is needed to make the next member of these homologous series.
  - 'Alk-' is a general term for the number of carbon atoms. The phrase 'alkyl group' is used to describe a combination of carbon and hydrogen atoms such as $-CH_3$ or $-C_2H_5$.
- Each carbon atom can form a maximum of four bonds. This ability of carbon atoms to form four bonds is responsible for the vast range of organic compounds.
  - The shape around each carbon atom in an alkane is tetrahedral.
  - The shape around the two carbon atoms in a double bond is flat (planar).
- Hydrogen atoms form a maximum of one bond each.
- The general formulae imply there could be an infinite number of carbon atoms in one alkane/alkene. (This is discussed further in the section on plastics, section 10.3.)
- The alkanes and alkenes are similar in that:
  - they contain hydrogen and carbon only
  - adding a $-CH_2-$ group produces the next alkane or alkene in the homologous series.
- The alkanes and alkenes differ in that:
  - all alkanes have only single covalent bonds – they are 'saturated'
  - all alkenes have one carbon–carbon double bond – they are 'unsaturated'.
- We use poly(ethene) (HDPE, High Density Poly(ethene), and LDPE, Low Density Poly(ethene)) every day. These are plastics that are made in great quantities around the world. (This is discussed further in the plastics activity later in the chapter.)

Properties of alkanes and alkenes

This class experiment illustrates differences in the properties between the two types of chemicals. To help students make sense of what they are doing, encourage them to make models of cyclohexane (Figure 10.1) and cyclohexene (Figure 10.2).

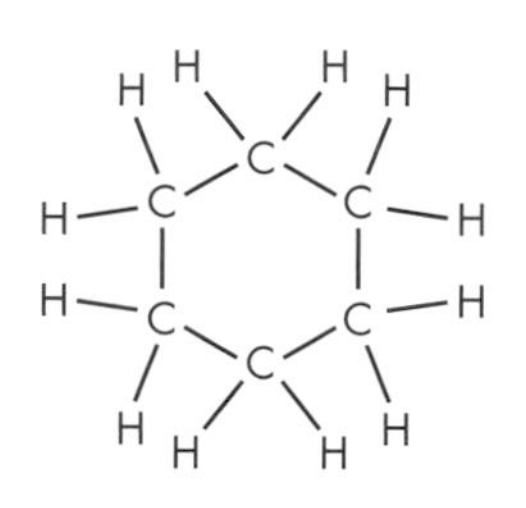

Figure 10.1
Structural formula of cyclohexane

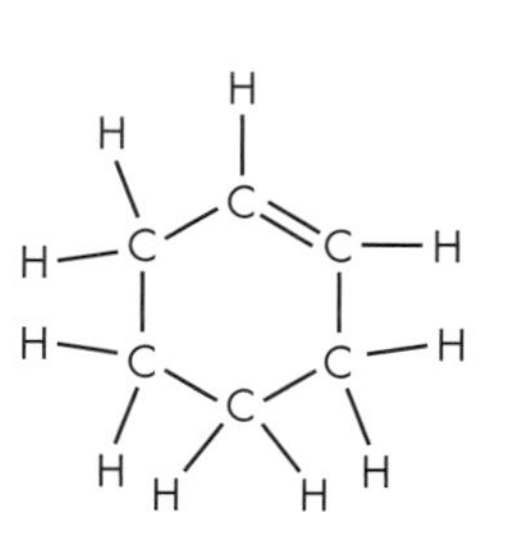

Figure 10.2
Structural formula of cyclohexene

Before carrying out the experiment, ask students:

- What does 'cyclo-' mean?
- How many carbon atoms correspond with the prefix 'hex-'?
- Will alkanes and alkenes react in the same ways or not? Explain.

PROCEDURE

Materials

- Molymod® set containing 6 carbon atoms and 14 hydrogen atoms, 2 bromine and 2 oxygen atoms
- about 2 cm^3 cyclohexane
- about 2 cm^3 cyclohexene
- two watch glasses
- a heatproof mat
- splints
- two dropping pipettes
- four test tubes, or smaller tubes, e.g. ignition tubes
- a test-tube rack, or a rack to take smaller tubes
- eye protection
- access to:
  - 0.2 $mol\,dm^{-3}$ acidified potassium manganate(VII) solution
  - 0.2 $mol\,dm^{-3}$ bromine water
  - a lighted Bunsen burner

Safety
All students should wear eye protection.

Method

Test the cyclohexane and cyclohexene as follows:

1. Set the watch glasses on the heatproof mat. Place about 5 drops of the cyclohexane on one watch glass and 5 drops of cyclohexene on the other. Light the samples.
2. Place 5 drops of each of cyclohexane and cyclohexene in two separate test tubes.
3. Take one pair of test tubes. Add a few drops of acidified potassium manganate(VII) solution to the cyclohexane in one tube and to the cyclohexene in the other tube. Shake the tubes gently.
4. Take the second pair of test tubes. Add a few drops of bromine water to the cyclohexane in one tube and to the cyclohexene in the other tube. Shake the tubes gently.

Observations

Table 10.3 Observations for tests on cyclohexane and cyclohexene

Chemical	Burning	Acidified potassium manganate(VII) solution	Bromine water
Cyclohexane	Yellow flame	No change	No change*
Cyclohexene	Yellow flame – more smoky than cyclohexane	Decolourises	Decolourises

*Note that bromine is soluble in cyclohexane so the colour may still exist but in the non-aqueous layer.

Questions to ask

- Were your predictions correct?
- Why does cyclohexene decolourise the two solutions?
- Why is no reaction seen with cyclohexane?
- Would we expect the same reactions with other alkanes and alkenes? (Use the molecular models to help explain the observations.) Explain why/why not.
- What does this tell us about the differences between alkanes and alkenes?

Discussion points

Only the cyclohexene reacts and so decolourises the potassium manganate(VII) solution and the bromine water. This is because of the carbon–carbon double bond in the molecule. All alkanes are saturated molecules, with only single covalent bonds between all atoms, so there is no 'spare' bond available to react under these conditions. Alkenes are unsaturated because of the double bond.

The reactions observed with cyclohexene are called addition reactions. This is because extra atoms add to the molecule at the double bond. This is often called addition across the double bond. The new molecules formed are: with potassium manganate(VII), cyclohexan-1,2-diol (Figure 10.3); with bromine water, 1,2-dibromocyclohexane (Figure 10.4).

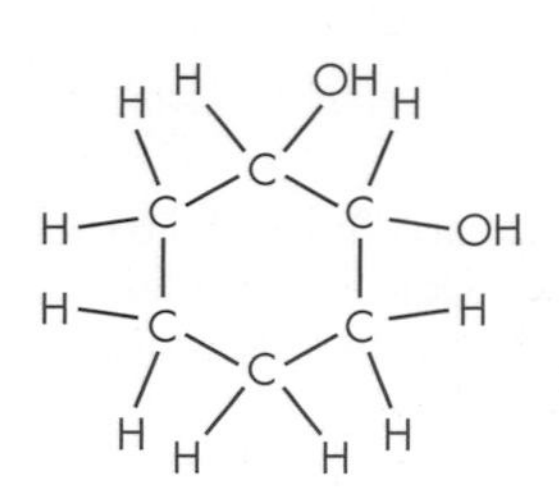

Figure 10.3
Structural formula of cyclohexan-1,2-diol. This is a member of the homologous series of alcohols.

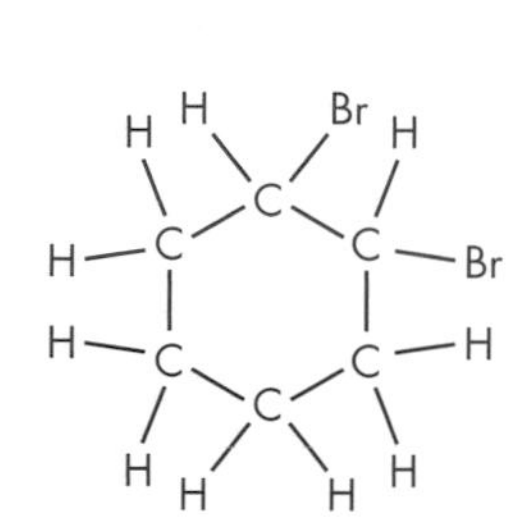

Figure 10.4
Structural formula of 1,2-dibromocyclohexane. This is a member of the homologous series of haloalkanes.

The reactions are typical of all alkenes. The two tests are referred to as tests for non-saturation.

Differences in the burning test are harder to distinguish. Alkanes and alkenes both combust easily. Alkenes tend to burn with a more smoky flame because of the lower ratio of carbon to hydrogen.

Cracking large hydrocarbon molecules

The first four alkanes are gases at room temperature and pressure. Natural gas is mainly methane. Other, larger alkanes are found in crude oil. The largest are in tar, or bitumen, used in making road surfaces. Enormous quantities of the smaller ones are used every day as fuels. Petrol comprises a mixture of alkanes of 4–12 carbon atoms long; diesel (named after the German engineer Rudolf Diesel, inventor of the diesel engine) is a mixture of alkanes 8–21 carbon atoms long; and aviation fuel (known as 'avgas', aviation gasoline or kerosene) is mainly octane (C_8H_{18}), but is a complex mixture of up to about 1000 chemicals.

To get enough of the smaller alkanes to meet worldwide demand, larger molecules are broken apart in a process called cracking. There are different types of cracking, each designed to produce specific types and sizes of molecules. The principle, however, is similar – breaking up a long-chain carbon atom molecule into smaller ones. The long-chain chemical is heated, vaporised and passed over a hot catalyst. The large molecules break into smaller ones, some of which are gases. The gases are collected over water and can be tested for unsaturation.

The following experiment, which can be done as a class experiment but is best done as a demonstration for students at this level, mimics the cracking process. The demonstration involves passing Vaseline® vapour over hot, broken porcelain and collecting the gas(es) produced. The products are tested. Care is required to ensure no blockage of the delivery tube occurs, while a non-return valve must be used to help avoid cold water sucking back into the hot equipment.

PROCEDURE

Materials

- a boiling tube and bung
- a delivery tube fitted with a non-return valve
- clamp stands, bosses and clamps
- a trough
- water
- up to six test tubes to collect gas, with corks to fit (rubber reacts with alkenes)
- a test-tube rack
- mineral wool
- liquid paraffin or fresh Vaseline
- some finely divided porcelain catalyst ('porous pot')
- a Bunsen burner
- to test the gas for unsaturation: access to acidified potassium manganate(VII) solution and/or bromine water
- Molymod® set containing about 20 carbon atoms and 42 hydrogen atoms – use to make a model of the paraffin molecule $C_{20}H_{42}$ present in Vaseline
- chemicals, pipettes and splints to test the gases produced (as used in the practical to demonstrate the properties of alkanes and alkenes, page 310)
- screens to protect the audience

Safety

All students should wear eye protection.

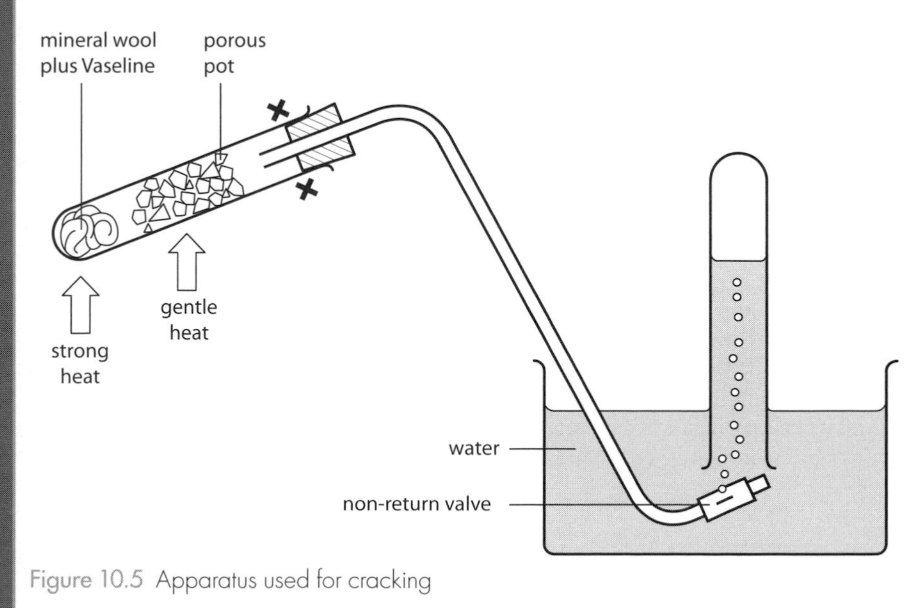

Figure 10.5 Apparatus used for cracking

Method

Set up the apparatus as shown in Figure 10.5. Here are some tips to help:

- Set the boiling tube at a slight angle downwards to help maintain a gap between the mineral wool and the porcelain catalyst.
- It is important that the Vaseline (paraffin) is fresh as this helps to ensure the cracking process occurs efficiently.
- Mix Vaseline into the mineral wool. Push the mineral wool/Vaseline mixture down to the end of the boiling tube using a spatula. The Vaseline could be melted very gently to help this.
- Clean grease away from the sides of the boiling tube using a paper towel.
- Fill most of the rest of the tube with porcelain catalyst. Plenty of catalyst is important – do not stint on this.
- Clamp the tube in position firmly at the neck, not over the catalyst.
- To make a non-return valve, take a piece of Bunsen tubing (or tubing that fits snugly over the delivery glass tubing) about 5 cm long. Cut one short slit (about 1 cm long) lengthways in the middle part of the tubing. Block one end tightly with a piece of solid glass. Fit the opposite end to the open end of the delivery tube. The valve will allow gas bubbles to escape through the slit, which will close immediately afterwards. The valve prevents water sucking back up the delivery tube and potentially shattering the hot boiling tube dramatically into smithereens.

1. Start by heating the porcelain catalyst with a strong, roaring blue Bunsen flame. It is important this gets very hot. Do not be tempted to start by heating the Vaseline. This creates the risk of blocking the delivery tube with cooled, long-chain Vaseline molecules. The point is not to vaporise the Vaseline before the catalyst is hot enough to crack the long-chain molecules.
2. When the porcelain is hot, heat the Vaseline gently. To do this, change the Bunsen flame to a gentle blue flame and waft it under the end of the boiling tube from a distance. The Vaseline will vaporise quite easily.
3. The vapour will start to flow over the hot catalyst. Move the Bunsen flame back to the catalyst and continue to heat it strongly. In this way, move between the Vaseline and catalyst, altering the Bunsen flame accordingly. When Vaseline vapour passes on to the hot catalyst this will start to turn black.
4. Initially, gas coming out of the delivery tube will be air so should not be collected. When the Vaseline molecules crack, the bubbles will start to change as small alkene and alkane molecules pass down the delivery tube. For example, ethene bubbles tend to be smaller than air bubbles and are produced at a rapid rate, often in a steady stream. When this occurs, place the upturned boiling tube over the slit in the non-return valve.
5. Collecting the gas involves careful co-ordination with the heating process. Have at least three water-filled test tubes lying in the trough, with corks nearby. When the gas needs to be collected, use one hand to hold the tube above the slit in the valve. The gas bubbles displace water from the tube. Try to keep an eye on the heating at the same time, but if necessary, stop heating and move the Bunsen away from the apparatus. Cork each tube as it fills with gas. It can be hard at first to tell which gas is air and which are alkanes/alkenes, so if in doubt have more tubes ready for gas collection rather than fewer.
6. When the process is complete, no further Vaseline will remain in the boiling tube. Stop heating and lift the delivery tube out of the water.
7. A lighted splint can be placed at the mouth of a test tube to test combustion; drops of acidified potassium manganate(VII) can be added to another tube; drops of bromine water added to a third tube.

Observations

Ask students to observe changes to the mineral wool, porcelain catalyst, gas bubbles and to the test liquids.

Questions to ask

Make a model of a long-chain alkane such as $C_{20}H_{42}$. This molecule, eikosane, is solid at room temperature. Molecules of around this length are present in Vaseline.

- Which molecules are in the gas?
- What happens to the Vaseline molecules when they are heated?
- Why does the porcelain turn black? What is the black stuff?
- Why must the porcelain be hot?
- Why is the non-return valve needed?
- Why do we collect the gas(es) over water?

Discussion points

- Gas molecules in the collecting tubes include ethene, ethane, propene and propane. Usually more alkenes than small alkanes are produced. The test results show that unsaturated molecules are present.
- The small molecules are produced when the larger Vaseline molecules break up. The porcelain must be hot because energy is needed to break bonds. This can be demonstrated using the Molymod model of the paraffin molecule.
- The porcelain turns black due to carbon deposits.
- The gases are insoluble in water so can be collected over water. The non-return valve stops water getting into the equipment. If cold water hits the hot tube it will shatter.

10.3 Plastics

Plastics are polymers. 'Poly' means many, thus a polymer molecule is a large molecule made by joining many small molecules together. The small molecules that go into a polymer are called 'monomers'. A plastic may have more than one monomer, for example, two monomers may alternate to create the polymer molecule, such as in nylon. This section looks at different polymers based on the simplest alkene, ethene (C_2H_4).

Testing poly(ethene)

Poly(ethene) is probably the world's most commonly made polymer. It was first made by accident in 1898 when a German chemist, Hans von Pechmann, was heating a substance called diazomethane under pressure. His assistants found a white, waxy substance in the reaction vessel. They were able to find out that it was made of chains containing $-CH_2-$. At that time, no one knew how to capitalise on the discovery. In 1933, two British chemists, Eric Fawcett and Reginald Gibson, working at ICI (Imperial Chemical

Industries) in Cheshire, came across the same white, waxy substance when they were heating ethene with another compound, called benzaldehyde, under high pressure. Eventually, chemists worked out how to produce the white solid, known as 'polyethylene', on demand. During the Second World War it was used secretly to insulate cables. The process of making poly(ethene) (its formal, systematic name) has developed since, making the plastic an extremely common and versatile substance.

The following class practical asks students to investigate the properties of different plastic substances by carrying out a number of tests.

PROCEDURE

Materials

- Molymod® set including 10 carbon atoms and 20 hydrogen atoms
- scissors
- tongs
- access to water (to test density)
- a set of masses on a hook (to test tensile strength)
- access to a Bunsen flame (to test for softening)
- plastic carrier bags
- samples of high-density poly(ethene) (HDPE) and low-density poly(ethene) (LDPE) – to find these, look for the plastic recycling symbols (shown in Figure 10.6) on bottles, bags and other plastic 'rubbish' put out for recycling.

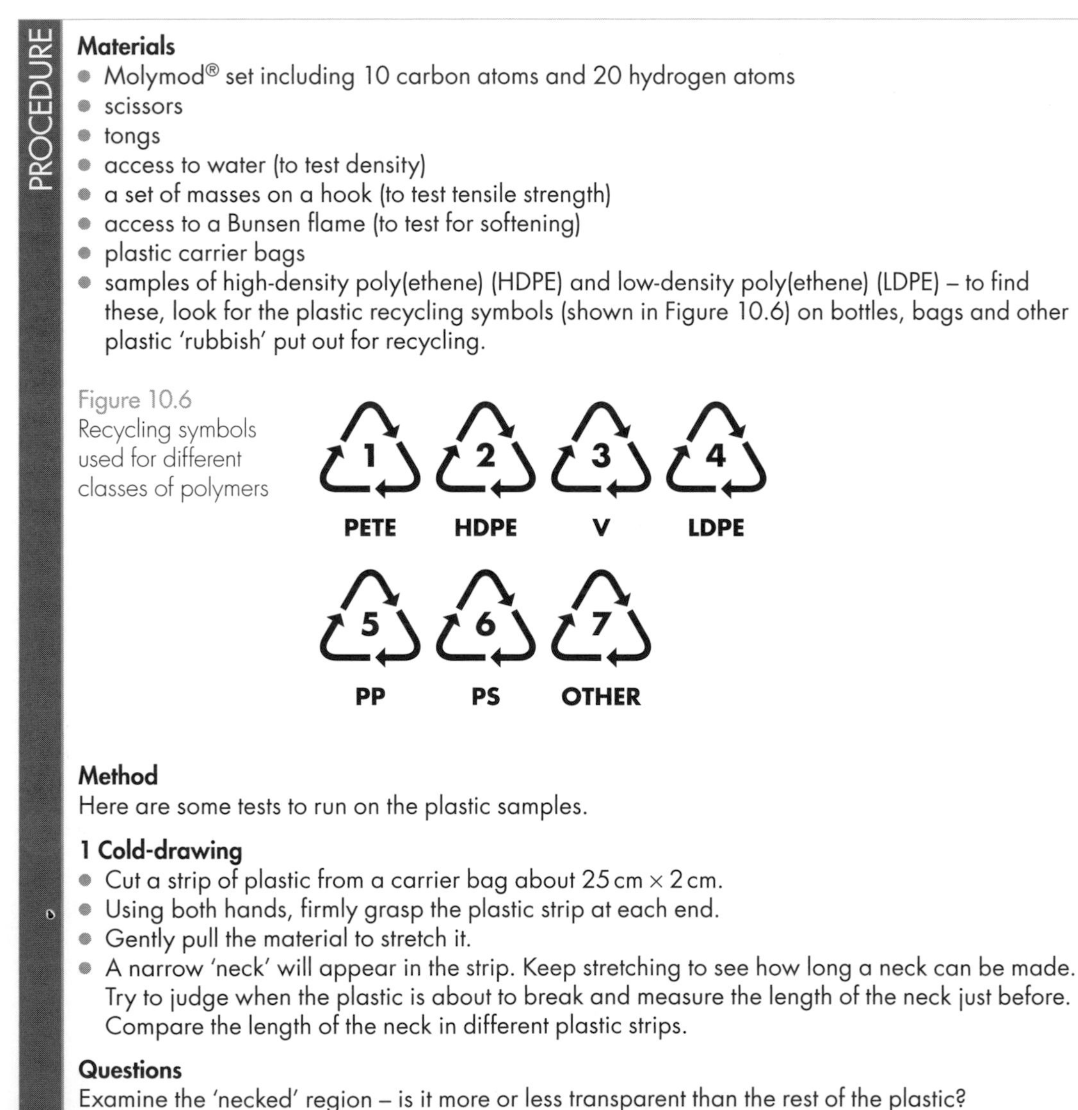

Figure 10.6 Recycling symbols used for different classes of polymers

Method

Here are some tests to run on the plastic samples.

1 Cold-drawing

- Cut a strip of plastic from a carrier bag about 25 cm × 2 cm.
- Using both hands, firmly grasp the plastic strip at each end.
- Gently pull the material to stretch it.
- A narrow 'neck' will appear in the strip. Keep stretching to see how long a neck can be made. Try to judge when the plastic is about to break and measure the length of the neck just before. Compare the length of the neck in different plastic strips.

Questions

Examine the 'necked' region – is it more or less transparent than the rest of the plastic?
Does the necked region itself seem stronger or weaker than the rest of the plastic?

Notes
The energy supplied by pulling the plastic strip created the necked region. Here, the plastic molecules have been pulled into alignment, creating a crystalline region in which they are packed in a regular arrangement.

The necked, crystalline region should be more opaque than the rest of the plastic. The necked area is usually stronger than the rest of the plastic. In the non-necked region, the plastic molecules are randomly organised.

Students could be asked to test a whole carrier bag: when heavy weights are placed in the bag, the region around the handles necks – this enhances the strength of the bag.

2 Tensile strength
Ask students to plan an experiment to compare the tensile strength of LDPE and HDPE. Basically this involves noting the length the plastic will stretch before breaking under a particular mass.

Questions
Which plastic is stronger? Why is this?

3 Softening
Ask students to plan an experiment to compare the extent to which the two plastics soften under heat.

Question
How can 'softness' be measured accurately?

4 Flexibility
Ask students to compare the flexibility of the two plastics. Students should consider how to test flexibility.

Question
In what situations are flexible and non-flexible plastics needed?

Discussion points

Poly(ethene) is formed by polymerisation of ethene. This means making many ethene molecules react together to form long chains of carbon atoms, each of which has two hydrogen atoms attached to it. A catalyst, is used to help speed up the reaction. A variety of catalysts are used in making poly(ethene), depending on the type required. HDPE is produced with the aid of a Zeigler–Natta catalyst (after the chemists who invented this type of catalyst), as this helps to produce a higher proportion of straight chains. LDPE is often produced using a complicated chemical process called 'free radical polymerisation'. The reaction can be represented as shown in Figure 10.7.

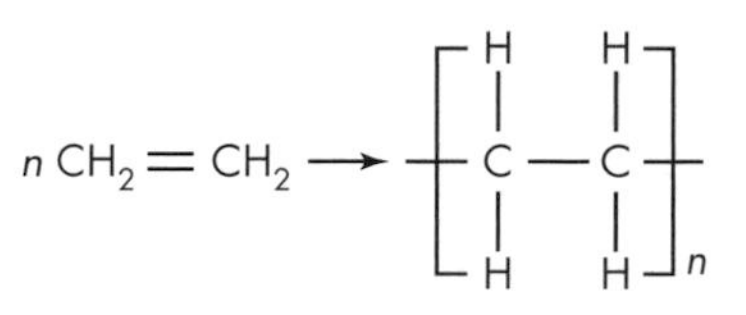

Figure 10.7 A reaction equation for the polymerisation of ethene

LDPE molecules are often branced, so do not pack together as easily, creating a less-dense structure. The straighter, less branched molecules in HDPE pack together relatively tightly, so HDPE will have higher tensile strength, be less soft and less flexible than LDPE.

Note that the costs of the two plastics vary. At the time of writing LDPE was trading at about £616 per metric tonne and HDPE at £726 per tonne. Students could discuss situations requiring plastics in which the additional cost of HDPE would be worthwhile.

Depolymerising poly(ethene)

To 'depolymerise' poly(ethene), carry out the cracking experiment described earlier (page 313), replacing Vaseline and mineral wool with pieces of poly(ethene).

Questions to ask

- What equation represents the depolymerisation of poly(ethene)?
- Is ethene the only product? How would we know?
- Write an equation showing how ethene reacts with bromine water.

Discussion

There is no guarantee that ethene is the only product from the reaction – the long-chain hydrocarbon molecules can break in various places. The equation for the reaction between ethene and bromine is:

$$C_2H_{4(g)} + Br_{2(aq)} \rightarrow CH_2BrCH_2Br$$

Structure and properties of ethene-based polymers

Ethene is the base molecule for a variety of polymers that make well known plastics. In this activity, samples of poly(ethene) (LDPE, HDPE), poly(propene) (PP), poly(chloroethene) (PVC, poly(vinylchloride)), poly(styrene) (PS) and poly(tetrafluoroethene) (PTFE) are collected and compared. Different plastics can be identified by the recycling numbers (as shown in Figure 10.6, page 316) located on the underside of each item.

PROCEDURE

Materials

- Molymod® set including 20 carbon atoms, 40 hydrogen atoms, 10 chlorine atoms, 20 fluorine atoms
- samples of poly(ethene) (LDPE, HDPE), poly(propene) (PP), poly(chloroethene) (PVC, poly(vinylchloride)), poly(styrene) (PS) and poly(tetrafluoroethene) (PTFE): plastic recycling numbers will identify the first four; PTFE is used commonly as a non-stick coating on cookware.

Relevant data about the plastics and their recycling numbers are shown in Table 10.4.

For each plastic, create a sheet of information which includes the plastic recycling number, monomer formula and name, plastic formula and name, and the T_m and T_g values (explained in 'Discussion points', page 320). Create an 'Other' sheet (officially this is recycling number 7) for non-ethene-based plastics with different numbers to those for the plastics in Table 10.4.

Table 10.4 Information and data about five different plastics based on the ethene monomer

Recycling number	2	3	4	5	6
Common name	High-density polyethylene	Polyvinylchloride	Low-density polyethylene	Polypropylene	Polystyrene
Abbreviation	HDPE	PVC or V	LDPE	PP	PS
Formal name	High-density poly(ethene)	Poly(chloroethene)	Low-density poly(ethene)	Poly(propene)	Poly(phenylethene)
Monomer formula and structure	C_2H_4 $H_2C{=}CH_2$	CH_2CHCl $H_2C{=}CHCl$	C_2H_4 $H_2C{=}CH_2$	CH_2CHCH_3 $H_2C{=}CHCH_3$	$CH_2CHC_6H_5$ $H_2C{=}CHC_6H_5$
Polymer formula and structure	$[-CH_2-CH_2-]_n$	$[-CH_2-CHCl-]_n$	$[-CH_2-CH_2-]_n$	$[-CH_2-CH(CH_3)-]_n$	$[-CH_2-CH(C_6H_5)-]_n$
T_m values (°C)	~130	~147	~110	~165	~150
T_g values (°C)	27	81	–125	–10	100

Method

1 Ask students to bring samples of plastics into school.
2 Place your sheets of information in a row along a bench or desk. Have a spare place for 'other' plastics.
3 Make Molymod models of the plastic monomers. Place these on the corresponding plastic information sheets.
4 Start by asking students to pile of their all plastic samples together.
5 Then ask students to sort the plastics into the identified locations using the recycling symbols on each sample.
6 Students can tabulate how each plastic is used, together with information such as the monomer and polymer formulae, and T_m, T_g. This could be extended to include research about costs of plastics and other uses.

Questions to ask

- In what ways do the monomers vary?
- In what ways are the monomers similar?
- How do the physical properties of the plastics vary?
- What are the different plastics used for?
- How do the properties and monomer structures relate to how a plastic is used?

Discussion points

- Plastics which do not have recycling numbers can be placed in 'Other'.
- Plastic with recycling number 1 may also be found commonly – this is poly(ethenebenzene-1,4-dicarboxylate), better known as PETE or PET which stands for poly(ethylene terephthalate). At GCSE level it is advisable to sort this as 'Other'. Note that PETE is used often for products which manufacturers want to be perceived as 'premium' or 'healthy' products, such as water or other drinks, and cleaning products such as shampoo, conditioner and shower gel. PET is an example of a condensation polymer, formed when two different monomers are reacted together in an alternating sequence. The term 'condensation' is used because a small molecule, often water (H_2O), is released (or 'condensed') whenever the two monomers bond together. Other well-known condensation polymers include Nylon and polyurethane.
- The T_m value is the melting temperature. Plastics normally melt over a range of temperatures. This is because the molecules are so long there is no precise temperature at which solid changes to liquid. The value given is mid-range. The symbol (~) that precedes the T_m figure in Table 10.4 represents the fact that the value is not precise.
- The T_g value is the glass transition temperature. At or below this temperature the plastic structure becomes glassy and crystalline (as discussed in the earlier practical, see page 317). The effect is to make the plastic brittle so its physical properties change. Below T_g, plastic may shatter more easily, but also has higher tensile strength.

- The plastics are all made from monomers based on ethene, C_2H_4.
- Poly(chloroethene) (PVC): changing one hydrogen atom in an ethene molecule to a chlorine atom creates chloroethene. This is the monomer for the plastic we call PVC. PVC can be used to make window frames, as 'uPVC' (unplasticised PVC), and plastic films such as cling film. In PVC the long molecules are kept apart by small molecules called 'plasticisers'. Without the plasticiser, the long chains pack firmly together creating a rigid structure. The ability to alter the physical properties using a plasticiser means that the polymer can be adapted to a wider range of uses. There is discussion about polymer stability, as these molecules can 'leak' out of the plastic structure. For example, plasticiser in cling film dissolves in fatty foods such as cheese or meat, tainting the flavour; plasticisers in medical products may leak out over time, or in high-temperature conditions. However, the overall benefits from using these plastics need to be weighed against any possible potential risk from leakage.
- Poly(propene) (PP): changing one hydrogen atom in an ethene molecule to a methyl group ($-CH_3$) creates propene. This is the monomer for the plastic poly(propene) or PP. This can be used for storing food: freezer containers are often made of polypropene. The T_g temperature means it will become crystalline in a domestic freezer which is normally set at around –18 °C. This tends to suit food products which need rigid protection at this temperature.
- Poly(styrene) (PS): changing one hydrogen atom in an ethene molecule to a benzyl group (C_6H_5-) creates styrene or phenylethene. This is the monomer for poly(styrene). The plastic is used for a wide range of packaging. It can be 'blown' to make low-density packing chips and is used in more dense forms as food trays, yoghurt pots and disposable cups.
- Poly(tetrafluoroethene) (PTFE): changing all four hydrogen atoms in an ethene molecule to fluorine atoms creates tetrafluoroethene. This is the monomer for polytetrafluoroethene or PTFE. This is better known as Teflon®, the coating used to make cookware non-stick. PTFE does not have its own recycling number, so would be placed in group 7 which is 'Other'.
- By sorting the plastics, students should recognise common features in how each material is used, as well as differences between the uses for the different plastics. Showing the different monomers will help connect the uses to the formulae.

Investigating a biodegradable polymer: poly(ethanol)

The plastics we make and use on a daily basis are very stable substances, meaning that they do not break down easily. The cracking experiment (described earlier) provides strong evidence for this – a lot of energy is required for a plastic molecule to break into smaller pieces. In fact, most plastics would take at least 100 years to biodegrade naturally in landfill sites. This, besides the ongoing depletion of finite (non-renewable) crude oil resources, is the main reason we need to recycle plastics. An alternative (but incomplete) solution is to use readily biodegradable plastics. An example is poly(ethanol) (colloquial name polyvinyl alcohol; note this is different from polyvinyl acetate used in PVA glue), which is a water-soluble polymer with the structure shown in Figure 10.8.

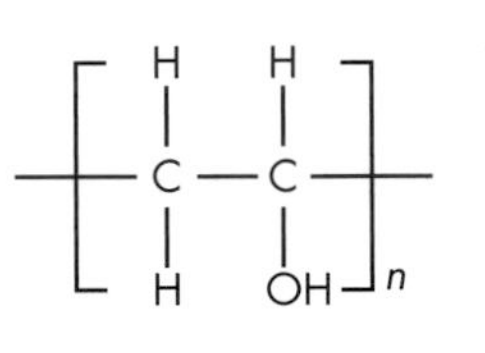

Figure 10.8 Structural formula of poly(ethanol)

Poly(ethanol) is used to make laundry bags to save nurses and carers from handling soiled items. Anglers use small poly(ethanol) bags to tie bait samples on fishing lines. Poly(ethanol) laundry and fish-bait bags dissolve at different water temperatures. The speed and temperatures at which they dissolve can be investigated. This activity is adapted from *Polymers Information and Activity Book* (details of which are given in 'Other resources' at the end of the chapter).

PROCEDURE

Materials

- samples of poly(ethanol) (an internet search on 'polyvinyl alcohol bags' produces websites from which laundry and fish-bait bags can be bought cheaply; at the time of writing, 50 fish-bait bags (10 cm × 10 cm) were available for £3.99 and 25 laundry bags (84 cm × 66 cm) for £11.70)
- water at different temperatures
- a glass rod
- a stopwatch

Safety

All students should wear eye protection.

Method

Give students two samples of polyethenol from different sources and ask them to investigate the properties of the plastics and any similarities and differences they find between them, writing a report on the findings. The tests may include those described earlier (in the section on 'Testing poly(ethene)', see pages 316–7). Students may be surprised to find that the plastics dissolve in water.

Students could also investigate or research other biodegradable plastics. The 'Other resources' section at the end of the chapter provides further details.

Questions to ask

- What differences and similarities in properties are found?
- What could these plastics be used for?
- In what ways are biodegradable plastics 'better' than non-biodegradable ones?
- What other biodegradable plastics are there?
- Why are biodegradable plastics only a partial solution to our recycling problem?

Discussion points

Students' investigations comparing the two types of polyethenol should show that the plastics differ in the water temperature at which they dissolve, but share similar properties such as tensile strength.

Other biodegradable plastic types are photosensitive and 'synthetic'. Photodegradable plastics are sensitive to UV light. Synthetic plastics have starch grains built into their structures. These are attacked by micro-organisms in soil. Companies are also producing 'compostable' plastics which degrade in the conditions common to composting plant materials, yielding carbon that can re-enter the carbon cycle. At the time of writing, such plastics remain expensive to produce and do not provide realistic alternatives for many of the broad range of normal plastic uses.

10.4 Natural products

Ethanol

Ethanol, CH_3CH_2OH, is commonly known as 'alcohol'. It is the second member of the homologous series of primary alcohols; highly toxic methanol (CH_3OH) is the first. The general formula for the primary alcohols is $C_nH_{2n+1}OH$. The characteristic part, or functional group, of the alcohols is –OH, called the hydroxyl group.

Ethanol is made by fermentation of sugars, such as glucose, by yeast in the following reaction:

$$C_6H_{12}O_{6(aq)} \rightarrow 2CH_3CH_2OH_{(aq)} + 2CO_{2(g)}$$

The source of sugar varies. Grape juice is used to make wine, using natural yeasts on grape skins. Allowing fermentation to stop naturally produces a dry wine, with low sugar content. Stopping fermentation at an earlier stage produces a sweeter wine, as more sugar remains in the product. Making beer by brewing cereal grains such as barley, wheat, rice or corn is one of the oldest known procedures common to many human societies. A culture of yeast is added to grain, while flavour is produced by adding other ingredients. In the UK hops, which give the beer a malty taste, are common. The malt-flavoured spread Marmite® is a brewing by-product.

The strength of a drink containing ethanol is measured in 'per cent proof'. The proof value is twice the volume per cent of ethanol. Thus, spirits, which are around 80% proof, contain 40% by volume of ethanol. The strength can be increased by distillation as shown in the practical on fractional distillation which is discussed later in this section (page 328).

Ethanol has possibilities as a renewable fuel. This can be discerned from the first experiment which illustrates its combustible qualities.

■ The non-burning £5 note

This demonstration experiment shows that ethanol is combustible. It is taken from *Classic Chemistry Demonstrations* (detailed in the 'Other resources' at the end of the chapter).

PROCEDURE

Materials

- a Bunsen burner
- tongs
- a heatproof mat
- three beakers, 250 cm^3
- about 75 cm^3 ethanol
- about 10 g sodium chloride – table salt (optional)
- water
- one currency note to 'burn' (Note: it may be best to avoid Australian and New Zealand notes as they contain large quantities of polymer so will not absorb liquid as efficiently as paper-based ones; as a result they may melt, but it would perhaps be interesting to find out!)
- two pieces of paper cut to resemble a currency note

Safety

All students should wear eye protection.

Method

Before the demonstration, prepare the beakers. Put about 50 cm^3 water in one, 50 cm^3 ethanol in the next and a mixture of 25 cm^3 water, 25 cm^3 ethanol in the third. Add a small spatula of sodium chloride to the ethanol and water mixture (see 'Discussion points' opposite).

1 Dim the lights so the flames can be seen more easily.
2 Light the Bunsen on the heatproof mat. Turn the flame to gentle blue.
3 Soak one of the paper pieces in the water. Pick it out with tongs and hold it in the lighted Bunsen flame. It will not ignite until the water has evaporated.

4 Soak the second paper piece in ethanol. Pick it out with tongs and hold it in the Bunsen flame. The ethanol and paper will ignite.
5 Take the currency note. Place it in the third beaker. Do not tell the audience anything about the liquid in the beaker. They can later be asked what they think it is, given the previous two tests.
6 Pick out the note with the tongs and hold in the Bunsen flame as before. This time, the ethanol ignites and burns away, but the note will not, due to absorption of water.

Questions to ask

- Will X, who so kindly donated the note, get his/her money back?
- What evidence/advice/comfort can they expect based on previous tests?
- Why will the note not burn?
- What is burning?
- What does this tell us about the properties of ethanol?

Discussion points

Ethanol has a higher ignition temperature (425 °C) than paper (230 °C). The water protects the paper from burning in the heat of the Bunsen flame.

Adding sodium chloride helps to make the ethanol–water mixture flame visible. A yellow-orange flame will be seen due to the presence of sodium ions. Without this, ethanol burns a faint blue colour.

Ethanol can be used as a fuel. Some cars have been adapted to run on bioethanol. However, grain supplies would need to be enormous to meet the potential demand.

■ The 'breathalyser' reaction

Students will be aware that driving with alcohol in the bloodstream is dangerous. Some European countries have a 'zero' limit – they will not tolerate any blood alcohol at all. In the UK, the limit is currently 0.8 g/l or 80 mg/cm^3. This is often called '80 mg' or '80 promille'. The UK breath alcohol limit is 35 mg/cm^3.

Breathalysers use the reaction in this demonstration as the basis for testing breath alcohol (again this is from *Classic Chemistry Demonstrations*, see 'Other resources').

PROCEDURE

Materials

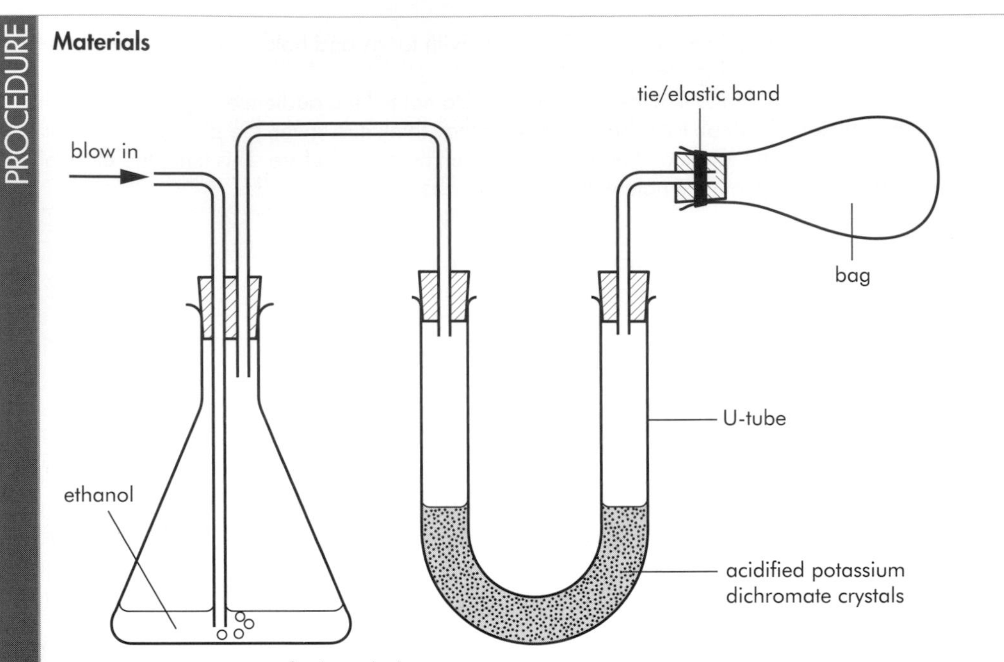

Figure 10.9 Apparatus for breathalyser reaction

- a U-tube (arm length about 10 cm), each arm fitted with a one-hole bung and glass tubing
- two conical flasks, 250 cm^3, each fitted with a two-hole bung and two pieces of tubing
- tubing to connect parts of the apparatus
- a plastic freezer bag, 'sandwich' size, or access to a filter pump
- cable tie or elastic band or about 15 cm string/thread
- one beaker, 100 cm^3
- 30 g potassium dichromate crystals
- 100 cm^3 ethanol
- 3 cm^3 2 $mol\,dm^{-3}$ sulfuric acid

Safety

All students should wear eye protection.

Method

Before the practical:

- Prepare the U-tube. Weigh out the potassium dichromate crystals in the beaker. Add 3 cm^3 sulfuric acid (ratio 10 g potassium dichromate : 1 cm^3 sulfuric acid). Mix thoroughly. The crystals will moisten, but will not dissolve. Pour the mixture into the U-tube. Tap the tube on the work surface to settle the crystals.
- Place ethanol in one of the flasks. Fit the bung and glass tubing (Figure 10.9). Ensure that the long end of the tubing is below the surface of the ethanol.
- Attach the plastic bag securely to the one-holed bung at the other end of the tubing.
- Before connecting up, blow air into the flask so the air becomes saturated with ethanol vapour. Make sure the plastic bag is empty and deflated. This will ensure that more ethanol vapour, rather than air, is in the system.

1 Connect the flask to the U-tube, and the bag to the other arm of the U-tube (Figure 10.9). Blow again into the flask, so ethanol vapour passes over the crystals and into the bag. The bag should inflate.
2 Observe what happens to the crystals. A brown colour should appear. This will turn green over time.
3 Once the bag is inflated, it can be squeezed gently to pass the vapour back over the crystals. Alternatively, in place of the bag, attach a filter pump. This will draw ethanol vapour over the crystals.

Questions to ask

- Why do the crystals change colour?
- How can this test be used to determine breath alcohol levels?

Discussion points

Original breathalysers used a bag and crystals similar to this set-up. Today, breathalysers use the same reaction, but in a more sophisticated form, linked to electronic determination of the depth of colour produced. They can be bought quite cheaply over the internet.

The reaction in the breathalyser is quite complicated, but can be summarised as:

$$2Cr_2O_7^{2-}{}_{(s)} + 3C_2H_5OH_{(g)} + 16H^+{}_{(aq)} \rightarrow 3CH_3COOH_{(l)} + 4Cr^{3+}{}_{(aq)} + 11H_2O_{(l)}$$

dichromate ions (orange); ethanol; ethanoic acid; chromium ions (green)

Estimates suggest that the breathalyser has saved many lives since its first use in the UK in 1968. Students could be asked to research the social history relating to this. Discussion could be linked with social issues related to alcohol consumption.

■ Fractional distillation of an ethanol/water mixture

The concentration of ethanol in water can be increased by fractional distillation. The principles behind this method are also used to prepare fractions from crude oil. Note that a 100% 'pure' sample of ethanol cannot be obtained, since ethanol and water form a constant boiling point mixture at about 95% ethanol.

PROCEDURE

Materials

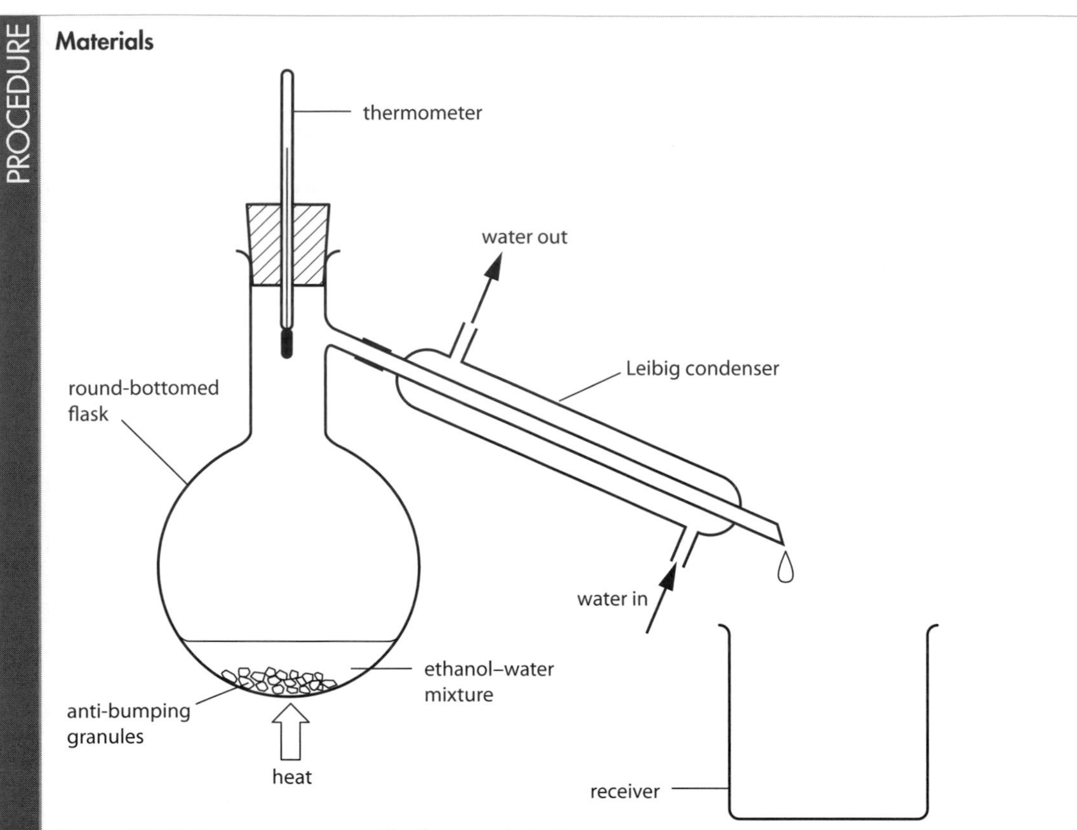

Figure 10.10 Apparatus required for fractional distillation

For this experiment, 'quick-fit' glassware is best. Otherwise the list below will suffice.

- a round-bottomed flask, 500 cm^3, with side arm
- a Leibig condenser, with rubber tubing
- a one-hole bung to take side arm and to fit the Leibig condenser
- a one-hole bung to take thermometer and to fit flask neck
- a −10 to 100 °C thermometer
- clamp stands, bosses and clamps
- an electric heater to heat the liquid in the flask (safest option) or a Bunsen burner
- heatproof mats
- access to cold water tap and sink
- a funnel
- about 150 cm^3 50 : 50 water : ethanol mixture or the product from a fermentation experiment
- a receiving vessel, such as a small beaker
- anti-bumping granules
- safety screens
- to test the original mixture and product: acidified potassium dichromate solution; watch glasses, lighted splint, dropping pipette, test tubes, test-tube rack

Safety

All students should wear eye protection. The demonstration should be carried out behind a safety screen. If using a Bunsen burner to boil ethanol, use a boiling water bath so that if the flask breaks it does not cause a fire.

Method

Before the demonstration, set up the apparatus using Figure 10.10 as a guide, leaving out the thermometer for the moment. Make the ethanol–water mixture or have ready a product from a fermentation reaction. Position the safety screen between the apparatus and audience.

1. Use the funnel to pour the distillation mixture into the flask.
2. Add one spatula of anti-bumping granules.
3. Fit the thermometer so that the bulb sits at the side-arm junction.
4. Start water flowing through the condenser (from bottom to top).
5. Set the receiving vessel in place.
6. Switch on the electric heater or start heating using a gentle blue flame. Note that the mixture will boil at around 80 °C.
7. Maintain a steady heat so the liquid simmers continuously.
8. Monitor drops of distillate appearing in the receiving vessel.
9. Continue collecting distillate until about 25 cm^3 is available for testing.
10. Stop heating the flask.

Questions to ask

- Why do we need anti-bumping granules?
- What is the temperature of the vapour going into the condenser?
- What does the temperature tell us about what the distillate might be?
- What will be left behind in the flask?
- How can we test the distillate?
- What would we need to do to get all the liquid in the flask to pass through the condenser?

Discussion points

The vapour will be at about 78 °C, the boiling point of ethanol. Thus, the distillate (the liquid collected in the receiving vessel) will be mainly ethanol. Liquid remaining in the flask is mainly water. To get the water to vaporise, the temperature would need to be increased to 100 °C.

The ethanol can be tested as suggested previously by burning and by reaction with an acidified solution of potassium dichromate (which turns green in the presence of ethanol). The results can be compared with tests on the distillation mixture. The mixture should not burn (as proved in the earlier demonstration 'The non-burning £5 note'), but pure ethanol combusts easily with a blue flame.

The use of fractional distillation in production of spirits such as whisky and brandy can be discussed.

Note also that the same process on a much larger scale can be used to separate out a complex mixture of liquids of different boiling points using a fractionating tower, effectively a very long, vertical arrangement of condensers. This is the principle used in separating crude oil fractions.

Carboxylic acids

Carboxylic acids form a homologous series. The general formula for the carboxylic acids is $C_nH_{2n+1}COOH$. These acids are called 'weak' acids. This is because not all the molecules ionise in solution:

$$CH_3COOH \leftrightarrows CH_3COO^-_{(aq)} + H^+_{(aq)}$$

The characteristic part, or functional group, of the carboxylic acids is –COOH, called the carboxyl group. This group has the structure shown in Figure 10.11.

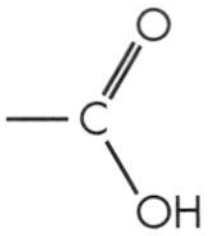

Figure 10.11 The carboxyl group

Many carboxylic acids are found in nature. For example, the simplest carboxylic acid is methanoic acid, HCOOH. This is the irritating chemical produced in red ant bites. Its common name, formic acid, comes from the Latin name for ant, 'formica'. Others include the following:

- ethanoic acid (CH_3COOH), better known as vinegar
- propionic acid (CH_3CH_2COOH) which gives Swiss cheese its flavour
- butyric acid ($CH_3CH_2CH_2COOH$) responsible for the taste of rancid butter
- carbonic acid (H_2CO_3) formed when carbon dioxide dissolves in water and used to add 'fizz' to fizzy drinks
- 2-hydroxypropane-1,2,3-tricarboxylic acid, better known as citric acid, found in fruit juices
- ascorbic acid or vitamin C (formal name: 2-(1,2-dihydroxyethyl3,4-dihydroxyfuran-3-one)
- oxalic acid (HOCOCOOH), a toxic dicarboxylic acid with two carboxyl groups; it is common at low concentrations in edible plants such as rhubarb, spinach and sorrel. One estimate suggests that 4 kg of spinach contains enough oxalic acid to poison a person.

Properties of carboxylic acids

The carboxyl group gives these compounds acidic properties. These can be demonstrated in a class experiment. This is drawn from the RSC's book *Classic Chemistry Experiments* (detailed in the 'Other resources' section at the end of the chapter).

PROCEDURE

Materials

- test tubes, about four per group for each acid tested
- a test-tube rack
- dropping pipettes
- a glass rod
- a 2 cm piece of magnesium ribbon
- access to:
  - $0.05\,mol\,dm^{-3}$ ethanoic acid solution, about $50\,cm^3$ per group
  - $0.4\,mol\,dm^{-3}$ sodium hydrogencarbonate solution, about $5\,cm^3$ per group
  - $0.4\,mol\,dm^{-3}$ sodium hydroxide solution, about $5\,cm^3$ per group
  - universal indicator solution (full range)

Safety

All students should wear eye protection.

Method

Ask students to test the properties of the ethanoic acid solution. To do this, put 1–2 cm depth acid in each of four test tubes. Then test the samples as follows.

1. Add a few drops of universal indicator solution. Note the colour. Keep this as a reference tube for tests 2 and 3.
2. Add a few drops of universal indicator solution. Note the colour. Add sodium hydrogencarbonate solution dropwise, noting any colour changes. Continue additions until no further colour change occurs.
3. Add a few drops of universal indicator solution. Note the colour. Then add sodium hydroxide solution dropwise, noting any colour changes. Continue additions until no further colour change occurs.
4. Add a small piece of magnesium ribbon to the ethanoic acid. Note what happens. Test the gas produced.

Questions to ask

- What colour changes occur?
- What substances were produced in the reactions with the alkalis?
- What gas was produced?
- What does this tell us about ethanoic acid?
- How do these reactions compare with other acids?
- What might we predict about other carboxylic acids?
- Write equations for the reactions.

Discussion points

Addition of sodium hydroxide, a strong alkali, produces an alkaline solution within a few drops. More sodium hydrogencarbonate is needed to produce the same colour as this is a weak alkali.

Equations for the reactions are:

with sodium hydrogencarbonate

$$CH_3COOH_{(aq)} + NaHCO_{3(aq)} \rightarrow CH_3COONa_{(aq)} + CO_{2(g)} + H_2O_{(l)}$$

with sodium hydroxide

$$CH_3COOH_{(aq)} + NaOH_{(aq)} \rightarrow CH_3COONa_{(aq)} + H_2O_{(l)}$$

with magnesium

$$2CH_3COOH_{(aq)} + Mg_{(s)} \rightarrow (CH_3COO)_2Mg_{(aq)} + H_{2(g)}$$

The salts produced are called sodium ethanoate and magnesium ethanoate. The CH_3COO^- ion is called the ethanoate ion. Conventionally the salts of carboxylic acids are written with the carboxylate ion first.

The test results show that ethanoic acid behaves in the same way as any mineral acid, such as hydrochloric, sulfuric and nitric acids. However, it is a weak acid, which means fewer hydrogen ions are in ethanoic acid solutions compared to other acids. We can predict that other carboxylic acids would react in the same ways.

Take this opportunity to discuss the use of ethanoic acid as vinegar and a preservative for foods such as onions, beetroot, eggs and in condiments such as piccalilli.

■ Vitamin C

Vitamin C, ascorbic acid, is essential in the body for growth and repair. It has a complicated structure and formula, summarised as $C_6H_8O_6$. Our bodies cannot make or store vitamin C, so we must take in a regular supply amounting to about 60 mg daily. Fortunately, vitamin C is easily available in fruit. The following experiment, based on that given in the Vitamin C chapter in the RSC book *Contemporary Chemistry for Schools and Colleges* (detailed in the 'Other resources' section at the end of the chapter) provides a reliable means of testing for and comparing vitamin C levels in different fruit juices. The experiment involves reacting ascorbic acid with iodine, according to the reaction:

$$C_6H_8O_6 \quad + \quad I_2 \quad \rightarrow \text{dehydroascorbic acid} + \quad 2I^-$$

ascorbic acid + iodine → dehydroascorbic acid + iodide ions

The product solution is colourless compared to the starting reaction mixture. The presence of iodine can be traced by addition of starch indicator to the mixture. The blue-black starch–iodine complex has disappeared at the end point of the reaction.

PROCEDURE

Materials

- $10\,cm^3$ vitamin C solution
- one drop per test of iodine solution
- $10\,cm^3$ starch solution
- $50\,cm^3$ water
- fruit juices to test – ensure the packaging is available
- dropping pipettes
- test tubes, sufficient for one per test
- a test-tube rack
- a $10 \times 10\,cm$ white card as background

Safety

All students should wear eye protection

Note

The vitamin C solution acts as the 'standard'. Use a vitamin C tablet containing a stated amount of vitamin C (such as 1000 mg) or dissolve a known mass of vitamin C powder in water to produce a standard solution. If preferred, a range of standard concentrations could be produced from this and the test results used to produce a 'standard curve' in which the number of drops added (*y* axis) is plotted against the known concentration of vitamin C (*x* axis). This graph can then be used to read off the values obtained for any fruit juices tested, provided they fall within the range of vitamin C standards.

Method

For each test:

1. Put $1\,cm^3$ starch solution in a clean, fresh test tube.
2. Add $5\,cm^3$ water and mix by gentle shaking.
3. Add one drop of iodine solution. Note the blue-black colour.
4. Place the white card behind the tube.
5. Add the vitamin C solution/fruit juice dropwise. Shake the tube after every five drops. Count the drops.
6. Keep adding until the blue-black colour disappears. This will produce a grey-white colour in the tube.
7. Note the number of drops added when no further colour change is produced.

Further investigations to try

- Compare storage conditions: prepare and test vitamin C levels in fruit juice samples from the same pack which have been heat treated, exposed to light, frozen, and stored outside a fridge.
- Investigate changes in vitamin C levels with increased storage time.
- Investigate changes in vitamin C levels with temperature.
- Compare fresh and packaged juices: prepare and test freshly squeezed fruit juices.
- Compare juices from different manufacturers: do branded and generic supermarket juices contain the same or different amounts? Compare labels/nutrition panels to find out what amounts of vitamin C manufacturers claim are present.

Questions to ask

- Why does the solution have to be added drop by drop?
- Which juice has the most vitamin C?
- Why is the vitamin C solution needed?
- Why is the starch solution needed?
- Use the results for the standard vitamin C solution to work out the vitamin C concentration in $1\,cm^3$ of the fruit juices.
- Which juice contains the most vitamin C?

- Which juice(s) give(s) the recommended adult daily allowance of vitamin C?
- Do the results match the packaging?
- How do vitamin C levels differ between packaged and fresh fruit juice?
- How do different conditions (temperature, freshness) affect the amount of vitamin C in a juice?
- How does exposure to light affect vitamin C levels? Would you recommend buying freshly squeezed juice that is in a transparent bottle?

Discussion points

Vitamin C levels in fresh fruit vary from around 4 mg/100 g in a pear to 1150 mg/100 g in rosehips. The highest levels found internationally are in the camu camu fruit, which records 2700 mg/100 g.

Typical numbers of juice drops to reach end-point are given in Table 10.5.

Table 10.5 Comparing vitamin C from various sources

Fruit juice/vitamin C tablet	Number of drops to end point
1000 mg vitamin C tablet in 100 cm^3 water	15
Lemon	30
Lime	35
Orange (day-old open carton)	40
Kiwi	15
Melon	20
Mango	30

Note that the smaller the number of drops, the higher the vitamin C content. This is because the amount of iodine is fixed beforehand. Students may need help to realise that the highest concentrations of vitamin C are found in the juices needing the smallest number of drops to react with all the iodine. The starch solution is needed to produce a consistent, visible end point. In Table 10.5, kiwi fruit are shown to have the highest concentration of vitamin C.

Changing conditions, such as temperature, light, storage time for example, deplete vitamin C levels. Freezing juice may, however, have less impact on vitamin C level.

Note that these results show how labile ascorbic acid molecules are. This helps to explain why we need a constant supply of the vitamin.

Students could investigate the vitamin C deficiency disease scurvy and why British sailors got the nickname 'Limeys' (this arose when the Royal Navy started issuing a daily dose of lemon or lime juice to sailors to prevent scurvy).

Esters

Esters are formed when carboxylic acids and alcohols react together. The sweet-smelling compounds associated with ripeness of fruit such as pears, bananas, melons and pineapples are esters. A classic, simple ester is ethyl ethanoate, produced in the reaction between ethanol and ethanoic acid:

$$\underset{\text{ethanol}}{CH_3CH_2OH_{(l)}} + \underset{\text{ethanoic acid}}{CH_3COOH_{(l)}} \rightarrow \underset{\text{ethyl ethanoate}}{CH_3COOCH_2CH_{3(l)}} + H_2O_{(l)}$$

Ester names comprise two parts: the first is from the alcohol and the second from the acid. Ester formulae are written the other way round, so the acid part of the name comes first while the alcohol section is at the end.

Reactions producing esters are generally slow. Accordingly, fruit takes time to ripen as ester concentrations build up gradually. In the esterification reaction detailed later in this section, a catalyst, hydrogen ions, is used to increase the reaction rate.

Strictly, esterification reactions are reversible reactions (see Chapter 5). The ester can break down, or hydrolyse, by reacting with water, and the alcohol and acid form as products. The two reactions, one making the ester and the other breaking down the ester, can theoretically occur indefinitely. Thus, if esters are needed in large quantities, chemists withdraw the ester product as it is made in order to increase yield by keeping the forward reaction, the one producing the ester, going and not giving the reverse reaction much chance to take place.

Besides being found in fruit, the fragrant nature of esters makes them useful in perfumes and other items requiring pleasant scents. Some chemical companies specialise in producing esters for these purposes. Their non-toxic nature makes esters suitable for use in artificial flavourings. Combinations of esters yield different scents and tastes, allowing chemists to 'fine tune' an ester mixture to specific requirements. For example, strawberry yoghurt can have a grassy or fruity smell and taste depending on the ester combination used to create the artificial flavour.

■ An esterification reaction

This reaction is probably best done as a demonstration as it involves use of concentrated acid, but, with an appropriate risk assessment, the small scale means that a class experiment is not impossible. Providing students with opportunities to handle 'difficult' chemicals is important for developing their skills.

PROCEDURE

Materials

- a test tube
- a beaker, 250 cm^3
- water
- a Bunsen burner
- a heatproof mat
- a gauze
- a tripod
- pipettes – preferably graduated pipettes measuring up to 5 cm^3
- 1 cm^3 glacial ethanoic acid
- 2 cm^3 ethanol
- about three drops of concentrated sulfuric acid
- an evaporating dish
- sodium chloride solution (approx. 1 mol dm^{-3})
- tongs

Method

1. Put 1 cm^3 glacial ethanoic acid in the test tube.
2. Add about three drops of concentrated sulfuric acid.
3. Shake the tube to mix the two acids.
4. Add 2 cm^3 ethanol.
5. Shake the tube again.
6. Put about 150 cm^3 cold water in the beaker.
7. Place the test tube containing the reaction mixture in the beaker of water.
8. Heat the beaker slowly.
9. Keep the temperature as high as possible, without the mixture in the test tube boiling, for about 20 minutes. Turn off the heat.
10. Get the evaporating dish ready by half-filling it with sodium chloride solution.
11. Pour the contents of the test tube into the evaporating dish.

The ester should rise to the surface of the liquid in the dish as an oily liquid with the pleasant smell of pear drops.

Reversing the reaction

1. To reverse the reaction, draw off the ester droplets with a pipette, placing the ester in a fresh test tube.
2. Add a few drops of a dilute mineral acid (hydrochloric, sulfuric, nitric) or aqueous hydroxide (sodium or potassium) and boil the mixture.
3. It is harder to test if the reverse reaction has occurred – some of the experiments described above could be adapted to help with this.

Questions to ask

- What is the smell?
- Where/when can similar chemicals to this one be found?
- What could this chemical be used for?
- Why is the acid needed (it does not appear in the equation for the reaction)?

Discussion points

The equation for the reaction is as given on the previous page. The acid acts as a catalyst to speed up the rate of reaction, otherwise the ester would take a long time to produce.

The ester is ethyl ethanoate. It has a smell similar to the ester used in making sweets called pear drops (the actual ester in pear drops is a more complicated molecule called '3-methyl-1-butyl ethanoate'). These can still be bought from pick 'n' mix sweet shops. (At the time of writing, a website is offering 250 g of pear drops for around £2. See 'Other resources' at the end of the chapter.)

Students can suggest different fruits that have similar smells. Some research (detailed in the 'Other resources' section) will reveal the esters present in a range of fruits. This could be extended by looking at sweet-tasting foods that lists 'artificial flavouring' in its ingredients – these are almost certainly combinations of esters.

Lipids: fats and oils

Biochemists use the term lipid to describe oils and fats. Lipids are esters.

Figure 10.12
Structure of a lipid

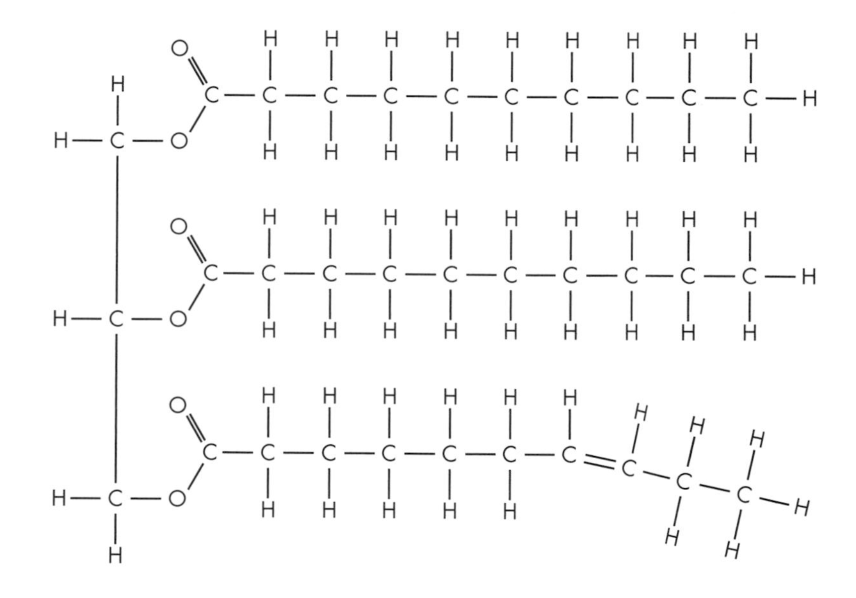

A lipid has a 'backbone' which is derived from the molecule glycerol. This has the shape of a capital letter 'E' (Figure 10.12). Three molecules are attached to the prongs of the E by ester linkages (–COO–). Each of these molecules is a fatty acid. The three fatty acids can be the same or different.

The molecular structure of the fatty acids varies. The notion of unsaturation applies to them, just as to the hydrocarbons discussed in section 10.2. Students may be aware that some fats, called unsaturated fats, are 'good' for us while others, saturated fats, are not. The next experiment (from *Contemporary Chemistry for Schools and Colleges*, detailed in the 'Other resources' section at the end of the chapter) uses the same principle as that shown in section 10.2 to test for saturation in fat molecules.

Fat molecules are found in liquid oils (vegetable, oil seed rape, olive, etc.) and in solid cooking fats such as margarine, lard, Trex®, etc. Margarine is hydrogenated vegetable fat, made by adding hydrogen across the double bonds of unsaturated fat molecules obtained from plants. Complete addition produces a saturated fat, comprising molecules with only single bonds. A 'partially hydrogenated' fat has a more complex structure, as some double bonds remain. Some of these double bonds are in a 'trans' arrangement, creating molecules called 'trans fats'. These are regarded widely as highly undesirable, as dietary trans fats have been linked to raised levels of heart disease. Some products, such as 'spreadable butters' vary, but are often a composite of natural and synthetic fats, designed to achieve a compromise between taste and physical properties.

This experiment investigates how edible fats and oils differ in flavour. The basic test requires assessment by flavour, allowing discussion of how we can 'measure' fairly a quality relying on personal opinion. This discussion can be extended to consider how food scientists 'measure' taste. Students could be asked to cook the chips 'blind', that is, with oils/fats identified only by a random number, not name. They can all vote on flavour, with the 'winning' and 'losing' oil/fats revealed when votes are counted. In trials, olive oil proved to be a favourite. People often prefer the taste of fats which are higher in saturated rather than unsaturated molecules. Discussing reasons for this will help lead to consideration of healthy eating issues.

This activity should be carried out in a food technology room. Student groups could each cook one or two bag of chips using different oils/fats, giving a good range across the class. Alternatively, the experiment could be done as a demonstration.

PROCEDURE

Materials

- knife
- potato peeler
- frying pan
- kitchen spatula to turn the pieces while frying
- balance or kitchen scales
- paper towels
- plate or small paper bags

For each bag of chips:

- 100 g potato pieces, peeled, rinsed and dried
- 10 g of oil or fat for frying (see Table 10.6 for suggestions)

Safety

Teachers are advised to consult a model risk assessment (e.g. CLEAPSS model risk assessments for D&T 3.019 Frying and Grilling) and be aware of how to deal with an oil or fat fire.

Method

1. Peel, then chop the potatoes into 1 cm cubes (this shape minimises the frying time and is easiest for sharing).
2. Rinse the chips to remove excess starch and dry them on kitchen paper.
3. Measure 100 g chips.
4. Heat the oil/fat in the frying pan on a medium heat. Do not overheat.
5. When the oil/fat is hot, test the temperature by dropping in one chip. If the oil/fat is ready, it will sizzle.
6. When the oil/fat is ready, add all the chips to the pan.
7. The chips will take about 5–7 minutes to cook. Use the spatula to ensure the chips cook evenly.
8. When the chips are golden brown all over, they are ready. Note that the shade will differ depending on the oil/fat used.
9. Tip the chips on to fresh kitchen paper to drain off any excess oil/fat.
10. Tip the chips on to a plate. Add a pinch of salt.
11. Note the appearance of the chips and record this in a results table.
12. Once the chips have cooled, taste them and compare the ones cooked in different oils/fats.

Questions to ask

- Which chips tasted the best?
- Which chips looked the best?
- Find out more about the oil/fat used to cook your favourite and least favourite chips. What differences are there between the molecules in the two oils/fats?

Further investigations

The experiment could be extended by carrying out these investigations:

- How does the variety of potato affect the flavour of the chip?
- Does the flavour differ if the chips are deep-fried, shallow-fried or cooked in an oven?
- What is the best temperature for cooking chips?
- How can we make 'healthy' chips?

Discuss with students why unsaturated fats are believed to be better for us than saturated ones. The answer lies in the double bonds present. These are reactive and allow the molecules to break up more easily in the body. Saturated molecules are harder to move around and dispose of as they break up less easily.

Table 10.6 shows how common fats and oils differ from each other in composition.

Table 10.6 Percentages of saturated and unsaturated molecules in common fats and oils

	Saturated			Unsaturated			
Fat or oil	Myristic acid C14 / %	Palmitic acid C16 / %	Stearic acid C18 / %	Oleic acid C18 / %	Linoleic acid C18 / %	Other polyunsaturated acids C18 and longer / %	Other polyunsaturated acids C14 and 16 / %
Beef fat (dripping)	6	27	14	50	3	–	–
Cocoa butter (vegetable fat)	–	24	35	38	2	–	–
Fish oil (sardine)	5	15	3	–	–	50	27
Cod liver oil	6	8	1	29	–	35	20
Corn oil	1	10	3	50	–	34	2
Olive oil	–	7	3	85	5	–	–
Peanut oil	–	8	3	56	–	33	–
Soya oil	–	10	2	29	–	58	–
Sunflower oil	–	6	2	25	–	67	–

Other resources

Books

Garforth, F. & Stancliffe, A. (1994). *Polymers Information and Activity Book*. York: University of York.

Hutchings, K. (2000). *Classic Chemistry Experiments*. London: Royal Society of Chemistry.

Kind, V. (2004). *Contemporary Chemistry for Schools and Colleges*. London: Royal Society of Chemistry.

Lister, T. (1995). *Classic Chemistry Demonstrations*. London: Royal Society of Chemistry.

Websites

Kind, V. (2004). *Beyond Appearances: Students' Misconceptions about Basic Chemical Ideas*
www.rsc.org/Education/Teachers/Resources/Books/Misconceptions.asp

Polyethenol is available from:
www.anglersnet.co.uk/cheap-fishing-tackle/pva/cat_11.html (fishing bait bags)
www.win-health.com/soluble-laundry-bags.html (large laundry bags)

Pear drops can be bought from:
www.aquarterof.co.uk/pear-drops-p-157.html (currently 250 g for £1.97)

Entertaining information about food additives, including esters, is available at:
www.chewonthis.org.uk/factory_food/additives_home.htm

A company selling high-impact liquid flavours is Stringer Flavours Limited, based in Tring, Hertfordshire, UK. Their website is:
www.stringer-flavour.com/default.asp
The company offers flavour samples from a long list.

Vitamin C is readily available in various forms (with zinc, timed release, powder). A typical range is available from:
www.justvitamins.co.uk/Vitamin-C/Default.aspx

This company also offers a product called 'rosehip 5000 mg' which would make a good, concentrated solution:
www.justvitamins.co.uk/products/Rosehip-5000mg-1160.aspx

11 Earth science

Elaine Wilson

11.1 Understanding how rocks are formed
- How does loose sediment 'stick' together to form a 'rock'?

11.2 Identifying rocks and their characteristics
- Minerals in granite
- Comparing granite and sandstone
- Identification of hand specimens of rocks

11.3 Weathering and the effects of physical, chemical and biological processes on rock formation
- The difference between weathering and erosion
- Physical weathering: the effect of heating and cooling a rock
- Chemical weathering: the effect of a weak acid on limestone
- Biological weathering

11.4 The sedimentary rock cycle: a simple model for rock formation

11.5 The rock cycle

11.6 Exploring plate tectonics
- Using real data to explain tectonic activity

11.7 Earth 'system' science: the case of climate change
- Calculating a carbon footprint
- Reducing your carbon footprint
- Researching the impact of climate change
- Using evidence to monitor carbon dioxide levels
- Encouraging group talk

Introduction

Earth science is an all-embracing term for the various branches of science related to the study of planet Earth. The chemistry of the Earth (geochemistry) is an important, multifaceted subject worthy of a discrete book in its own right. Consequently it will be difficult to offer here a comprehensive coverage of the subject that is relevant to chemistry teachers. Nonetheless the core ideas which have been shown to cause most conceptual difficulty have been included as a starting point. The chapter will provide a short section setting out background subject knowledge together with recent research on barriers to student learning, followed by suggested teaching sequences and approaches.

Background

The relevance of chemistry to Earth science lies in the key concept of differentiation, that is, the process by which the elements become distributed throughout the Earth. This principle encompasses the origin of minerals, rocks and fossils.

When the Earth and other planets accreted around 4.5–4.6 billion years ago, the mixture of elements remaining reflected the cosmic abundances. Through a series of complex chemical reactions the Earth warmed up and differentiation of the constituent elements took place.

Elements such as oxygen, silicon, magnesium, iron, aluminium and calcium were displaced into the mantle during formation of the iron core. During the formation of the crust a number of compounds known as minerals were formed. Minerals are the inorganic building blocks from which rocks were formed. Although there are more than 3000 known minerals, only about 20 are very common and only nine constitute 95% of the crust. There are two categories of silicate-based minerals:

1. Heavy silicate minerals, containing heavy elements such as iron, magnesium, calcium and aluminium. These heavy silicates are known as biotite mica, amphibole, pyroxenes, olivine and feldspar.
2. Light silicate minerals, containing lighter minerals such as silica (silicon dioxide) and oxygen, aluminium and potassium, form the minerals called quartz, muscovite mica, orthoclase feldspar and plagioclase feldspars.

Rocks are made up of mixtures of these nine common minerals, as summarised in Table 11.1.

Table 11.1 Silicate minerals comprising rocks

Silicate minerals including heavier elements	Silicate minerals comprising only lighter elements
Biotite mica Amphibole Pyroxenes Olivine Feldspar	Quartz Muscovite mica Orthoclase feldspar Plagioclase feldspars

There are three types of rocks: igneous, sedimentary and metamorphic.

Igneous rocks are formed by the cooling and crystallisation of a silicate melt. The molten rock material from which igneous rocks form is called magma. Magma is molten silicate material and may include already formed crystals and dissolved gases. The name magma applies to silicate melts within the Earth's crust; when magmas reach the surface they are referred to as lava. The principal constituents of magma are oxygen, silicon, aluminium, calcium, sodium, potassium, iron and magnesium. The viscosity of the magma is largely controlled by the silica (SiO_2) content and the melting point is controlled by the water content. Silica is the most abundant component and ranges in abundance from 35 to 75%.

The recent discovery of plate tectonics provides a grand unifying theory which helps explain how rocks are formed. Recent evidence has shown how the different types of igneous rocks are formed by the partial melting of parts of the crust at different depths. For example, basalts are formed in the upper mantle, granites are formed in the deep mantle and andesites are formed in subduction zones. Each type of rock can be identified by mineral composition and texture. Igneous rocks formed from lava, that is magma that reaches the surface, tend to cool quickly and form small mineral crystals, whereas rocks formed from magma deeper in the crust tend to cool down more slowly and form larger crystals. So it is the chemical composition and the cooling temperature of the original magma which determine when minerals crystallise and hence which type of igneous rock is formed.

Sedimentary rocks are the product of surface processes such as weathering, erosion, rain, stream flow, wind, wave action and ocean circulation. The starting materials for sedimentary rocks are the rocks outcropping on the continents. Processes of physical and chemical weathering break down these source materials into small fragments of rock such as gravel, sand or silt. Sedimentary rocks form when these initial sediments solidify by cementation and compaction. The most significant feature of the majority of sedimentary rocks is that they form layers or strata in chronological order. Consequently sediments can provide a preserved record of former climates and landscapes. Frequently, fossil remains of animals and plants that lived during these time periods are found preserved in their respective sediment units.

Metamorphic rocks are formed when the original texture, composition and mineralogy of igneous, sedimentary and existing metamorphic rocks have been changed by conditions of high pressure and extreme high temperatures. This metamorphosis is often brought about in the environment most commonly encountered at the centre of mountain belts such as the Alps or the Andes. These metamorphic rocks and associated igneous rocks formed deep in the crust make up about 85% of the continental crust.

Previous knowledge and experience

Studies of students' prior understanding of geology show that it is not obvious that rocks are actually materials from the Earth's crust. Indeed some students are only likely to identify large, jagged samples as rocks while smaller ones are described as 'stones'. Rock material is not often seen as 'natural' material. Students sometimes regard cut and polished samples as 'man made' while rock is a material straight from a quarry.

Furthermore students do not understand readily that rock material is made up of compounds called 'minerals' and frequently use the words 'rock' and 'mineral' interchangeably.

Choosing a route

A sensible starting point to help explain the changing Earth story to young learners is by introducing surface processes, such as the formation of rocks, and then moving on to look at the evidence to support the theory of plate tectonics. Using this approach mirrors the way that geologists have come to understand surface processes and how more recently they have been able to offer an explanation for the causes of the spectacular events witnessed on the surface of the Earth.

11.1 Understanding how rocks are formed

Understanding how rocks are formed is a more helpful starting point than describing the features of rocks. This approach is more engaging, and is more likely to help students understand the processes involved in the formation of the changing Earth.

■ How does loose sediment 'stick' together to form a 'rock'?

Students may believe that sediments become hard rocks simply by being compacted together. This is only the case with fine-grained sediments, such as clays, where electrostatic attraction helps to hold the particles together. Sands and coarser-grained sediments require some form of natural 'cement' to bind the particles together. Ask a student to take a handful of sand and see if it is possible to make a 'rock' by squeezing it as hard as possible. They will find that they cannot do it.

The chemicals that form the necessary cement come from groundwater, which is present in most sediments. In practice, the cementing agent is usually either silica (SiO_2), calcium carbonate ($CaCO_3$) or various compounds of iron. Most naturally occurring rocks are more strongly cemented than those which can be made in a school classroom.

The idea of particles being bound together can be tested in the laboratory, however, using a variety of 'cements'. Before carrying out each part of the activity, ask students to predict how strong they think each 'rock' will be when it has dried.

PROCEDURE

Materials

- some sand
- water
- an old plastic cup/similar small pot
- scissors
- various cements – salt, sugar, plaster of Paris, etc.
- an old 20 cm^3 syringe
- a hacksaw

Safety

Students should wear eye protection when testing their 'rocks'.

Method

1. Dampen some sand with water and pack it tightly into the bottom of an old plastic cup or small pot.
2. Cut away the plastic carefully and leave the sand pellet to dry.
3. Repeat this several times, but mix the sand beforehand with any suitable 'cements' that come to hand. Use a ratio of about one part of cement to four parts of damp sand.
4. If an old 20 cm^3 syringe is available, the nozzle end can be cut off with a hacksaw and the syringe used several times to make a more uniform series of 'rock' pellets than can be achieved with the plastic cup.
5. Ask the students to devise a fair way to test the strength of their rocks after they have hardened. When the rocks have become hard (which may take a day or so), students can then be invited to test them to destruction, to see which were made with the strongest cements. If possible, give them a piece of real sandstone to test as well. If they plan a series of tests, they should try the least destructive one first!

Discussion points
Many sedimentary rocks were once loose sediments that were cemented by natural cements in a very similar way. Natural cements were deposited by fluids flowing through the spaces between the grains.

See the 'Making a rock' section on www.earthlearningidea.com for full details and technical instructions. (Search the website for the appropriate activity and there will be a pdf file to download.)

11.2 Identifying rocks and their characteristics

Students' experience of small rocks, or of pieces of exposed rock, does not help them towards an awareness of a continuous rock 'crust' making up the surface of the Earth. Moreover, in everyday speech the words 'rock' and 'mineral' do not have the precise meanings that they do for the geologist. It will be important for students to understand that the word rock refers to the material which makes up that part of the Earth beneath our feet and that soil, sand or water are only quite a shallow covering over underlying rock. Opportunities to look at cliffs, quarries and caves to see exposed rock faces will help with this. Good photographs and videos can be a substitute for first-hand experience.

Collect some good-sized hand samples or borrow sets of rocks from your local university geology department. As a last resort, use the virtual rock kit found at:

www.earthscienceeducation.com/virtual_rock_kit/index.htm

Minerals in granite

Ask students to examine a sample of granite with a hand lens. They will find that it is made up of lots of little pieces of different-coloured crystals called grains. Each different type of grain is a different mineral. In most granite there are four different minerals, probably grey, pink, white and black.

Students could be asked to complete a table, stating the colour, lustre and size of grains in several samples of minerals.

Ask students working in groups to estimate the amount of each mineral in the sample. Which one is the most common?

Comparing granite and sandstone

Collect examples of different types of rock from your local area and from further afield. Take your students through this investigation sequence – using the clues in the rocks to find out how they formed.

Begin with two rocks, one made from sediment with obvious grains (a sedimentary rock such as sandstone), and the other a crystalline igneous rock with big crystals (such as a granite).

Ask the students to work in groups of three. Two students will each choose one of the two rocks and describe it carefully to another student. The third person will record key words and phrases used. These descriptions are then reported back to the rest of the class. This will identify key properties of rocks, namely: their colour, that they are made of 'bits' and that the surfaces feel rough.

Ask the students to predict what will happen to the masses (weights) of the two rocks after they have been put into water. When they have agreed a prediction, they should watch very carefully as both rocks are put into water together and left for about a minute. They will clearly see bubbles of water rising from the sandstone, but many fewer from the granite. Questions to ask about the sandstone: Where on the rock are most of the bubbles coming from? Why do they come from here? What does this tell you about the rock? Why is the sandstone different from the granite?

They should realise that: most of the bubbles rise from the top of the rock; they do this because the air in the spaces (pores) in the rock rises, allowing water to flow into the bottom; this shows that the rock is quite porous and that the spaces are connected (the rock is permeable). The granite has no connected spaces, so air and water cannot flow through. (For more information, visit the 'Rock detective' section of: www.earthlearningidea.com)

Identification of hand specimens of rocks

It is helpful to have large hand specimens available in the school and to devise a key for the specimens. A good starting point is to take photographs of one good example of an igneous, a sedimentary and a metamorphic rock, and label these so that the students can simply match the rock to the photograph and describe the distinctive features of each category. Then progress on to using a key to identify unknown rock specimens. An example is shown in Figure 11.1 on the opposite page. Keys work well when good hand specimens are available which match the description on the particular key.

11.3 Weathering and the effects of physical, chemical and biological processes on rock formation

The difference between weathering and erosion

When rocks are exposed to the effects of the weather, small pieces break off from the surface. Rock material is also removed when chemical interactions occur. These processes are called weathering. Weathering and erosion are often confused. Students may not readily accept that erosion is a separate process which involves the transport of the weathered fragments away from the parent rock.

Students' thinking about the sediments formed on the sea bed might help them to note that they have been transported a long way and that changes in the size, shape and composition of the mixture of fragments have taken place on the journey. It could be useful to introduce students to a laboratory-based stream table, to mimic, in miniature, the behaviour of real streams in transporting rock debris and depositing it where the 'stream' enters the 'sea'. Weathered rock falls under gravity and it is carried away by moving water, air and ice. This is called erosion.

The first step for students is to recognise that sedimentary rocks can only form from rocks which already exist and which are exposed to weathering at the Earth's surface. Only then will students be able to come to terms with the idea that weathered rock fragments are transported away from the parent rock and deposited as sediments on the sea bed. The timescale involved in weathering, erosion and sedimentation presents additional problems when students are trying to understand rock formation.

Figure 11.1
Key to some rocks commonly used for ornamental purposes

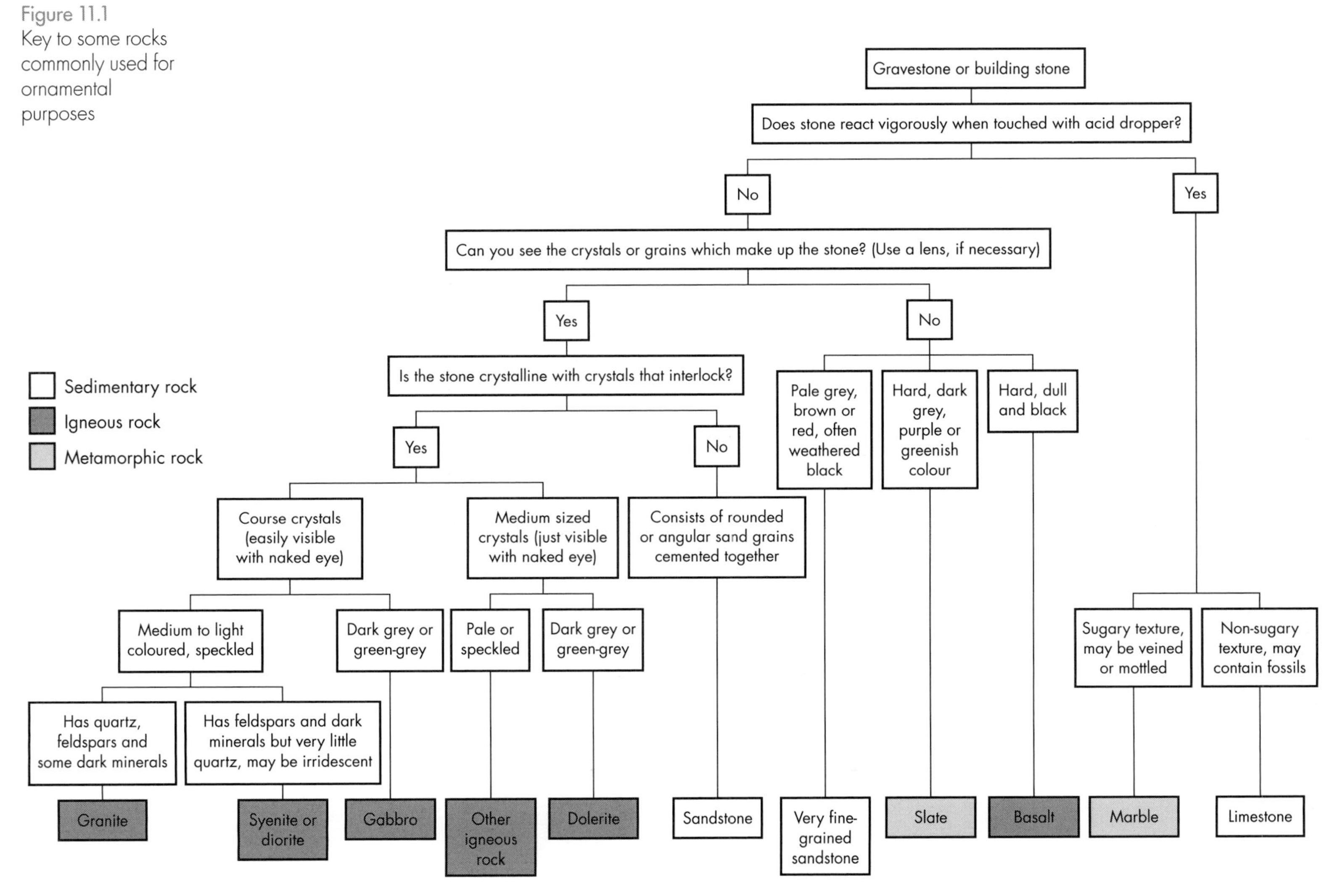

Physical weathering: the effect of heating and cooling a rock

PROCEDURE

Materials
- small rock samples of granite, chalk, limestone, slate and sandstone
- a Bunsen burner
- tongs
- a beaker of cold water

Safety
All students should wear eye protection. Students with long hair should tie it back.

Method
1 Heat a granite chip in the hottest part of a roaring flame for up to 1 minute.
2 Then drop the rock chip into the beaker of cold water.
3 Remove the rock sample and dry it off.
4 Repeat this heating and cooling five more times for the granite sample. Record what happens in a results table. Repeat this process for the other rock samples.

Questions
Which sample was the most easily weathered? How do you know this? Explain what happens to the particles during heating and cooling. Explain why some types of rock are weathered more than others by heating and cooling.

Note
This activity has been adapted from the Earth Science Education Unit's 'The dynamic rock cycle' workshop (www.earthscienceeducation.com).

Chemical weathering: the effect of a weak acid on limestone

When carbon dioxide from the air dissolves in rainwater it forms carbonic acid (a weak acid):

carbon dioxide + water → carbonic acid

If this weak acid rain falls on limestone, the limestone reacts and is weathered:

limestone (calcium carbonate) + carbonic acid → calcium hydrogen carbonate

PROCEDURE

Materials

- two boiling tubes, one of which has a bung
- deionised/distilled water
- universal indicator solution
- a clean straw
- powdered limestone
- a spatula

Method

1 Half-fill a boiling tube with deionised/distilled water.
2 Add six drops of universal indicator. Record the colour of the solution and corresponding pH (as detailed in Chapter 6).
3 Repeat this with a second tube of water and indicator.
4 Put a clean straw into one of the boiling tubes. Blow gently for about 30 seconds. Record any colour change and note the corresponding pH. Discard the used straw in the bin.
5 Add a spatula of powdered limestone.
6 Put a bung in the tube and shake it for 10 seconds.
7 Record any colour change or other signs of a reaction. Compare the colour with the other boiling tube. Record the results.

Questions

What happened to the water and the limestone powder? Explain why it happened. Where might this happen in the real world? Limestone buildings and statues in cities are weathered more quickly than those in the countryside. Why do you think this happens?

Note

This activity has been adapted from the Earth Science Education Unit's 'The dynamic rock cycle' workshop (www.earthscienceeducation.com).

Biological weathering

It is not just the weather that causes weathering of rocks. Plants and animals can also break down rocks. The photographs in Figure 11.2 show examples of rocks being weathered by a plant or an animal.

For each situation shown, ask students to:

1 identify the environment being weathered
2 identify the organism doing the weathering
3 explain how the organism causes the weathering
4 explain what happens to the products of the weathered rock.

Figure 11.2 Examples of biological weathering

11.4 The sedimentary rock cycle: a simple model for rock formation

Having introduced students to the formation of sedimentary rocks, the next logical step is to link the sediments, sedimentary products and weathering processes together into a cycle. A helpful way of doing this involves students working in pairs to carry out a card sort. Put together a series of product cards and process cards like those detailed on the next two pages.

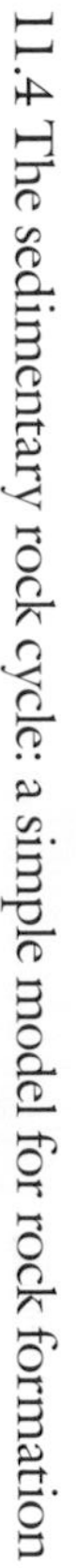

Figure 11.3
Examples of product cards:
a sedimentary rocks at the surface;
b weathered rocks;
c mobile sediments;
d layers of sediments;
e sedimentary rocks

Ask students to arrange the product cards in order and link them to the process cards given here.

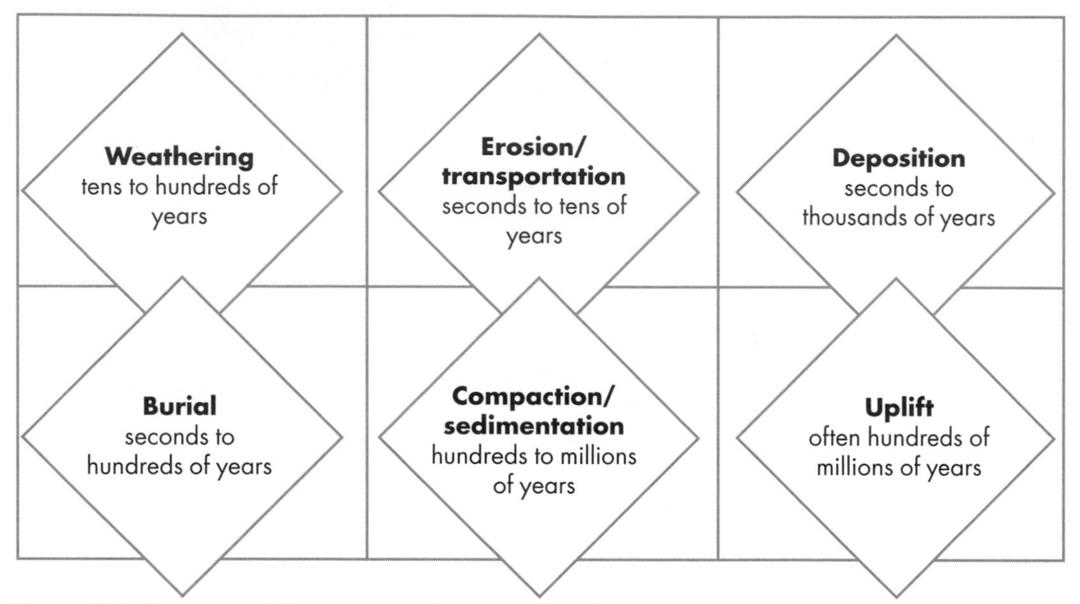

Figure 11.4 Process cards for use in a sedimentry rock cycle

■ Consolidating activity

Ask students to complete an imaginative recount of the life story of grains in the sedimentary rock cycle, entitled: 'Rocky and Sandy's story: the Grains' journey'. They should draw on the processes identified in the previous activity that are involved in the changes that a sediment grain undergoes. These include: the sediments being transported, deposition of mobile sediments, layering of sediments building up, sediment being compacted and cemented, uplift causing the sedimentary rock to be exposed at the surface and finally weathering causing the breakdown of sedimentary rock.

(Adapted from the Key Stage 3 National Strategy: Strengthening teaching and learning of geological changes in Key Stage 3 Science.)

11.5 The rock cycle

The next step is to widen the cyclical process to include igneous and metamorphic rocks. Diagrams are included in many school textbooks; there is an animated rock cycle at

www.geolsoc.org.uk/rockcycle and other resources are recommended at the end of the chapter. Ask students to label one of the two diagrams (Figure 11.5 or 11.6): either adding process labels to Figure 11.5, which shows the products, or adding products to the processes diagram in Figure 11.6.

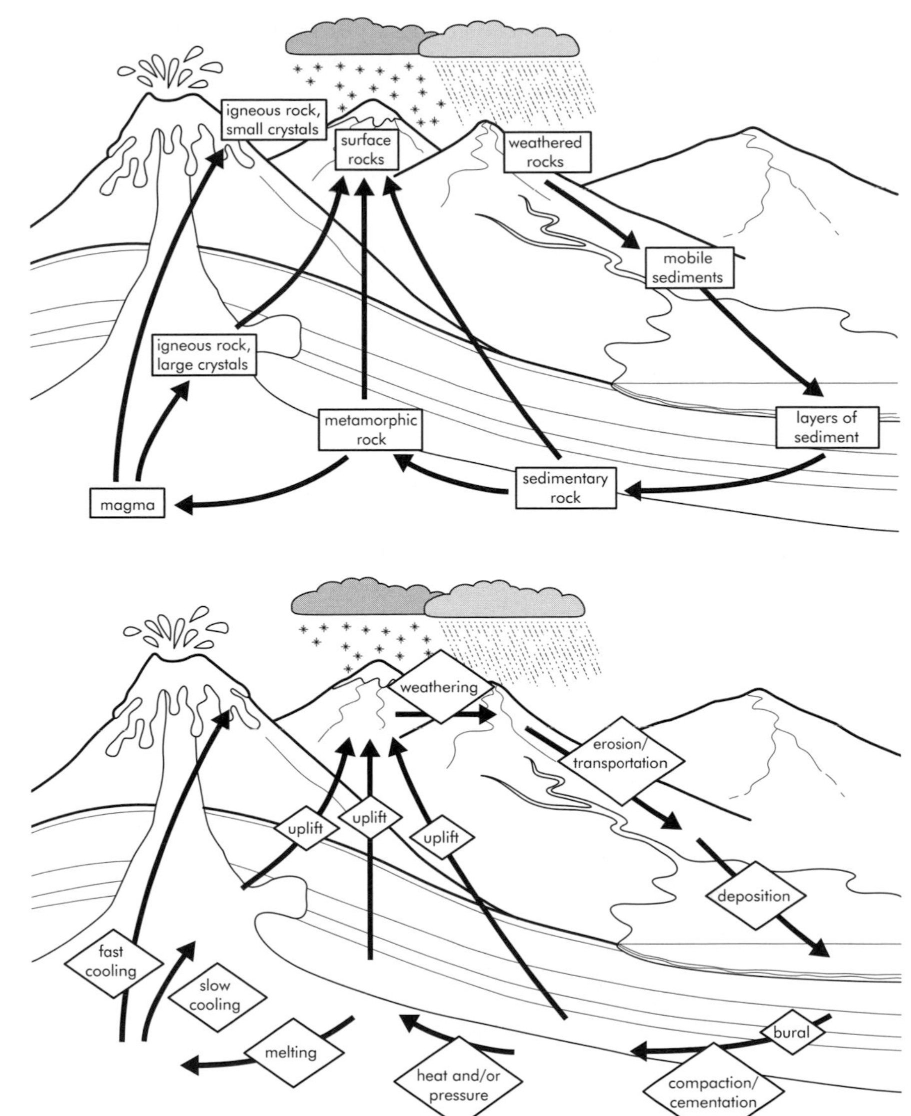

Figure 11.5 Products of the rock cycle

Figure 11.6 Processes involved in the rock cycle

11.6 Exploring plate tectonics

This theory of plate tectonics is a relatively recent way of explaining the causes and driving force behind the multitude of surface processes that geologists have been studying for many hundreds of years. In the last 30 years, plate tectonics has revolutionised our ideas about the structure of the Earth. Previously, no single theory had been able to account for all the observations and catastrophic phenomena witnessed on the Earth's surface. Plate tectonics explains the 'jigsaw' fit of the continents, the appearance of particular fossil traces found only in West Africa and the Atlantic coast of South America, the pattern of distribution of volcanoes and earthquakes on the Earth, and the drift of continents over very long periods of time.

In a nutshell, plate tectonics is based on recent evidence which suggests that the crust of the Earth is cracked into six large plates with a number of smaller plates of different sizes. These plates compose the outer solid rigid layer of the Earth (the lithosphere) and they move across the Earth's surface as giant slabs of rock floating on the semi-fluid layer of the mantle. They are probably carried by giant convection currents in the mantle.

Two distinct types of plates have been identified: oceanic plates and continental plates.

- Oceanic plates are found at the bottom of the oceans and are made from the same material as the upper mantle of the Earth which has been super cooled into solid rock of 10–16 km thickness.
- Continental plates and islands are made up of a mixture of lighter materials embedded within the same super-cooled rock as oceanic crust material. As the continental plates move, the land masses that are embedded in them are carried passively along.

See a diagram at http://csmres.jmu.edu/geollab/fichter/PlateTect/synopis.html

The plates move relative to one another in three ways:

1. Divergent plate boundaries: Oceanic crust is formed when plates move apart from each other and magma (molten rock) emerges between the plate margins forming new super-cooled crustal material. This is happening along the mid-ocean ridges such as that which occurs in the middle of the Atlantic. As the magma reaches the cold sea water, it is cooled suddenly to form solid rock and new ocean floor is formed on both sides of the gap between the plates. Volcanic islands are frequently found over the mid-ocean ridges. For example, Iceland is situated on the mid-Atlantic ridge and it has numerous volcanoes.
2. Convergent plate boundaries: When plates collide, one plate can slide down underneath another plate. As this happens, the rock making up the plate which is sliding underneath melts and, because it is less dense, it rises through the crust forming volcanoes. The lighter continental crust on the plate which is sliding underneath is folded up, forming mountain ranges. Such regions, called 'subduction zones', are characterised by the existence of deep ocean trenches, earthquakes and volcanoes.
3. Transform plate boundaries: When plates slide past one another and rub together in the process, the friction causes pressure to build up and when this pressure is released the plates move very

suddenly relative to one another causing earthquakes. This process is happening now along the San Andreas Fault off the coast of California.

Using real data to explain tectonic activity

This theory is highly relevant to students and you can draw on recent events in the news. At the time of writing there has been considerable seismic activity in New Zealand and Japan, and air travel has been disrupted by volcanic activity in Iceland. There are lots of excellent images available which can be used to illustrate both the surface effects and what is happening underneath the surface. It is helpful, however, to take time to explain the principles. The activity which follows will enable students to use actual data and evidence to make the links for themselves. The purpose of the exercise is to use a range of data to identify patterns and trends at plate boundaries and to explain what is happening at these boundaries.

■ Part 1: exploring trends and patterns

Students work in groups of three and take on a different 'expert' role. The three areas of expertise are: a) seismologist, b) volcanologist, c) geochronologist. Each expert should then join other experts in the class and convene specialist groups who will focus on one of the areas: seismological, volcanological or geochronological evidence. They use maps provided and links to other sources to help explain what story the data is telling.

Expert group 1: seismology data

Download a seismology map such as the one given at http://terra.rice.edu/plateboundary/downloads.html (choose the 'Seismology Map' in your preferred format). Each dot on this map represents the epicentre of an earthquake. The epicentre of an earthquake is the surface location which is directly over the hypocentre of the earthquake. The hypocentre is the location of the break in the rocks, whose relative movement caused the earthquake. Look too at the United States Geological Survey website for the most recent earthquake activity: http://earthquake.usgs.gov/earthquakes/recenteqsww/

Where are the active areas for earthquakes? Where is the most recent earthquake? What magnitude was this?

Expert group 2: volcanology data

Consult a volcanology map, for example on the US volcano watch website (http://volcanoes.usgs.gov) or the Global Volcanism Program (www.volcano.si.edu/reports/usgs). Where are the main volcanoes located? Where was the most recent eruption? Where do geologists anticipate the next eruption will happen?

Expert group 3: geochronology data

Geologists use a wide variety of geochronological tools and methods to estimate the age of rocks and sediments. Using measurements of radioactive isotopes, geologists are able to date all geological materials, from billions of years to the present time.

Download a seafloor age map such as the one from http://terra.rice.edu/plateboundary/downloads.html (choose the 'Seafloor Age Map' in your preferred format). The colours on the map represent the age of the oceanic crust. There is a bar on the right-hand side of the map that shows the age in millions of years. At the bottom of the scale, red represents very young oceanic crust, formed less than 9.7 million years ago. The oldest oceanic crust is represented in blue and is between 154 and 180 million years old. Note that all oceanic crust on Earth is younger than about 180 million years. This is actually surprising given that the Earth was formed about 4600 million years ago and there are areas of the continents where the rocks are known to be almost 4000 million years old. The oceans are young because old ocean crust becomes quite dense and tends to be subducted back into the mantle. Thus it does not remain on the surface of the Earth very long. Continental crust is relatively quite light, both when it is young and when it is old, so it tends to stay at the surface, nearly forever. These data were obtained from Dietmar Muller, a scientist at the University of Sydney in Australia.

Ask students to note where the oldest rocks are located and where the youngest rocks are being formed.

■ Part 2: putting all sources of evidence together

Specialist experts should then return to their original group of three and discuss the key findings from each specialism. Each expert should make a brief presentation to the rest of their group about their scientific speciality's data and classification scheme and describe the patterns emerging.

■ Part 3: explaining recent plate movement

A plate boundary map will be used to bring all three sources of data together. Download a plate boundary map such as the one from http://terra.rice.edu/plateboundary/downloads.html (choose the 'Plate Boundary Map' in your preferred format).

Allocate each group of three a plate boundary and ask them to describe in detail what is happening and the direction of movement of the plates, using the combined expert evidence collected in Part 1.

Students can be asked to describe what they think is happening at the plate boundaries below Iceland, Japan, New Zealand and off the Californian coast. Each group of three can present their ideas to the whole class so that all students can share their findings for the three plate boundary activities using very recent global events as examples.

A final consolidating activity is to make a model using a tennis ball. Full instructions are available at: http://volcanoes.usgs.gov/about/edu/dynamicplanet/ballglobe/index.php

11.7 Earth 'system' science: the case of climate change

In the mid-1980s, NASA began a systems approach to cataloguing the elements of the Earth's environment, their linkages, dependencies and fluxes. As a result, an important paradigm has emerged in the geosciences, analogous to the plate tectonics revolution of the 1960s and 1970s. This paradigm, called 'Earth system science' or simply the 'Earth system', acknowledges that there is an important synergy and constant interaction between the atmosphere (air), hydrosphere (water, including oceans, rivers, ice), biosphere (life), lithosphere or geosphere (solid rock) and human influences (called the anthroposphere). Consequently the Earth system is often represented by interlinking and interacting 'spheres' of processes and phenomena.

The atmosphere, hydrosphere, biosphere and geosphere form the simplest collection, with increasing input from the anthroposphere because of human activity. The most recent thinking in this area has come to the conclusion that it is unhelpful to reduce the system to a series of discrete spheres because in reality no part of the Earth system can be considered in isolation from any other part.

Earth system science embraces chemistry, physics, biology, mathematics and applied sciences, transcending disciplinary boundaries to treat the Earth as an integrated system. It seeks a deeper understanding of the physical, chemical, biological and human interactions that determine the past, current and future states of the Earth. Earth system science provides a physical basis for understanding the world in which we live and upon which humankind seeks to achieve sustainability.

The most recent research evidence from NASA and the UK-based Natural Environment Research Council (NERC) starts from the premise that Earth is the only planet we know of that sustains life. From collective knowledge so far the evidence very strongly suggests that life on Earth is critically dependent on the abundance of water in all three phases – liquid, vapour and ice. Furthermore, carbon exists in a variety of forms and is the very basis of life; in the atmosphere, carbon is fully oxidised as carbon dioxide. Other compounds of carbon, such as methane and particulate carbon in the form of soot, are also present in the atmosphere, and these add to the greenhouse effect which helps to make the Earth habitable. We also have strong evidence to suggest that the Earth's atmosphere and electromagnetic field protect the planet from harmful radiation while allowing useful radiation to reach the surface and sustain life. Finally, the Earth exists within the Sun's zone of habitation and, with the Moon, maintains the precise orbital inclination needed to produce our seasons.

> 'These remarkable factors have contributed to Earth maintaining a temperature range conducive to the evolution of life for billions of years. The great circulation systems of Earth – water, carbon and the nutrients – replenish what life needs and help regulate the climate system. Earth is a dynamic planet; the continents, atmosphere, oceans, ice and life ever changing, ever interacting in myriad ways. These complex and interconnected processes comprise the Earth system, which forms the basis of the scientific research and space observation that we refer to as Earth system science.'
>
> Source: NASA, 2011

Previous knowledge and experience

The big question at the time of writing which will challenge students is the controversy surrounding the debates about global warming. While there is considerable scientific evidence available, students may well be confused by the strong messages being made by substantial numbers of non-scientists who argue that global warming is not an issue. The key point being disputed surrounds the causes of increased global average air temperature, especially since the mid-twentieth century. The debates focus on whether this warming trend is unprecedented or within normal climatic variations and whether humankind has contributed significantly to it.

Active teaching approaches

■ Calculating a carbon footprint

Ask students to keep track of their use of power sources for various purposes (heating/cooling, electricity, transport, etc.) for a week. They can use the online calculator at www.carbonfootprint.com/calculator.aspx to estimate their yearly footprint and compare it to UK and world averages.

■ Reducing your carbon footprint

Divide the class into groups of three and hand out a mixed-up set of cards (made using the information given here) to each group. The cards should have the text from the left column of Table 11.2 on one side and the explanatory text from the right column on the other side. Ask students to sort them into the following groups:

- appliances
- travel
- home insulation
- heating and lighting.

Table 11.2 Text for the cards

Install double glazing.	Double glazing works by trapping air between two panes of glass, creating an insulating barrier that reduces heat loss, noise and condensation. Double glazing cuts heating through windows by half.
Switch to energy-efficient light bulbs throughout your home.	This will save energy and cut electricity bills.
Fill your kettle with only as much water as you need.	Boiling more water than you need for a cup of tea wastes time and transfers more energy than is needed.
Lower your thermostat by 1 degree.	This should reduce the heating bill by about 10%.
Turn off the light when you leave a room.	This will transfer less energy and cut your electricity bills.
When you can, take the train instead of flying.	Aeroplanes release more than twice as much carbon dioxide as trains.
Turn your appliances completely off (rather than leaving them on standby).	Leaving appliances on standby transfers energy without doing anything useful. It is possible to save 5–10% on the electricity bill.
Unplug mobile phone and other chargers when you are not using them.	Chargers transfer energy whenever they are plugged in.
Turn down your TV display brightness.	Manufacturers set them to their brightest levels for display in showrooms. Turning it down will save electricity.
Defrost your freezer at least once a year.	This will keep it working as efficiently as possible.
Install a hot water jacket.	Insulating the hot water cylinder is one of the simplest and easiest ways to reduce energy transfer and save money. Fitting a 'jacket' around the cylinder will cut 'heat' loss by over 75%.
Use public transport for local trips where possible.	Using public transport instead of a car typically reduces journey-related emissions by half.
Install loft insulation.	Without loft insulation you could be losing as much as 15% of heating costs through your roof. Loft insulation acts as a blanket, reducing heating of the loft space from the house below.
Walk or cycle for journeys less than 1 mile.	A car journey of less than 1 mile is the most inefficient type of journey as the engine does not have time to warm up properly and hence burns more fuel.

Ask students to sort the cards from easiest to hardest based on the effort required to achieve them. Lead a discussion about which of the suggestions from the card-sorting game would be easiest to realise. Why would others be more difficult? Ask students to re-sort the cards from least to most expensive to show how costly they would be to do. Is this the same order as before? Then ask them to pull out all those that they do already. How could they work with their families to implement some of the remaining suggestions at home?

■ Researching the impact of climate change

Use Google Earth® to see how climate change could affect the planet. Ask students to create a presentation to show some examples of the impact of global warming around the world.

Start Google Earth then look at the list of layers in the 'Layers' panel which is in the bottom-left corner. Layers show information placed on top of the Google Earth satellite images of the Earth. For example, they can show the borders of countries and the names of cities, or buttons which link to photos of places.

Experiment with the different parts of the Layers panel. You can switch different layers on and off by clicking on the relevant boxes to tick or untick them. See what happens to the Google Earth view. Use the panel to find out more about climate change: click the plus icon next to Global Awareness in the Layers list; this will expand to show more layers. Find the UNEP icon – this stands for United Nations Environment Programme. Click the tick box next to it to switch on this layer. The UNEP 'Atlas of Our Changing Environment' shows photos of places around the world where the environment is changing. These places can be located by looking for the square blue and black UNEP icons as you explore Google Earth.

There are different reasons why these places are changing. For example, some are urban areas becoming larger with increasing population size. However, others may be caused by climate change.

Retreating glaciers

Click on the UNEP icon over Greenland to get information on the retreating Helheim glacier. This contains two photos to compare. Click on Iceland's UNEP icon to see an overlay in a box and get an overlay photo to show iceberg calving. Click on Further information to get more detailed information and pictures. This can also be viewed on a Google map on the UNEP website: na.unep.net/

■ Using evidence to monitor carbon dioxide levels

Carbon dioxide (CO_2) is a greenhouse gas that is strongly correlated with global temperatures. It is believed that the more CO_2 there is in the atmosphere, the warmer Earth's atmosphere and surface become, because of the ability of CO_2 to absorb Earth's outgoing infrared radiation. In this exercise, students will investigate the CO_2 data set from Mauna Loa, Hawaii, which dates from the late 1950s, and then estimate the rate at which atmospheric CO_2 has been increasing recently. This will help them to understand:

1 the basis for conclusions about how quickly CO_2 concentrations are rising globally
2 estimates for how long it will take for global CO_2 concentrations to double at today's rate of increase
3 the effects of looking at only a portion of a data set in terms of describing its properties.

Background information

1 CO_2 is a trace gas, meaning that the atmosphere has only a small amount of CO_2 compared to other gases (nitrogen and oxygen make up 99% of the atmosphere). Because CO_2 is a trace gas, adding CO_2, for example by burning fossil fuels, changes its atmospheric concentration more than if we were adding a gas like oxygen, since there is already so much more oxygen in the atmosphere.
2 The units for CO_2 concentration are ppm or parts per million. A value of 1 ppm for gas X in the atmosphere would mean that for a volume having one million total units of all gases, a single unit of the million units would be of gas X.
3 The slope of a line in a graph is the ratio of the rise (the y axis) over the run (the x axis) and has units of the rise axis over the run axis. For example, if a graph has time (hours) as the x axis and distance (miles) as the y axis, then a line will have a slope of miles/hour.
4 The best line through a set of data points can be defined in several ways, but in some sense (even by eye) has to minimise the difference between the data points and the line.
5 The amount of time needed for global CO_2 concentrations to double is an important input into climate change models for predictions of future climate.

Task 1

For Mauna Loa, the complete monthly CO_2 data set for the seven years between 2003 and 2009 consists of 84 data points. Each pair of students will plot on graph paper eight data points from the 84 monthly Mauna Loa CO_2 values (given in Table 11.3, where each row represents eight randomly selected monthly values).

Table 11.3 Monthly Mauna Loa atmospheric CO_2 concentrations in parts per million (ppm). G = group number, D1 = date (M/Y) of first observation, V1 = value of CO_2 concentration on that date (Data source: bluemoon.ucsd.edu/)

G	D1	V1	D2	V2	D3	V3	D4	V4	D5	V5	D6	V6	D7	V7	D8	V8
1	02/03	375.6	12/03	375.7	01/04	376.8	04/04	380.5	05/06	384.9	06/06	384.1	02/08	385.9	11/08	384.1
2	02/03	375.6	03/03	376.1	01/05	378.3	09/06	378.8	01/09	386.7	04/09	389.5	09/09	384.6	10/09	384.3
3	05/03	378.4	06/03	378.1	10/03	373.0	04/04	380.5	08/07	381.8	03/08	385.9	08/09	386.1	10/09	384.3
4	01/03	374.7	06/06	384.1	10/06	379.1	12/08	385.2	01/09	386.7	02/09	387.2	05/09	390.2	12/09	387.4
5	10/03	373.0	03/04	378.4	09/05	376.4	07/06	382.4	03/08	385.9	10/08	382.8	03/09	388.6	06/09	389.6
6	03/04	378.4	11/04	375.8	09/05	376.4	02/06	382.0	12/07	383.7	03/08	385.9	01/09	386.7	08/09	386.1
7	07/03	376.6	10/03	373.0	09/04	374.1	07/05	380.7	10/05	376.8	06/07	385.9	08/08	384.2	12/08	385.2
8	03/03	376.1	07/04	377.8	06/06	384.1	07/07	384.4	07/08	386.3	10/08	382.8	11/08	384.1	06/09	389.6
9	12/03	375.7	06/04	379.6	10/04	374.2	10/05	376.8	08/06	380.5	04/08	386.8	10/08	382.8	06/09	389.6
10	01/06	381.4	05/06	384.9	06/06	384.1	03/07	384.3	03/08	385.9	05/08	388.5	08/09	386.1	12/09	387.4
11	03/03	376.1	05/03	378.4	08/03	374.5	02/06	382.0	11/06	380.2	08/07	381.8	01/08	385.0	08/08	384.2
12	02/03	375.6	10/03	373.0	12/03	375.7	07/04	377.8	08/05	378.7	09/08	383.0	03/09	388.6	06/09	389.6
13	04/03	377.7	09/03	373.0	03/04	378.4	07/05	380.7	10/06	379.1	07/07	384.4	07/09	388.0	10/09	384.3
14	04/03	377.7	06/03	378.1	01/04	376.8	06/04	379.6	12/05	380.0	04/06	384.4	09/09	384.6	12/09	387.4
15	01/03	374.7	10/06	379.1	01/07	382.5	12/07	383.7	01/08	385.0	06/08	387.9	10/08	382.8	06/09	389.6
16	09/03	373.0	08/04	375.8	11/04	375.8	06/07	385.9	05/08	388.5	08/08	384.2	06/09	389.6	12/09	387.4
17	09/03	373.0	07/04	377.8	10/04	374.2	03/05	380.4	04/06	384.4	01/09	386.7	02/09	387.2	06/09	389.6
18	03/03	376.1	04/04	380.5	06/04	379.6	08/05	378.7	12/06	381.7	06/07	385.9	04/08	386.8	07/08	386.3
19	01/04	376.8	02/04	377.4	04/04	380.5	06/04	379.6	08/04	375.8	06/07	385.9	08/07	381.8	12/07	383.7
20	03/03	376.1	10/03	373.0	06/06	384.1	02/07	383.7	06/07	385.9	11/07	382.4	03/08	385.9	10/09	384.3
21	07/03	376.6	01/05	378.3	12/05	380.0	06/06	384.1	12/06	381.7	06/07	385.9	07/09	388.0	11/09	386.0
22	08/03	374.5	08/03	374.5	06/05	382.1	04/06	384.4	03/07	384.3	08/07	381.8	09/07	380.9	09/09	384.6
23	05/03	378.4	09/03	373.0	01/05	378.3	07/06	382.4	02/07	383.7	05/08	388.5	08/08	384.2	05/09	390.2
24	07/04	377.8	12/04	377.4	04/05	382.1	09/05	376.4	03/06	382.6	07/06	382.4	02/08	385.9	05/09	390.2
25	12/03	375.7	01/05	378.3	09/05	376.4	04/07	386.2	05/07	386.4	12/07	383.7	10/09	384.3	12/09	387.4
26	04/03	377.7	05/03	378.4	04/04	380.5	05/04	380.6	09/06	378.8	10/07	380.9	11/08	384.1	03/09	388.6
27	10/03	373.0	11/03	374.4	10/04	374.2	06/05	382.1	11/05	378.3	01/07	382.5	08/09	386.1	12/09	387.4
28	06/03	378.1	02/04	377.4	06/04	379.6	08/04	375.8	08/05	378.7	05/07	386.4	08/07	381.8	10/09	384.3
29	05/03	378.4	08/05	378.7	10/05	376.8	05/07	386.4	04/08	386.8	07/08	386.3	10/08	382.8	04/09	389.5
30	08/03	374.5	05/05	382.2	07/05	380.7	03/06	382.6	08/06	380.5	11/06	380.2	07/07	384.4	09/08	383.0
31	05/03	378.4	09/03	373.0	04/04	380.5	03/06	382.6	01/07	382.5	02/07	383.7	04/07	386.2	11/08	384.1

Each pair of students should use the first (earliest) data point of their chosen eight values to estimate the percentage of the atmosphere made up of the trace gas CO_2. (For example, 300 ppm = 300/1 000 000 = 0.0003; to get a percentage, $0.0003 \times 100 = 0.03\%$.)

Now ask students to estimate the slope of the data to two significant digits (for example, slope = 4.1). To do this, they must draw a line of best fit through their eight data points and then measure the rise (change in y axis) over the run (change in x axis) of this line. The slope should have units of ppm/year. Students should then compare their group's estimate of the slope with slopes estimated by other nearby groups. (Remember that each team is using a different eight random samples.)

Task 2

In this part, students will estimate how long, based on the rate of increase their group determined from their data set (in other words, the slope determined from their graph), it will take for atmospheric CO_2 concentrations to double from the value at the beginning of their data. For example, if the beginning value was 300 ppm and the rate of increase was 3 ppm/year, it would take 300/3 = 100 years to increase by 300 ppm (in other words, double).

Show the class the plot of all the monthly data which the student pairs have been working on (Figure 11.7). Ask them to describe the pattern now evident in the complete data set. How often does the pattern repeat? What might account for this pattern?

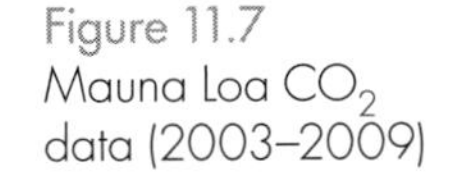
Figure 11.7 Mauna Loa CO_2 data (2003–2009)

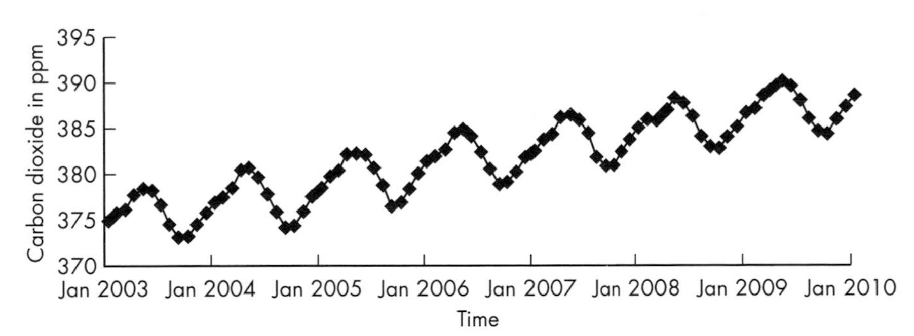

It is worth noting that the overall rate for the last 10 years is higher than the rate for the 10 years before, and that the rate may continue to increase in the future. Ask the students whether, if the rate increases in the future, the time needed to double the atmospheric CO_2 concentration will increase, decrease or stay the same.

Ask students to reflect on what they have learnt in this exercise. Individually, in their own words, they should write a one- to two-paragraph summary about what they have learnt from this entire exercise about the CO_2 data set at Mauna Loa.

This activity was adapted from: http://serc.carleton.edu/NAGTWorkshops/climatechange/recommended.html

■ Encouraging group talk

Climate Talk is a Democs kit which can be downloaded freely from: www.neweconomics.org/publications/climate-talk
The kit provides scenarios and stimulus material to set up group discussion about this controversial area of the curriculum. The purpose is to help young people to understand the science of climate change, the threat that it poses and what they can do to help.

Other resources

Books

Park, G. (2010). *Introducing Geology: A Guide to the World of Rocks*. Edinburgh: Dunedin Academic Press.

van Andel, T. (1994). *New Views on an Old Planet*. Cambridge: Cambridge University Press.

Webster, D. (1987). *Understanding Geology: Pupil's and Work Book*. Edinburgh: Oliver & Boyd, Pearson Education.

Websites

Rocks and rock formation:
www.earthlearningidea.com
www.earthscienceeducation.com

Plate tectonics:
http://volcanoes.usgs.gov/about/edu/dynamicplanet/index.php
http://mineralsciences.si.edu/tdpmap/

Climate change:
www.carbonfootprint.com
http://climate.nasa.gov/widget
http://climate.nasa.gov
www.esrl.noaa.gov/gmd/ccgg/trends
www.neweconomics.org/publications/climate-talk
http://serc.carleton.edu/NAGTWorkshops/climatechange/recommended.html
www.teachclimatechange.org
http://globe.gov

12 Chemistry in the secondary curriculum

Keith S. Taber

This handbook is not intended as a prescriptive document which tells teachers what should be taught in secondary chemistry, but rather has looked at key issues in teaching some of the core (and often problematic) topics which have commonly been included in secondary school courses. Just as the book is not designed to be prescriptive, we have also avoided the temptation to follow some particular curriculum formulation or examination specification and so to seek to 'cover' the content of any specific government document or examination authority. We hope that the guidance in this book will continue to be useful despite future specific changes in curriculum or examination courses. There are continuing important debates about what science should be taught during the secondary school years, and what chemistry deserves its place – especially where chemistry is a compulsory subject for all students.

The notion of 'pupil voice' or 'student voice' is taken seriously in many schools today, recognising that children should be involved in the major decisions that influence their lives. It has also been found that providing young people with some elements of choice in what they study, or allowing them to select particular activities to meet specified learning goals, can be highly motivating for many students. This always needs to be balanced against their status as minors, that is, as young people recognised as not yet in a position to make fully informed and considered decisions about what is in their best interests. There is then a difficult balance to be reached in judging how much input students themselves should have in decisions about what they will study. Teachers are often given the power to make some of these decisions for the young people in their care, and this brings responsibilities to make those decisions carefully and after due consideration.

Chemistry for enthusiasts

Such debates inevitably draw upon different ideas about the purposes of formal education, and so the justification for setting out the particular things that we expect all young people to meet during their schooling. Quite rightly, the idea that the school curriculum must include all the chemistry likely to be useful for the minority going on to study post-secondary chemistry, and that this should be taught to all students (perhaps with an implication that most will fail to engage with, or make sense of, much of the material, aiding the selection of those most suitable for progression), is no longer seen as viable or appropriate.

Yet, despite this, it is clear that part of the purpose of a curriculum is to offer learners a flavour of different subjects – both to help them decide whether a particular subject is of sufficient interest to consider studying it further, and to provide a basic understanding of the core ideas suitable for progression. This suggests that (i) all students should meet enough (representative) chemistry to support judgements about whether this is a subject they are interested in selecting for further study in some form after completing secondary education, but (ii) there should be curriculum differentiation that allows different groups of students to study more or less of the subject. This can meet the needs of those who are keen to take further courses, as well as those who have had their taster, and decided this is not a subject they wish to spend too much time working on.

All students should meet some chemical topics throughout the secondary years, but with some variety in the types of courses that can be offered to suit the needs for different student groups. Having a common core to the science curriculum, which can be supplemented by alternative optional elements, allows school departments and teachers to find the right curriculum topics for groups of students identified in the local teaching and learning context. Making these important decisions should be informed by consideration of the different purposes of education.

Chemistry for public understanding

It is widely accepted that a key role of school is to prepare young people for their place in adult society, where they will be: making decisions about consumer habits; acting (or not) on health and medical advice; engaging in civic life (perhaps by supporting particular organisations and pressure groups); and voting for their political representatives in consideration of mooted policies about such matters as environmental protection. Science, including chemistry, clearly has a major role here, as a great many of the decisions adults will face rely on understanding and critiquing arguments based upon (or claiming to be based upon) scientific evidence.

It is essential, therefore, that young people understand something of the nature of scientific processes; and of the nature of scientific models and theories, and how they relate to evidence. Of all the science subjects, chemistry offers exceptional opportunities to appreciate the relationships between evidence based on observation of the natural world, model-building as a means of making sense of evidence, and theoretical knowledge as a means to build up general explanations and make fertile predictions where solid evidence is not yet available. Chemistry, after all, is largely based upon an extensive set of theoretical models relating to entities (molecules, ions, electrons, bonds) which cannot be directly observed, but which support a coherent and extensive understanding of the material world. That is, a coherent and extensive understanding which allows us to create new substances and materials that quite possibly have never previously existed in the Universe, designed to meet particular needs (and increasingly to do so in ways that are sustainable).

In particular, all students should be supported to become citizens who can appreciate both the strength of science as a means of producing robust and reliable public knowledge, and the inherent limitations of science. All scientific knowledge should be considered provisional in the sense that the scientific attitude is to always be prepared to take another look when there is reason to do so. Often the case seems to have been made: but science always admits appeals based on convincing new evidence. This is an essential issue for teachers when, understandably, most of the chemistry met in school is well established and seems beyond question, yet much of the chemistry and other science which is part of public discourse seems to be the subject of uncertainty and disagreement. This can be understood when students appreciate something of the processes by which ideas – initially someone's unsubstantiated imaginings – are tested and developed, and slowly come to be seen as worthy of being considered sound (yet never quite certain) scientific knowledge. Clearly a very important part of school science must be focused on teaching something of the essential nature of science and its processes. This is the theme of the companion handbook on *Teaching Secondary How Science Works*, but the authors of the present volume have offered suggestions for where these themes can be emphasised in chemistry topics.

However, whilst encouraging a focus on teaching about the processes of science, it is important not to lose sight of the products of those processes: the models and theories themselves. Teaching about key concept areas in chemistry (particle theory, the periodic table, types of reaction, etc.) is important in its own right. For if students are to be helped to become citizens who understand public debates relating to chemistry (ozone depletion, safe storage of nuclear waste, pollution from combustion of fossil fuels, etc.) then they will need a sufficient level of literacy in terms of the core conceptual ideas of chemistry: what a molecule is; what happens in a chemical reaction; how chemical changes may be influenced by changing conditions; etc.

It is clearly not possible to set out with confidence a canon of agreed ideas that comprise what future citizens will need to know to engage with public discourse about socio-scientific issues; but there is a good case for thinking that many of the key topic areas covered in the present book are at least respectable candidates. The more chemistry we can teach, and the more sophisticated understanding we can offer, the more likely we are to prepare future citizens for this aspect of their future lives. Yet, of course, this always has to be balanced against all the other competing areas of knowledge that will have claims on a place in curriculum.

Chemistry for culture – and the ecosystem

Certainly chemistry teaching should be part of a balanced education. One perspective, the notion of a liberal education, would suggest that all young people should be offered the tools to appreciate the wider aspects of the culture of their society. The scientist and novelist, C. P. Snow, famously referred to the 'two cultures' and bemoaned that whilst a person knowing little of art and literature might be considered somewhat uncultured, ignorance of such scientific topics as thermodynamics seemed to be much better tolerated (sometimes almost celebrated) in polite society. In the twenty-first century such an attitude should not be acceptable. Modern life is highly dependent on science and the technologies it underpins. Moreover, human actions in the world have led to widespread extinctions, immense loss of habitat and resources, and potentially threaten the global environment as a suitable home for many living species – certainly including *homo sapiens* on the scale of current and projected world populations.

Science has not caused these problems – they are the result of human (individual and collective) decision-making and actions – but science has provided the tools for much of the damage. Science also offers the means by which we can hopefully find ways to better live in balance with our environment, and so facilitate the survival of the ecosystem in something like its current form. Science tells us that the Earth and its biota have faced a number of previous major changes, and so there is good reason to suspect that even if we cannot protect the environment from the implications of human activity, it will lurch into a new stable state, and life will probably carry on. However, that new stable state will not be the Earth we know, and many major groups of animals and plants will probably not survive the dramatic change as a new equilibrium begins to take shape, providing an ecology that will allow new species to evolve and dominate. Quite possibly, no humans will be around to see what that new world will be like.

Some would argue this is pessimistic, and the effect of human activity is exaggerated; but there is a broad scientific perspective that we have put excessive strain on the natural ecosystem, to such an extent that we may well be forcing the system past some turning point – beyond which it is likely to shift to a new state less suitable for human habitation. If this view is correct (and it seems a brave or foolhardy gamble to guess otherwise) then only people, working together across national boundaries, can avoid such a scenario. Assuming the human will and commitment is there, it will depend upon science to provide the essential know-how and technologies.

In this context, science cannot be seen as an optional add-on for being cultured: rather science must play a key part in a liberal education to create societies that are empowered to bring about positive change. A key part, but certainly not the only component: because the people who will be charged with this work will also require a knowledge of ethics, of politics and diplomacy, of human nature, of rhetoric, of economics, of history and international relations. These future leaders will rely upon the understanding and support of society in general. To assure humankind's future, we need science to be seen as a part of our common culture: a part that is closely interlinked with other important areas of human knowledge and experience. This could be seen as a strong argument for secondary science courses that include careful study of socio-scientific issues.

Chemistry for workers

Chemistry can therefore contribute to the knowledge needed by individual citizens, and can form part of the general education that informs the cultural context for societies to work together for the good of people worldwide. But chemical knowledge will have stiff competition for the available space in a core curriculum suitable for all secondary learners.

There will be more scope for the elective chemistry that some students may take on top of such a core 'entitlement' curriculum. School is, in part, preparation for the world of work, and some may argue for vocational chemistry to support those looking to follow particular career paths: chemistry for nursing, or for working with animals, or for hairdressers and beauticians, or for technician work. Here there is an important argument about the distinction between education as part of schooling, and training for particular work. Schooling should provide the basic conceptual framework upon which more specific vocational training can be built. School chemistry may well have a role in vocational preparation in that sense, but arguably schools should not be taking up the task of training up workers for specific jobs. It seems likely that the basic chemistry needed as a background for vocational courses is likely to be in line with that needed by citizens and cultured members of society: understanding of basic concepts that support further, more specific, learning in whatever topic and context individuals may later meet in their professional or civic lives.

It could be argued that much the same is true of the future chemist. After all, if schooling provides the basic conceptual background for the learner who might later need to learn more specific chemistry for a particular job, then why should it matter if that job will be as a hairdresser, a school laboratory technician, an industrial process chemist or an academic research chemist? Traditionally, universities have expected undergraduate students to arrive with strong background knowledge from college level ('sixth-form') courses that, in turn, required a broad prior knowledge of chemistry from secondary school study. However, if schools provide a sufficient background in basic principles to allow the identification of those who are fascinated by chemistry and have aptitude for the subject, then it should not be beyond college and university courses to fill in any required specifics. This may require adjustment of post-compulsory curricula, but as long as secondary education includes a sound background in the fundamental concepts of the subject, such an adjustment should be relatively easily accommodated in further and higher education.

So, arguably, a great deal of the detailed material that might be included in secondary courses is not essential for citizenship, for a liberal education, nor for progression to higher study of the subject: this material does not have to be widely learnt to safeguard democracy, culture or even the chemical sciences. This is surely true. For example, it is hard to argue that a detailed study of trends and properties in several groups of the periodic table (as many students may have studied some decades ago) – whilst it may provide a useful context for learning and reinforcing key concepts – should actually be considered in any sense essential to a sound education. Providing the basic chemical concepts are well understood by the end of secondary school, such detailed surveys of particular chemistry can always be undertaken later by those with a good reason to study such material. The same applies to the different homologous series of organic compounds: their chemistry is interesting to some students, and certainly of value for some purposes, but has limited claim on being a core part of school learning.

From such a perspective there are good grounds to see much traditional upper secondary school chemistry as a luxury that has limited claim on curriculum time. This has been shown by the success of courses that teach chemistry through the everyday contexts of major areas of application (transport, food production, fabrics, etc.), and those that focus on socio-scientific issues (global warming, pollution, sustainable development). These courses can draw upon the specific chemistry that is most relevant, given that school science can only ever consider a small proportion of the ever-increasing chemical knowledge base. That said, there is a strong argument for suggesting that time is spent teaching chemical basics prior to setting out on such courses. So one approach would see chemistry in lower secondary science (i.e. for 11–14 year olds) setting out the basic conceptual foundations of the subject, and providing a basis for supporting (and being reinforced through) later contextually based courses, or courses focusing on social issues that are likely to interest and engage many upper secondary students (i.e. 14–16 year olds).

Chemistry for the 'gifted' learner

For many students of secondary age, a study of chemistry motivated by problems and issues of obvious importance and relevance is more likely to maintain their attention than a more theoretical ('traditional') type of course which is organised around the internal logic of the subject itself. However, we should be aware that different groups of students have different needs. There certainly are some students of secondary age who will be fascinated by the disciplinary structure of chemistry and will be strongly engaged by exploring the development, coherence, and limits of its theoretical apparatus. For these students, an approach largely (if not exclusively) developed through studying the traditional themes and topics of academic chemistry will often be appropriate.

Such students are likely to be those labelled with terms such as 'gifted' (although this is a term that is not well defined, and risks creating an artificial distinction when a wide range of secondary students can show exceptional abilities in some aspects of chemistry). These are the students who thrive on a more intellectually demanding fare, and for whom the abstract and theoretical nature of chemistry as a science is not a turn-off, but a welcome opportunity to challenge their thinking. These learners will not usually be stimulated by a science curriculum consisting of a sequence of many discrete units – especially when they are studied superficially, largely taught as unrelated topics, and learnt in terms of rote phrases appreciated by examiners. Rather, they will be better served by opportunities for extended engagement with underlying patterns and a focus on integrating concepts.

Chemistry, when taught to fit their needs, offers a great deal for these students. It is a subject of great complexity, where evidence reveals patterns that offer clues to the nature of the natural world. It is a subject based on model-building and theory construction, but always answerable to testing against nature. That is, it is a subject that demands high levels of thinking skills, and repays careful and extended engagement with the subject matter.

This type of course also offers considerable opportunities for building in the application of mathematics, something which some commentators (such as SCORE, the umbrella organisation Science COmmunity Representing Education) feel is largely missing from many secondary courses. Chemistry offers much scope for careful quantitative practical work, as well as many opportunities for data handling and mathematical modelling. As one example, patterns in ionisation energies are normally only studied post-secondary, but offer some excellent opportunities for graphing complex data that can be linked with learning about the periodic table (and basic models of atomic structure).

So there is room for offering *some* students a chemistry course that is more traditional in terms of being primarily based around the structure of the discipline (periods, groups, classes of organic compound) to reflect a subject that has a strong theoretical aspect as well as having many important real-world applications. However, even for these students, it should be remembered that there is little value in attempting to offer a comprehensive survey of chemistry, as it is far too vast a subject to do that well in the available time, and whatever particular chemistry is later needed can always be studied in post-secondary courses. Indeed, this group of learners will benefit most from the opportunity to study a small range of topics from different areas of chemistry, but in depth, and through learning approaches that engage higher-level thinking skills and offer opportunities for extended laboratory work, problem-solving and authentic projects that are then reported to peers or wider audiences. Such a course will better meet the intellectual needs of this group of students, and – by showing something of the nature of chemistry as a science – may encourage many to pursue the subject further.

Last words

Ultimately, then, there are good reasons to consider chemistry a key part of the science education of all students, and a range of approaches can be taken to develop diverse courses in secondary chemistry suitable for different groups of students. It should be an aim to allow *all* students to appreciate the fundamental principles of the subject: to understand something of the nature of chemical substances and reactions, including our submicroscopic theoretical models – so providing a foundation for later learning (whether in formal courses, or through informal learning in response to personal interests and needs).

We should not be precious about wishing to teach our favourite specific areas of chemistry when these do not seem relevant to our students, but should be open to context-based approaches, highlighting the nature of science, and addressing chemistry-related socio-scientific issues which are often better able to engage student interest. Difficult decisions will need to be made over the value of having a distinct 'chemistry' strand within science education – as some groups of students may be better served by a more integrated approach to science, within which the disciplinary sources of different topics and ideas are identified. Some students, if a minority, will gain more from approaches which highlight the theoretical structure of chemistry as a scientific discipline, and require deep engagement with some of the many challenging concepts the subject offers. Most students will appreciate some element of choice, both in terms of having an input into the kind of science courses they follow and in being given a selection of alternative activities on some occasions. Most students will also benefit from some variety in their learning of chemistry, regardless of whether their course primarily has a focus on applications, science in society or disciplinary structure. Luckily for the chemistry teacher, as the present handbook demonstrates, chemistry is a subject that offers a great deal of variety in both its content and the ways it can be taught.

Other resources

Books

A companion volume in this series focuses on teaching about the nature of science: Kind, V., and Kind, P. M. (2008). *Teaching Secondary How Science Works*. London: Hodder Education.

Emsley J. (2010). *A Healthy, Wealthy, Sustainable World*. RSC Publishing. This book offers a very readable account of how chemistry is developing the materials we need to maintain and improve our living standards in more sustainable ways.

Advice on teaching the most able students in science classes is given in Taber, K. S. (Ed.). (2007). *Science Education for Gifted Learners*. London: Routledge.

Websites

The role of chemistry in a sustainable world

A downloadable article on this topic may be found at:
www.rsc.org/Education/EiC/issues/2011September/healthywealthsustainableworld.asp

Teaching science in context

A report on context-based science teaching is available from the University of York's website: Bennett, J. (2005). *Bringing science to life: the research evidence on teaching science in context*. York: University of York, Department of Educational Studies. Available at:
www.york.ac.uk/media/educationalstudies/documents/research/Contextsbooklet.pdf

Teaching socio-scientific issues

SATIS (Science and Technology in Society) is a series of resources first produced by the Association for Science Education in the 1980s and 1990s. Many of these have been updated, and are freely available from the website: www.satisrevisited.co.uk/

Teaching chemistry to the gifted

An article on how chemistry is a suitable subject for challenging the most able is freely available on the web: Taber, K. S. (2010). Challenging gifted learners: general principles for science educators; and exemplification in the context of teaching chemistry. *Science Education International*, **21**(1), 5–30. Available at:
www.icaseonline.net/sei/march2010/p2.pdf

Mathematics in chemistry and general support for chemistry teachers

The magazine *Education In Chemistry* (published by the Royal Society of Chemistry) has presented a series on teaching mathematics for chemistry. It also publishes a wide range of news, reviews and articles of relevance to teaching chemistry at school (as well as college and undergraduate) level. The magazine website is at:
www.rsc.org/education/eic

Chemistry Education Research and Practice is a free-access research journal published by the Royal Society of Chemistry, which publishes a wide range of articles about learning and teaching in chemistry. Articles can be accessed through the website at:
http://pubs.rsc.org/en/journals/journalissues/rp

Index